J. Ross Mackay
175 - 2803 West 41st Avenue
Vancouver, BC V6N 4B4
Canada

11 June 2004

Wind as a Geomorphic Agent in Cold Climates

Wind erosion and deposition are extremely important factors in cold climates because of the open space and scarce vegetation. Aeolian processes connected with sand drift in Polar environments are similar to those in deserts, but in cold environments frost and snow also play an important role. Snow, and its transport by wind, make this white seasonal desert quite different from its low-latitude analogues.

The Arctic is characterized by strangely eroded rocks, special wind-formed lakes, large sand dunes and immense loess deposits that owe their formation to aeolian processes controlled by snow cover and frost formation. *Wind as a Geomorphic Agent in Cold Climates* presents a detailed description and explanation of these wind-generated Polar landforms. It includes numerous illustrations and photographs that will assist the reader in identifying and interpreting these features: both modern-day and those preserved in the geological record from earlier glacial periods. Unlike most aeolian geomorphology texts, which do not cover the high-latitude regions north and south of 60°, this volume brings together important information from scattered literature sources concerning these cold environments.

The book provides an important introduction to this area of geocryology and will form a useful reference for graduate students and researchers in a variety of fields, including geomorphology, geology and environmental science.

MATTI SEPPÄLÄ received a Ph.D. from the University of Turku, Finland, in 1971 and subsequently took up a position at the University of Oulu. In 1978 he moved to the University of Helsinki where he is now Professor in Physical Geography. He has travelled widely in order to conduct numerous Arctic field campaigns and to collaborate with colleagues around the world. He has worked as a research fellow at the Universities of Uppsala, Heidelberg, Montréal, Cambridge and Durham. Professor Seppälä's research interests include cold environment geomorphology, geomorphological mapping, cartography, climamorphology, aerial photo-interpretation, permafrost studies and aeolian processes. He has contributed to several book publications and has served on the editorial boards of *Géographie physique et quaternaire* and *Permafrost and Periglacial Processes*. Professor Seppälä is a fellow of the Arctic Institute of North America and of the Canadian Geological Association, and is actively involved in a number of international organizations relating to Polar research. He is the recipient of various honorary awards including the Erskine Fellowship, University of Canterbury, Christchurch, New Zealand; and the Ragnar Hult Medal of the Geographical Society of Finland.

Studies in Polar Research

This series of publications reflects the growth of research activity in and about the Polar regions, and provides a means of synthesising the results. Coverage is international and interdisciplinary: the books are relatively short and fully illustrated. Most are surveys of the present state of knowledge in a given subject rather than research reports, conference proceedings or collected papers. The scope of the series is wide and includes studies in all the biological, physical and social sciences.

Other titles available in this series:

Wind as a Geomorphic Agent in Cold Climates

Matti Seppälä
University of Helsinki

CAMBRIDGE
UNIVERSITY PRESS

PUBLISHED BY THE PRESS SYNDICATE OF THE UNIVERSITY OF CAMBRIDGE
The Pitt Building, Trumpington Street, Cambridge, United Kingdom

CAMBRIDGE UNIVERSITY PRESS
The Edinburgh Building, Cambridge CB2 2RU, UK
40 West 20th Street, New York, NY 10011-4211, USA
477 Williamstown Road, Port Melbourne, VIC 3207, Australia
Ruiz de Alarcón 13, 28014 Madrid, Spain
Dock House, The Waterfront, Cape Town 8001, South Africa

http://www.cambridge.org

First published 2004

Printed in the United Kingdom at the University Press, Cambridge

Typeface Times 11/14 pt   *System* LATEX 2_ε [TB]

A catalogue record for this book is available from the British Library

Library of Congress Cataloguing in Publication data

Seppälä, Matti.
Wind as a geomorphic agent in cold climates / Matti Seppälä.
   p.   cm. – (Studies in polar research)
Includes bibliographical references and index.
ISBN 0 521 56406 9
1. Eolian processes–Polar Regions.   I. Title.   II. Series.
QE597.S46   2004   551.3′7′0911–dc22   2003066159

ISBN 0 521 56406 9 hardback

358 p.

Contents

The colour plates are situated between pages 141 and 142.

Acknowledgements

I am deeply grateful for the assistance and guidance of many people who have helped me during the compilation of this book. It is based on almost 40 years of study and unfortunately I am not able to list all those people who have helped me during field studies in the Arctic and Antarctica during that time. They have been mentioned in the original studies published earlier within different scientific forums and publications. All these friends I remember with great pleasure and I acknowledge their indirect assistance in the writing of this work. You cannot forget events such as one's tent blowing away in a thunder storm or being covered in a snow storm. The help of friends in dealing with these difficult conditions has been invaluable – without them I could not have collected the necessary data for my work.

I am very grateful for the help and encouragement of Professor Peter Williams, Ottawa, from the initial planning of this book, during its compilation, through to his final checking and revision of the English text and comments on its content. I also sincerely appreciate the useful suggestions on the chapter about thaw lakes made by Professor J. Ross Mackay, Vancouver. In all circumstances the author and only he is responsible for the contents of the book.

The following colleagues kindly agreed to place their original photographs at my disposal: P. Havas (Plate 1), W. Karlén (Plate 13), J. Käyhkö (Plates 2, 8, 10 and 12), D. Kobayashi (Fig. 12.1), E. Koster (Figs. 6.12, 8.17, 8.18), D. Krinsley (Plate 4), R. Larsson (Fig. 6.20), P. Lintinen (Plate 14), J. Ross Mackay (Fig. 5.7), K. Moriwaki (5.1), K. Nenonen (7.4), O. Salvigsen (8.6), H. Tabuchi (1.3), and W. Tobiasson (5.3). Original study material was placed at my disposal by B. Holmgren (Fig. 2.28) and I. S. Evans (Figs. 3.11, 13.7).

Many copyright holders have kindly given permission to modify and use copyright material. They have been listed separately. Although every effort has been made to trace and contact copyright holders, this has not always been successful and some enquiries remained unanswered at the time of going to press. I apologize for any apparent negligence and it is hoped that any further acknowledgements may be added in future editions.

Sometimes writing this book seemed to be an endless task but my family remained patient throughout, even when my wife, Mrs Raija Seppälä, and our children had to move to new places and attend school in a foreign language.

I am grateful for the opportunity to work at the library at the Scott Polar Research Institute (SPRI), Cambridge, UK where the librarians always kindly helped me in finding the most difficult reference material. The libraries of the University of Cambridge played a vital role in the preparation of this work.

Cartographers Mrs Kirsti Lehto and Mr Jari Patanen made the final drawings of the figures and Ms Leena Heiskanen printed most of the photographs, all at the Department of Geography, University of Helsinki.

Dr David Walton (editor-in-chief of this series of books), Dr Maria Murphy and Dr Susan Francis (in-house editors at Cambridge University Press) have been of great help during the production of this book. Mrs Frances Nex (copy-editor) caught with sharp eye unbelievably many slips in my presentation, and that improved the final result. I have enjoyed working with them.

This work would not have been possible without the financial support of the Academy of Finland (fellowship for a senior scientist); the University of Helsinki (Visiting Fellowship to Clare Hall College, Cambridge); and Hatfield College, University of Durham, UK (research fellowship). Lastly, I would like to thank Professor Paavo Talman, the Head of the Department of Geography, University of Helsinki, for his patience during the long writing process of this book, which has necessitated several periods of absence away from my normal work at the department.

Copyright holders who have kindly given permission to use their figures
A. A. Balkema Publishers, Rotterdam (2.10), American Geophysical Union, Washington, DC (12.5), American Meteorological Society, Boston (3.1), American Society of Agricultural Engineers, St Joseph, MI (11.3), Annales Botanici Fennici, Helsinki (5.2), Association of American Geographers, Blackwell Publishing, Oxford (1.1 and 3.12), Association francaise pour l'étude du Quaternaire (French Quaternary Association), Meudon, France (8.16), Cambridge University Press Cambridge, UK (3.9), Catena Verlag GMBH, Reiskirchen, Germany (4.11), Dansk Geologisk Forening, Copenhagen (4.1), Paul M. B. Föhn, Davos, Switzerland (13.2), Elsevier (3.6, 4.18 and 8.19), Erdkunde, Förderverein wissenschaftliche Geographie, Bonn, Germany (16.2), F.-K. Holtmeier, Münster, Germany (9.2), Fennia, Geographical Society of Finland (4.14, 4.17, 6.1, 6.9, 6.21, 6.25, 6.26, 8.7 and 8.8), Geografiska Annaler (4.7, 8.12, 8.13, 8.14 and 13.6), Geografiska Föreningen i Göteborg, Sweden (9.1), Géographie physique et Quaternaire, Montréal (10.3), Geological Society of Finland (4.15 and 4.16), Hokkaido University Press, Sapporo, Japan (4.12), International Glaciological Society, Cambridge, UK (11.1 and 14.6), J. R. L. Allen, Reading, UK (Table 4.4), John Wiley & Sons, Chichester (4.6 and 15.1), John Wiley & Sons, Hoboken, NJ (2.16 and 2.17), Jonas Åkerman, Lund, Sweden (5.8 and 6.17), Jukka Käyhkö, Turku, Finland (6.19), Kevo Subarctic Research Station, University of Turku, Finland (2.4 and 3.14), MIT Press, Cambridge, MA, USA. (11.2 and 11.4), National Institute of Polar Research, Tokyo, Japan (12.4), New Zealand Journal of Science and Technology, The Royal Society of New Zealand, Wellington (3.13), New Zealand Journal of Geology and Geophysics, The Royal Society of New Zealand, Wellington (4.4), Nordic Council of Ministers, Copenhagen (6.11), Quaternary Research, Elsevier Science, Oxford (10.1), Revue Canadienne de Géographie, Département de Géographie, Université de Montréal (7.4), Royal Meteorological Society, Reading, UK (3.2 and 3.3), Soil Science, Lippincott Williams & Wilkins, Baltimore (8.5), Taylor & Francis AS, Solli, Norway (15.4), The American Geographical Society, New York (2.1 and 2.23), The Geological Society of America, Boulder, Colorado (8.3), The Geological Society Publishing House, London, UK (8.15), The Royal Canadian Geographical Society, Vanier (2.11), The University of Chicago Press (7.1 and 7.3), Topographic Service of FDF, Finland (6.3), The University of Arizona Press (2.3).

1

Introduction

Sand dune formation has been studied for a long time (e.g. Sokolów 1894) and several excellent text books on that topic have been published recently (e.g. Greeley & Iversen 1985; Pye & Tsoar 1990; Nordstrom *et al.* 1990). Large sand seas and very high sand dunes in the deserts in low latitudes are such fascinating features that they have attracted students much more than cold environment features. Detailed surveys of aeolian landforms in tropical, subtropical and temperate environments are numerous. A very extensive literature exists on deserts, their morphology and climate (e.g. Cooke & Warren 1975; Mabbutt 1977; McKee 1979; Thomas 1989, 1997; Cooke *et al.* 1993; Abrahams & Parsons 1994) and coastal environments (e.g. Bakker *et al.* 1990; Carter *et al.* 1992).

Most presentations of aeolian landforms are limited to latitudes lower than 60°, with most close to the equator. In these studies the Polar world has been totally excluded. However, we find in the Polar regions real deserts with very limited or no vegetation. People seldom think of the regions with dry snow as white deserts (e.g. Giaever 1955). We can call Polar regions outside glaciers seasonal Polar deserts, without liquid water, with no vegetation, where snow cover melts away and with wind drifting granular material in an unlimited way. These seasonal cold deserts are limited to a zone where temperature in the winter is frequently fluctuating around 0 °C which means liquid water and hard pan in the snow cover. Occasionally wind there is also able to drift and pack snow especially when new. This book examines cold environments with respect to wind and its geomorphic impact in high latitudes. Much of what is presented here can be applied also in Alpine regions and in cold regions at lower latitudes (consider, for example, Mongolia). The topic has been studied earlier, for example by Hörner (1926) and Samuelsson (1926), but apparently since then there has been no review published of the geomorphic effects of wind in cold environments.

Extremely strong winds are well-known both from the Arctic and Antarctica. In South Victoria Land, Antarctica and on both sides of the peninsula of West Antarctica, 'where rock is exposed denudation is powerful as a result of summer insolation, frost weathering, and strong winds. The violent wind acts as a formative agency on rock and snow surfaces. It transports snow, dust, and sand; creates small heaps of coarse detrital material like those in stony deserts; it polishes boulders and bed rock; it furrows the snow and shapes it into sharp-edged sastrugi' (Nordenskjöld & Mecking 1928: 287).

Hobbs (1931) called attention to the (according to him) erroneous conception regarding the conditions outside the Pleistocene continental glaciers of North America and Europe. He had observed in Greenland fierce storms which blew outward from the ice of continental glaciers and the restriction of drainage to a brief warmer season during which meltwater issued from beneath the glacier. In 1942 Hobbs wrote: 'As soon as the brief warm season has come to an end, the agent of extraglacial transportation is no longer running water but strong wind currents directed radially outward from the glacier, or nearly at right angles to the glacier front'. Wind takes over from the thaw water the work of transportation so totally that Hobbs (1942) considered wind as the dominant transportation agent within extra marginal zones to continental glaciers. He had found how 'dust, sand, and smaller pebbles are lifted and carried away, leaving behind an armour of pebble pavement to protect the material below, and in every way this pavement is similar to the *sêrir* of low-latitude deserts'.

Thus an overall presentation of aeolian geomorphology from the cold regions was missing. Hamelin and Cook (1967: 129) wrote in their glossary of periglacial phenomena: 'The place of wind action in periglacial geomorphology is not fully understood. It does, however, contribute to both depositional and erosional features. Lack of vegetation in periglacial areas greatly increases the efficacy of the wind. The accumulative or depositional aspect is by far the most important. Some loess is of periglacial origin and is deposited by wind action. . . . Abrasion is performed by wind, both by blowing sand and snow.'

However, the importance of wind as a geomorphic agent in cold climates has been known a long time (e.g. Sapper 1909; Enquist 1916; Samuelsson 1926); the High Arctic deserts are rather seldom mentioned in literature (e.g. Passarge 1921; Fristrup 1952a, 1952b, 1952–53; Smiley & Zumberge 1974). In 1993 it was still possible to publish a book about *Canada's Cold Environments* (French & Slaymaker 1993) without mentioning the term wind erosion or deflation, although strong winds are pointed out. Tricart (1970) credits deflation with a greater transporting role than running water in periglacial areas. Periglacial environments are characterized by strong wind action. Climatic dryness and the sparse plant cover are only partly important for the aeolian processes, but above all the frost and melting glacial ice produce large quantities of unconsolidated sediments favourable for wind transport (Troll 1948). Peltier (1950: 221) believes that wind has a secondary role that increases in the later part of the periglacial cycle.

Maximum wind action occurs in morphogenetic regions with low annual precipitation and high or low mean annual temperatures (Peltier 1950: Fig. 6; Thornbury 1954: 60–65). In cold morphogenetic environments Peltier (1950: Fig. 7) considers glacial, periglacial, boreal and partly savana (sic! term used in unusual context) regions, which have from moderate to strong wind action (Fig. 1.1).

In the existing literature of the twentieth century we can find certain centres of periglacial aeolian studies. First a strong school of this aeolian geomorphology was established in Uppsala, Sweden (Enquist 1916, 1932; Hörner 1926; Högbom 1923; Samuelsson, 1921, 1925, 1926). The papers were published in a local journal: *Bulletin of the Geological Institution of the University of Upsala.* Then the centre of gravity moved to the central European sand belt and to Poland where many ancient late Pleistocene and Holocene sand dunes of cold climate origin appear (e.g. Dylikowa 1958; Galon 1958; Kobendza & Kobendza 1958;

Fig. 1.1 Morphoclimatic regions and the activity of aeolian processes according to mean annual air temperature and precipitation compiled after Peltier (1950: Figs. 6 and 7). © Association of American Geographers, Blackwell Publishing.

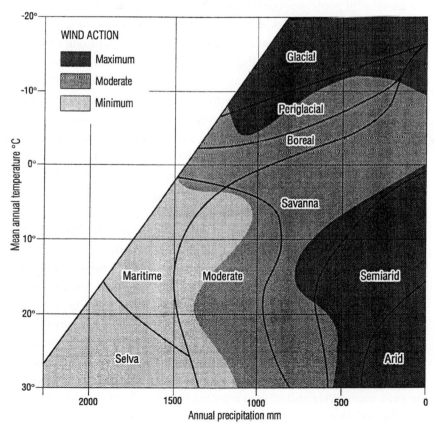

Mrózek 1958; Pernarowski 1959, 1966; Stankowski 1963; Wasylikowa 1964; Urbaniac 1969; Kozarski 1991). In Lódz the *Biuletyn Peryglacjalny*, founded by Jan Dylik, was published. Poser (1948, 1950), by continuing the old German tradition in late glacial sand dune studies (Sokolów 1894; Solger 1910), established in Göttingen, Germany, a study centre of periglacial processes in general and aeolian processes, too. Several studies were published in *Abhandlungen der Akademie der Wissenschaften* in Göttingen in the 1970s and 1980s. At present we have no very well specialized, leading cold environment research group studying cold environment aeolian processes but the interest seems to be split into several places (Seppälä 1975b; Niessen *et al.* 1984; Koster 1988, 1995). Recently a new journal in cold environment studies has been founded: *Permafrost and Periglacial Processes*, which contains some aeolian geomorphology, too.

Gary *et al.* (1972: 232) in the *Glossary of Geology* defined *aeolian* (=*eolian*) as 'pertaining to the wind; especially said of rocks, soils, and deposits (such as loess, dune sand, and some volcanic tuffs) whose constituents were transported (blown) and laid down by atmospheric currents, or of landforms produced or eroded by the wind, or of sedimentary structures (such as ripple marks) made by the wind, or of geologic processes (such as erosion and deposition) accomplished by the wind'. Nevertheless, wind transport is a major part of sedimentology and perhaps it has received less attention than it deserves (Pettijohn *et al.* 1987: 310).

I would like to expand further the use of the term *aeolian* to the secondary effects of wind in cold environment. Then we are speaking of snow drift and accumulation, glacier formation, permafrost formation, deflation features which could later be lake basins, and deposition of mineral material together with snow, so called *niveo-haeolian* deposits. Deflation of snow cover can increase the thickness of permafrost, and form patterned ground which has connections with wind activity. Regions with glaciers and some other areas in middle latitudes have been included in this presentation as a part of the cold climate aeolian geomorphology.

Aeolian processes are environmentally controlled. The basic factors are the latitude which determines the amount of solar radiation and relief which both strongly affect climate (Fig. 1.2). Dominating meteorological factors from the aeolian point of view are temperature and precipitation which both affect the main limiting factors: vegetation and moisture of the soil surface and also soil formation. The special controlling factors of aeolian processes in the cold environments are frost, ice, snow, icing and meltwaters. Large quantities of meltwaters in spring and summer partly prevent the aeolian drift of mineral material and partly produce favourable sand surfaces to be deformed by wind. Most of the year snow cover and ground frost in certain parts of Polar regions can totally stop the deflation of the ground, but as contrast in some other regions the winter activity is the predominating one. As a result of favourable climatic factors wind can transport in cold environments sand, silt and dust, in some cases also small stones and especially large amounts of snow. Particle drift causes abrasion on rock surfaces and deflation forms. Material is deposited. Sand dune formation is only a small portion of the total aeolian activity in the cold environment. Snow deposition is probably much more important for landforms. A secondary effect of the thinning of snow cover is deep frost formation. Many of these processes and events have some kind of feedback on aeolian processes, for example, wind action drifts snow forming glaciers, which then affect wind pattern (Fig. 1.2).

In extreme conditions as in the Antarctic cold deserts, in the absence of contemporary fluvial and other geological and biological processes capable of removing weathered material from rock surface, aeolian abrasion acts as a controlling factor in the rate and magnitude of erosion (Malin 1992: 27).

Wind plays an important role in cold climate geomorphology. Tricart (1970: 143) enumerates the main reasons for this: (1) high frequency of strong winds and storms in the Arctic; (2) low precipitation; (3) reduction of the vegetation cover by the cold, which is increased (4) by the wind blowing away the snow and exposing the ground surface, leading to the destruction of the vegetation; and to (5) the concentration of water in the ground by freezing which tends to dry

Fig. 1.2 Flowage diagram of the aeolian processes and formations combined with other limiting and supporting factors in general.

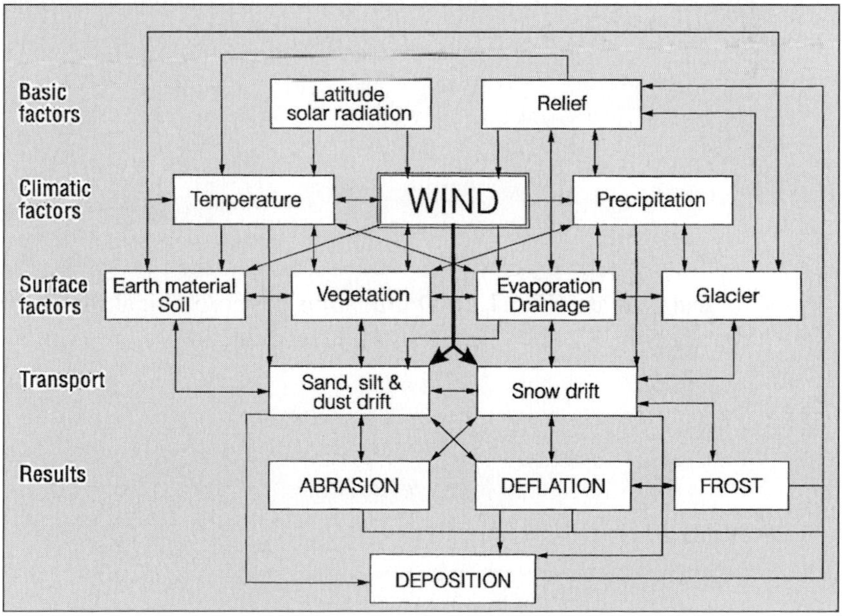

out the surface layer of silty soils, and to the production of abundant fine debris, especially silt, by frost weathering. This material is suitable for long distance transport by wind.

Wind is everywhere and in the circumpolar environments and on high mountains it is strong. There its force is not much limited by high vegetation like forests. When considering the importance of aeolian processes in cold environments we have to keep in mind how vast areas of the earth are covered by snow permanently or seasonally. Dry snow is relatively light material and easily drifted. Snow as such does not form geomorphic features but when it is accumulated in big amounts and forming glaciers and snow patches or melting, then it produces erosional and depositional landforms of nival origin and wind drift is the primary reason for them. In certain places snow cover is thin and in other places thick and the controlling factor is wind. Blowing snow itself is a major problem in cold environments, e.g. Canada (French & Slaymaker 1993: 9). Accumulation and redistribution of snow in the mountains has considerable effects on snow transport, glacier dynamics, snow drifts and avalanche danger (Dyunin & Kotlyakov 1980).

Fig. 1.3 Wind piled sea ice blocks on the shore of the northern Gulf of Bothnia, Hailuoto, Finland. Photo by H. Tabuchi.

We should consider the winter conditions much more when studying the periglacial phenomena and the frost action on landforms. In the literature we can find a lot of knowledge of snow and of its physical characteristics which should be integrated into the geomorphic processes. In the summer time the periglacial phenomena are mainly in rest phase and melting. For example, data of freezing on the active layer and the impact of accumulating snow are limited. This book partly tries to open that gate a little.

In this book the stress of wind on sea ice (see Sverdrup 1957) and coastal processes is largely excluded, although cold environments give several characteristic features on beaches, too, such as rafted ice walls pressed up on the shores by wind (e.g. Alestalo & Häikiö 1976) (Fig. 1.3), ice-pushed boulders (Philip 1990) and boulder barricades on shores (e.g. Rosen 1979; Dionne 1994), permafrost degradation by wave action on Arctic coasts, buried ice blocks melting in beach sand, drifting of icebergs, ice scouring and plucking of rock surfaces on tidal flats (e.g. Dionne & Quessy 1995), etc.

By knowing the present-day wind action in cold environments we have a better chance of understanding also the ancient aeolian features found in the vicinity of the glaciated regions. The palaeoenvironment reconstructions deal mainly with temperatures and humidity, interpreted through plant remains, especially pollens, and through soil formation. Some wind regimes relating to the aeolian landforms are described in the literature but the winter conditions are almost totally neglected, with the exception of the evidence of permafrost. Occurrence of snow is considered mainly for niveo-aeolian deposits. It seems to be obvious that today's aeolian processes in cold regions are considerably weaker and scattered compared with the Pleistocene activity in North America, central Europe, Asia, Argentina and New Zealand when the glaciers reached their maximum stage.

2

Delimitation and characterization of cold environments

Polunin (1951: 308) pointed out the confusing matter of giving a precise definition or delimitation of the Arctic as a region. However, we should try to define the cold environment in the sense that cold has some impact on aeolian processes. We can easily find large regions where cold is limiting vegetation so totally that it provides no protection from the influence of wind. This applies in Antarctica and great parts of the Arctic. But in Arctic and Subarctic regions we also find totally scrub-, moss- and lichen-covered areas where the aeolian processes are limited even though it is cold enough for deep permafrost formation. On the other hand the cold season could be important for part of the year and give opportunity for wind action in environments which are not even in the Subarctic climatic zone. An example is the prairies where very cold, continental winters dominate with thin snow cover and extremely strong winds without limiting factors. In North America this type of seasonally cold aeolian environment is found as far south as Wyoming, for example. At present the aeolian processes are not extremely strong in Wyoming but we can find geomorphic evidence proving that wind has had a strong influence: ventifacts (Sharp 1949), periglacial soil wedge forms (Mears 1987), sand dunes, and mushroom rocks. Drifted snow even wears rock surfaces there locally. It means that if the climate gets somewhat more severe then the conditions are really favourable for aeolian processes again. We do not consider in depth the low latitude cold-supported aeolian regions, nor the mainly mountainous areas in low latitudes which have similar characteristics to high latitudes in spite of the strong solar radiation.

Polar regions in general are cold. The Polar Circles (66°33'03" N and S; 2606 km from the poles) are simple criteria for defining cold regions lying poleward of the circles and they are governed by the presence of "midnight sun" or absence of "midwinter sun". The main reasons why the Polar regions are cold are: (1) Earth is a sphere which means that the sun shining in Polar regions is always at a low angle and does not bring much warmth to Polar areas; (2) the axis of the Earth is tilted in an oblique angle (23°27') to its ecliptic, the plane on which the Earth circles the sun, such that the Polar areas lose the sun for part of the year and they do not receive direct solar heat the year round. At the poles the day is six months long and the sunless winter night lasts six months. (3) Snow covers the ground most of the time and with its very high albedo it reflects most of the incoming solar energy back into space so the warming effect of the sun is lost. (4) Polar air is very clean without many dust particles and it contains very little water vapour. This means very small storage of the outgoing long wave heat radiation.

The Antarctic Circle encloses a very cold region. Unfortunately such a single criterion in the case of the Arctic Circle is useless for the purpose of our delimitation. Because of its size, diversity, complexity, and the random and scattered manner, the Arctic is difficult to define. The experience and terms used to describe it have become part of our everyday language without a precise recognition of their meaning. For example, the 'limit' for cold for a person from middle or low latitudes means a different temperature than for a person living in the high latitudes. Cold is a relative term. Different plants grow in different temperatures. For many plants the growing starts when the temperature rises above +5 °C but some grow under snow (e.g. Salisbury 1985). Snowdrop (*Galanthus nivalis*), for example, blooms through the snow.

Good (1955) used Washburn's (1953: 271) definition of the Subarctic: the circumpolar belt of dominant coniferous forest (taiga) with the mean annual air temperature (MAAT) below 0 °C and where the average temperature of the warmest month is more than +10 °C, but with less than four months above this temperature. As a result the Subarctic region covers, for example, the whole of Finland except the south coast, and Sweden from the north almost to Stockholm (Fig. 2.1). From the aeolian geomorphological point of view this large Subarctic is less interesting. Thick forests stabilize snow and soils very effectively. Our opinion is that the Subarctic region is a much narrower zone close to the boreal forest limit, in which we can find at least sporadic permafrost (Fig. 2.2).

2.1   Climate

"Arctic" is usually defined with regard to environmental characteristics rather than simply to the Arctic Circle (66° 33′ N). Polunin (1951: 309) defined the Arctic in the following terms: treeless, with the winters largely dark and cold, with high windchill and less than 50 days between spring and fall frosts, the subsoil mostly permanently frozen with frost-heaving important; annual precipitation normally below 500 mm and "largely in the form of snow which drifts and is packed tightly by the wind"; the soils generally moist in summer but the air of low absolute humidity, and with sheltered salt and fresh water frozen in the winter. One of the definitions is determined by the Nordenskjöld formula $[w = 9 - 0.1k]$ where w is the mean temperature of the warmest month and k is the mean temperature of the coldest month over a cycle of years. Nordenskjöld proposed that the southern limit of the Arctic should be the line along which the warmest month has a mean temperature equal to w in his formula. He also suggested that this line comes closer to the Arctic tree line than Köppen's 10 °C line (Fig. 2.2).

Hopkins (1959) made an attempt to show that there is a faithful correspondence between a temperature parameter (cumulative summer warmth) and the forest-tundra boundary in Alaska. He discovered a good correlation between the vegetation at the station and the number of degree-days above $+10$ °C and secondly the mean temperature of the coldest month (Hopkins 1959: 216).

Fig. 2.1 Arctic regions limited with the MMT isotherm 10 °C of the warmest month and Subarctic regions limited with MMT not higher than 10 °C for more than four months of the year according to the definitions by Washburn (1953). Source: Good (1955: Fig. 1). © The American Geographical Society.

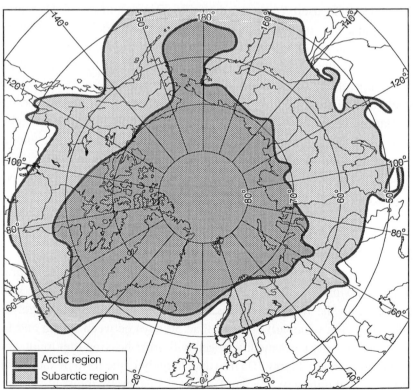

Cold is a climatic term but climatic borders are defined by means of statistical countings and therefore they are more like zones of transitions where the conditions fluctuate back and forth year-to-year, than exact limits. The meteorological stations are far from each other and the intermediate regions have to be covered with interpolations and often neglect the local topography or waterbodies or vegetation characteristics. Still we use, for example, the magic +10 °C isotherm of the warmest month from Köppen, as a boundary of the Arctic which has an average annual temperature of 0 °C (Washburn 1953: 269) because it has been drawn close to so many other limits which are relevant for the cold areas (Fig. 2.2). This limit is widely accepted but perhaps insufficient. It omits totally the long cold winter. Such omission allows certain places, e.g. Iceland, with very small annual range of air temperature and often very humid, to become Polar in climate (Baird 1964: 4) (Fig. 2.3). Climate is a combination of all the different physical factors of the environment. The +10 °C isotherm of the warmest month

Fig. 2.2 Southern limits of continuous, sporadic and mountainous permafrost (Hegginbottom 1993), 10 °C isotherm of the warmest month (*Encyclopedia of Geomorphology, Arctic Regions*) and Nordenskjöld's Arctic limit.

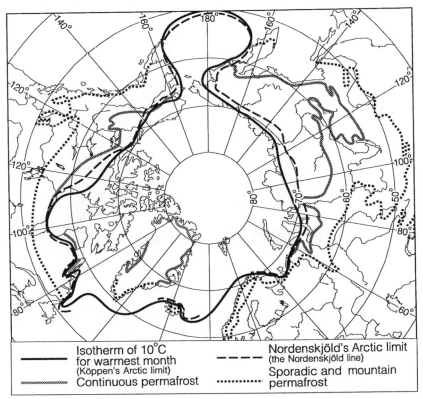

| Isotherm of 10°C for warmest month (Köppen's Arctic limit) | Nordenskjöld's Arctic limit (the Nordenskjöld line) |
| Continuous permafrost | Sporadic and mountain permafrost |

as a border of the Arctic is a very broad generalization. Inside this limit we
can find glaciers, but also sand dunes and close by patterned ground. In fact all
these Arctic phenomena are controlled by different local climatic factors. The
more we learn to know about the local climates and microclimatic phenomena
the more sceptical we become to the general climatological presentations. The
Earth's surface seems to be divided into small mosaic-like patterns with their
own local climatological characteristics which are rather difficult to present on
maps on a continental scale. These mosaic-like climatic patterns resemble the
vegetation patterns close to the Arctic forest limit (cf. Hustich 1948: Fig. 1;
1953: Fig. 1; 1966: Fig. 1). Washburn (1953: 269) pointed out that the boreal
forest limit tends to extend northward as forest fingers along major waterways.
The border of the cold environment can be drawn by means of some reasonable
period with air temperatures below 0 °C. Cold means that we have freezing and
thawing of water. The lack of trees and the dryness of the snow make the snow
easily transported.

Fig. 2.3 Arctic deserts, mean July 10 °C isotherm and mean annual precipitation, after
Péwé (1974: Fig. 3.1). © 1974 The Arizona Board of Regents. Reproduced by
permission of the University of Arizona Press.

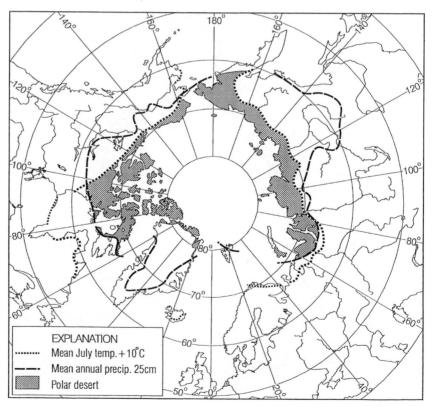

Tricart (1970: 19–27) divided the frost climates into three types:

(1) Dry climates with severe winters
Very low winter temperatures, short summers with very variable temperatures, large daily temperature range. Annual precipitation some 200 mm increasing towards the south as the summers get warmer.

(2) Humid climates with severe winters
(2a) The Arctic type corresponds to part of Köppen's ET climates. The wettest part has no definite seasons. Summers are marked by 3–4 months above 0 °C, mean annual air temperature below 0 °C, and humidity is higher than in type (1). Annual precipitation is normally in excess of 300 mm. An appreciable snow cover reduces the importance of deflation. Fog can last for weeks.

(2b) The mountain type. Monthly temperatures show variations similar to the Arctic type. The summer maxima are similar but winters are less cold. Relief and aspect have an important impact on temperature distribution. Temperature inversions are characteristic. Valley bottoms have often a greater number of freeze-thaw cycles.

(3) Climates with a small annual range
(3a) Island climates in high latitudes
(3b) Mountain climates of low latitudes.

We are especially concentrating on the first two types. For our purposes Tricart's classification considering both temperature and precipitation is better than the somewhat similar French's (1976: 6) temperature-based division into (1) High Arctic climates in Polar latitudes with extremely weak diurnal pattern, and strong seasonal pattern; small daily and large annual temperature range. (2) Continental climates in Subarctic latitudes with weak diurnal pattern, strong seasonal pattern and extreme annual temperature range. (3) Alpine climates in middle latitude mountains with well-developed diurnal and seasonal patterns. (4) Climates of low annual temperature range on subpolar islands and low latitude mountains.

An example of a Subarctic temperature regime in rather high latitudes is shown by the thermoisopleth diagram of mean hourly temperatures (Fig. 2.4) at Kevo (69°45′ N), Subarctic Finnish Lapland, where the diurnal pattern especially in the winter time is very weak but the summer temperatures relatively high (+15 °C, maximum +30 °C). The further north we go the longer is the period with small diurnal temperature range due to the Arctic night.

2.1.1 *Temperatures in the Arctic*
In cold environments the mean monthly air temperature stays below freezing for several months and in the Arctic often for the greater part of the year

(8–10 months). Mean annual air temperature (MAAT) is misleading if we do not also consider the annual temperature range. A MAAT of 0 °C with a very small annual range is a great limiting factor for vegetation as on the islands outside Antarctica, but it may also give good conditions for the growth of boreal forests if the differences of the mean temperatures of the coldest and warmest months range from 20 to 35 °C and the summer temperatures are well above 10 °C (not the case on the Antarctic islands). Temperatures in the Arctic are higher than in Antarctica and the main reason may be that the Arctic is dominated by an ocean (5.5 mill. sq. miles or 14.2 mill. sq. km) in its centre and the land masses (2.530 mill. sq. miles or 6.55 mill. sq. km) are largely not covered by glaciers (Sater 1968: 23–25). Earth's surface gets more heated during the summer. This brings fairly warm and moist air masses and rain to the Arctic. Sea water also warms the air even through the thick ice cover in the middle of the Arctic. This type of connection to warming Earth or sea water is missing in the southern Polar area.

Summer temperatures in the Arctic are surprisingly uniform (Fig. 2.5). The difference between the mean temperatures at the North Pole and on the Siberian coast is just about 10 °C. The summer is coldest (< -12 °C) in central Greenland. The highest summer temperatures (about $+16$ °C) occur in eastern Siberia and central Alaska.

Fig. 2.4 Thermoisopleth diagram constructed from monthly means of diurnal temperatures at 02, 08, 14 and 20 o'clock, Kevo, northernmost Finland. (Seppälä 1976a, Fig. 4.)

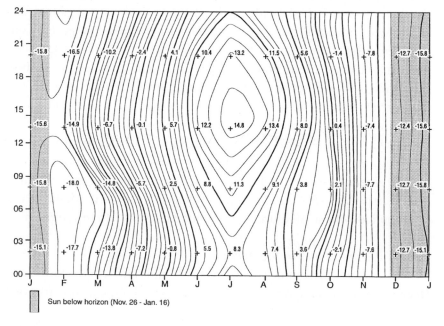

Sun below horizon (Nov. 26 - Jan. 16)

Fig. 2.5 Mean air temperature in northern Polar regions in January and July. Modified after the *Polar Regions Atlas* (1978).

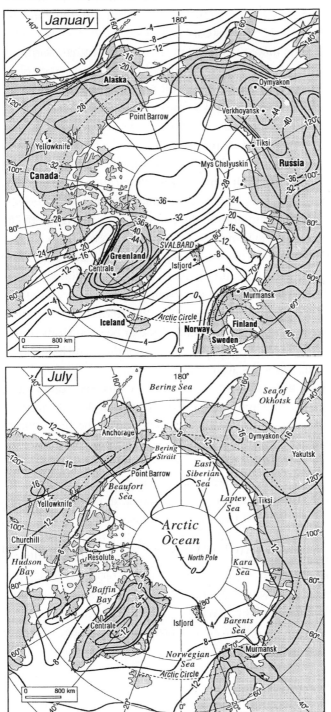

The ocean has a strong impact on the air temperature in the Arctic also during the winter as a result of the increased heat flux to the atmosphere from the sea ice (Fig. 2.5). According to Zhang *et al.* (1996) the Arctic coastal zone in northern Alaska (MAAT about $-12\,°C$) has cool summers and relatively warm winters. The Arctic inland zone more than 120 km from the coast has warmer summers and the coldest winters in the region although the MAATs are very close to each other. The Arctic foothills zone (800–1500 m a.s.l.) has the warmest winters of northern Alaska, due to the weaker temperature inversion, but its summer temperature is slightly lower than that of the Arctic inland zone.

The minimum temperatures of the northern hemisphere are recorded in the Subarctic eastern Siberia where the summers are reasonably warm. January or February is the coldest month in most areas in the northern hemisphere. Central Greenland and eastern Siberia with a mean monthly temperature (MMT) of less than $-44\,°C$ are the coldest regions in the winter (Fig. 2.5). Northern islands and the Keewatin region (central Canadian Arctic) are the coldest in North America (MMT $<-32\,°C$) (Fig. 2.5). In Eurasia the winter isotherms may be more or less parallel with the coasts, or alternatively parallel to latitude, getting colder southwards. In North America they are often perpendicular to the coast. This indicates that the cold air masses penetrate southwards as tongues from the North Pole. The annual range of mean temperatures can be seen from some selected weather stations in the Arctic regions (Fig. 2.6). For three and half to four months the mean air temperature stays above zero centigrade.

Alison (according to Markov *et al.* 1970: 31) called the Arctic "a huge desiccator". In winter, the Subarctic mimics the Arctic, while in summer it mimics the temperate zone. Arctic air masses are dominant in the Subarctic in winter, while air of temperate latitudes is dominant in summer. For instance, most of Greenland lies in the Arctic, but its southern coast belongs to the Subarctic. Atlantic air masses of the temperate zone characterize it in summer.

Williams (1961: 346) suggested $<+3\,°C$ mean annual air temperature as a rough limit for periglacial solifluction and patterned ground development. French (1976: 5) followed this definition closely and made a further subdivision of the periglacial domain by the $-2\,°C$ mean annual air temperature. In regions with MAAT between $-2\,°C$ and $+3\,°C$ the frost-action processes occur but do not necessarily dominate, and with MAAT $< -2\,°C$ frost-action processes dominate. Koster (1988: 69) agrees with French to define the cold-climate regions as having a MAAT less than $+3\,°C$ as well as all areas with snow climate, where the coldest mean monthly temperature is $< -3\,°C$.

2.1.2   *Climate of Antarctica*

In general Antarctica has very uniform climatic conditions but even there differences in orography result in differences in climate. In Antarctica both

Fig. 2.6 Mean monthly air temperatures and precipitation of selected weather stations in the Arctic (Walter & Leith 1960). Mean annual temperature in °C; mean annual sum-total precipitation in mm; mean daily minimum of the coldest month; absolute minimum temperature; altitude above sea level. Scales 0, −10, −20, −30, 10, 20 °C; 0, 20, 40, 60, 80, 100 mm.

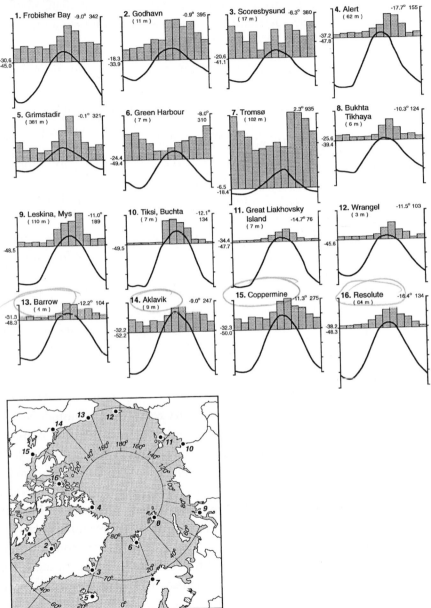

summer and winter are colder than in the Arctic (Fig. 2.5 & 2.7). The climate there is neither continental nor maritime, it is glacial (Nordenskjöld & Mecking 1928: 10). Antarctica is the highest continent on average which means that it is chilled by its high elevation, too.

Sverdrup (1938: 11) distinguished three types of Polar climate: marine, coastal and continental. The basic climatic zones for East Antarctica of V. A. Bugaev (according to Solopov 1967: 95–96) are:

1. Intra-continental high mountain zone, with altitudes more than 3.0–3.2 km above sea level. Mean annual air temperature (MAAT) −50 to −60 °C. Annual precipitation less than 80 mm. Winds are weak.
2. Ice slope zone, having a width of 700–800 km and altitudes from 0.3–0.5 to 3.0–3.2 km a.s.l. MAAT ranges from −20 to −50 °C. Higher precipitation than in the intra-continental zone. Very strong katabatic winds due to gravity and snow drifting occurs.
3. Coastal zone, with altitudes less than 0.3–0.5 km a.s.l. MAAT from −10 to −20 °C. In summer the air temperature is close to 0 °C, annual precipitation about 400 mm. Fairly strong winds and frequently occurring drifting snow. Oases differ from the ice shelves in their meteorological regimes.
4. Near coastal drifting ice zone extending from the Antarctic coast to the ice edge. Cyclonic activity causes special climatic conditions. High precipitation. Low-level cloudiness and fogs are almost constant. The air temperature in summer is close to zero. In winter the temperature is slightly lower than at the coastal zone.

For the South Pole the mean annual air temperature is −49.2 °C and the isotherms seems to follow the contours very well. At 80° S latitude MAAT ranges between −27 °C and −57 °C. MAAT at the coast of Antarctica is about −10 °C or less except on the northern part of the Antarctic Peninsula. 0 °C MAAT is located far outside Antarctica on the ocean about 60° S (Weyant 1967: plate 1). Therefore the circumpolar north is much warmer than Antarctica. This can be seen from the monthly mean temperatures, too. The mean July temperature at the North Pole is about 0 °C (Fig. 2.5) and at the South Pole the January mean temperature is about −30 °C (Fig. 2.7). The astronomic definition and the statistical mean temperatures of different latitudes on the northern hemisphere are very schematic in many ways. The ocean currents and winds moderate the temperature differences between the high and low latitudes. For example, the Hudson Bay region 55° N is very Arctic in character compared with coastal areas of North Norway some 70° N, however, the latter is some 1700 km more north (Figs. 2.5 & 2.6).

Fig. 2.7 Mean July and January air temperatures in Antarctica. Modified after the *Polar Regions Atlas* (1978).

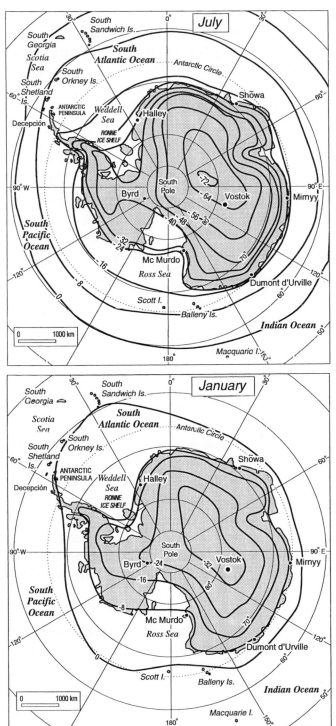

2.1.3   Precipitation

In general precipitation is low in cold climates. Most of the annual precipitation (total often less than 250 mm/a) for example in the Canadian Arctic is in the form of snow (Baird 1964: Fig. 23). In eastern Canadian Arctic and on the coastal areas of central and southern Greenland the annual precipitation is rather high, 350–400 mm (Fig. 2.6). Due to snowfall the precipitation may have been underestimated up to 40 or 50 per cent. The errors arise because snow does not lodge in really windy sites (Hare & Ritchie 1972: 362). As an example of the difficulties in precipitation recording we shall take isoplethic maps of snowfall and precipitation in Canada and Alaska (Hare & Hay 1974: Figs. 15 & 18) (Fig. 2.8). In the regions with mean annual snowfall of 100 cm the total annual precipitation is 10 or 20 cm (Arctic Islands and northern Alaska). The density of snow in the calculations is taken as 0.1 g cm^{-3} (Hare & Hay 1974: 93) which is a very low value because in the Arctic the density of snow is often 0.25–0.35 (Hare & Hay 1974: Fig. 19). Most of the total precipitation in cold environments often is in the form of snow which means that the values of total precipitation indicated for northern Canada and Alaska are "very approximate" as the authors considered for the isopleths of the snowfall (Hare & Hay 1974: Fig. 18). According to Zhang *et al.* (1996) snowfall forms about 40 to more than 50% of the total precipitation in northern Alaska. About one third of precipitation in northern Lapland is snowfall, too (Seppälä 1976a: 6).

In west Greenland the summer rains are dominating and winter is drier while on the east coast of Greenland the situation is opposite: winter is a more rainy season than the summer (Fig. 2.6). In Iceland north of Vatnajökull and in Svalbard rainfall is less than 320 mm. In north Greenland (in Peary Land 53 mm; Walter & Leith 1960) and Ellesmere Island and on some islands of the east Siberian coast the precipitation is very low (<100 mm) (Fig. 2.6). On the coast of the Beaufort Sea (e.g. Aklavik and Coppermine) the precipitation is more than twice as high as that in Barrow or Resolute (Fig. 2.6). The Norwegian coast is an exception in the high latitudes. Temperature is high in the winter time and precipitation is often three times higher than on the coast of Greenland due to the Gulf Stream. Often in the Arctic the maximum precipitation takes place during the short summer but in Tromsø, Norway, for example, the most rainy seasons are late autumn and winter (Fig. 2.6).

In Subarctic areas with reasonable aeolian activity as in Lapland the most rainy months are in the summer time (often July) (Seppälä 1976a).

In Antarctica practically all precipitation is in the form of snow. Precipitation in Antarctica is not measured as it falls, but rather as its total accumulation over a year. Snow accumulation is due to new snow, hoarfrost and drift snow. Snow storms in the Arctic and Antarctica occur even when no precipitation takes place. During a strong wind there is a lot of drift snow in the air hindering the visibility. Drift snow makes it extremely difficult to measure the total precipitation. The accurate measurement of fresh snowfall is quite impossible. The snow gauges are of dubious performance. The gauges may underestimate annual total precipitation, on average, by about 75%, ranging from 20% to 180% (Zhang *et al.* 1996: 513).

Fig. 2.8 Mean annual snowfall and precipitation (in cm), 1931–60, in Alaska and Canada. Source: Hare & Hay (1974: Figs. 15 & 18).

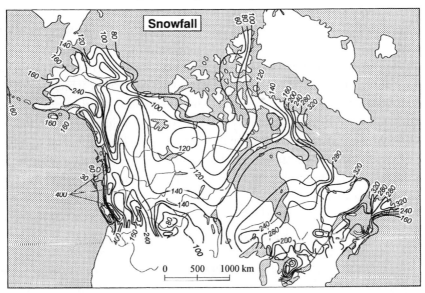

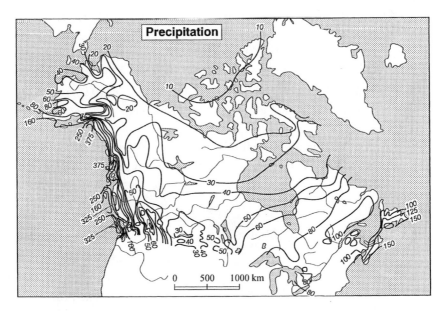

Decrease in snow surface height is a result of sublimation, ablation, deflation, and settling, which means that also the density of snow should be measured to find out the net accumulation. Therefore the direct amount of precipitation is not such an interesting matter in Antarctica. Instead the annual budget of snow is measured on the glaciers and ice shelves with snow gauges. It is difficult to know how representative these snow site measurements are regionally and in time (Loewe 1954: 9). For example, close to the coast of Antarctica in Adélie Land the approximation for the mean annual precipitation is about 30 cm of water (Loewe 1954: 8–9).

The oases areas of Antarctica (Fig. 2.9) have low precipitation and high sublimation which keeps them icefree. The dry-valley area close to McMurdo Sound in an environment otherwise dominated by snow and ice is due to several factors: (a) a greatly reduced inflow of glacier ice; (b) topography which makes low precipitation and hinders the access of windblown snow; and (c) greatly increased local temperature and ablation beside bare rock (Gunn & Warren 1962: 14). Miotke (1983) described one of the dry valleys, Taylor Valley, as colder than Siberia and very arid: "drier than the Sahara". The Vanda meteorological station recorded the annual precipitation under 100 mm and in 1970 the total was only 7 mm (Miotke 1983: 23).

Fig. 2.9 Antarctic ice sheet (white), ice shelves, oases and main nunataks. Source: Solopov (1967).

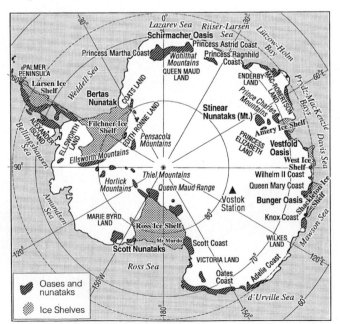

2.1.4 Humidity

Low temperatures during the winter make the environment very dry. Air has very low absolute humidity and plants above the snow surface are suffering frost drying because liquid water is missing. We can find for example some 1000 year old low table-like Juniper bushes in Finnish Lapland (Kallio *et al.* 1971), which have a thick stem but are less than 1 m high because all the branches growing above the maximum snow surface will dry. Also the height of many other bushes (willows, dwarf birches etc.) indicate the depth of snow cover (e.g. Clark *et al.* 1985).

Humidity is especially important when considering the aeolian activity. Low temperatures limit the water content of air and even in the Arctic deserts (Fig. 2.3) we find high *relative* humidity in the winter time which then minimizes evaporation. In the winter time the relative humidity can be in general above 100% (Fig. 2.10) but the absolute water content is very small. We have to keep in mind

Fig. 2.10 Mean relative air humidity (%) in Antarctica in July and in the western Arctic in January. Source: Zav'yalova (1987 Fig. 2).

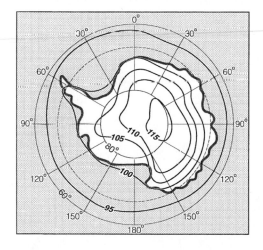

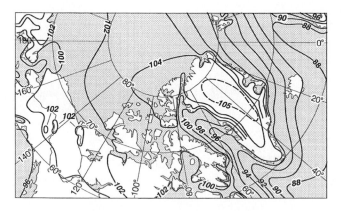

that already at 70° latitude the Polar night lasts some 54 days and towards the poles it gets still longer being 6 months at the poles. The lack of straight solar radiation has a great influence on the low evaporation, too.

Relative humidity in the middle of winter (July) is well above 100% in most parts of Antarctica except the eastern coastal regions (Fig. 2.10).

Looking for example at the relation of water deficiency to water excess in northern Canada we find that most regions are classified humid or subhumid (Fig. 2.11). The driest regions are found in western Canada in the rain shadow of the Rocky Mountains and in the Northwest Territories.

Precipitable water in the entire column of air over Canada and Alaska has been computed by Hay from radiosonde data (Hare & Hay 1974: 84). The precipitable water in the winter time should be considered when thinking of the humidity and possible sublimation. In Arctic Canada and northern Alaska the precipitable water in January is normally less than 4 mm (Fig. 2.12). In summer time the precipitable water in the entire column of air over Canada and Alaska is higher, ranging from 12 mm in northernmost Arctic to above 22 mm in coastal regions and the Great Lakes district (Hare & Hay 1974: 84–85).

Fig. 2.11 Moisture regions in Canada (Sanderson 1950; Hustich 1951: map).

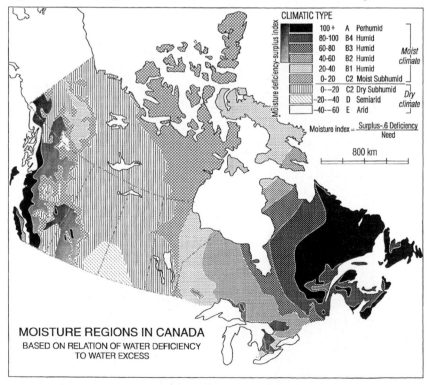

2.1.5 Mountain climate

Polar regions are not the only cold areas; non-polar mountain summits and also some mid-continental lowlands have very low temperatures in the winters. The geographical factors affecting mountain climates are latitude, continentality, altitude and the topography. Barry (1992) in his excellent book gives a covering overview on these features. It is not our task to repeat too many of these facts. The higher the latitude the lower the snow line occurs in general and this indicates the coldness in the mountains. Wind velocity over mountains is related to the topography rather than altitude. However, in middle and high latitudes there is an increase of wind speed with height due to the global westerly winds. In some locations, terrain configuration may increase wind speeds near the surface above those in the adjacent free air (Barry 1992: 60). Air temperature drops about 0.6 °C/100 m in the free atmosphere when the altitude increases. This does not mean that the temperature gradient is the same on mountains because the slopes cause uneven heating by radiation and uplifting of warm air. Gravity also slides dense air down the slope resulting in inversion especially in the winter time.

Troll (1944) delimited the mountain periglacial zone by the upper forest limit and the lower snow limit. In many cases the wind action is very strong in the upper forests, indicated by *Krummholz* formation (i.e. wind shaped trees)

Fig. 2.12 Distribution of the mean depth of precipitable water (mm) over Canada and Alaska in January. Source: Hare & Hay (1974: Fig. 13).

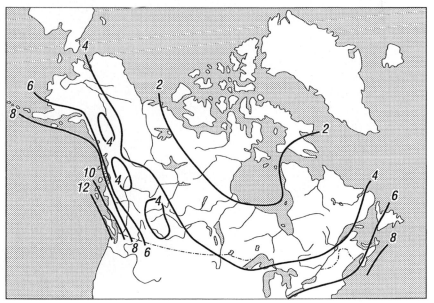

(Fig. 2.13). Seasonal snow accumulation (e.g. Clark *et al.* 1985) and surely the upper snow and glacial zone are important from the aeolian geomorphic point of view. When trying to delimit the cold regions with characteristic natural phenomena we can use features with clear limits or with less precise limitations which change from year to year. As indicators the regions of glaciers, annual snow cover, permafrost, seasonal frost and forest limit can be mapped.

2.2   Seasons

In the Arctic winters are cold and summers cool. The climate of the Arctic is not easily characterized in simple terms. Conventional seasonal names can be applied only in some restricted sense and their duration is irregular in space and in time, both within a year and from one year to the next according to Barry & Hare (1974: 17). They summarized seasons in land areas of the Arctic as follows:

Winter (6–8 months)   snow cover, some bare rock
Spring (1 month)   unstable snow cover, some melt ponds in south, bare rock

Fig. 2.13 Wind eroded trees (*Krummholz*), mountain hemlocks (*Tsuga mertensiana*), on the slope of Flattop Mtn., Anchorage, Alaska. All photographs by the author if not otherwise mentioned.

Summer (2–3 months)   wet and dry tundra, lakes, bare rock, snowbeds, glaciers
                      and ice caps
Autumn (1–2 months)   snow cover, lakes freezing, some bare rock

Spring especially can be very short. Snow melts fast, rivers are released from ice cover in a few days or even in hours and flowers are blooming very soon. Summer can be 'cancelled' and plants not produce flowers and seeds every year because of unfavourable conditions. Plants are mainly perennial. Snowing is possible also during the so-called summer. The moisture in the ground in early summer is very characteristic for the Arctic. Melting snow, increasing precipitation during the warm season and low evapotranspiration are the reasons for wetness.

2.3   Seasonal snow cover

In cold environments a large part of the precipitation is in the form of snow. It covers the ground for several months of each year. Annual snow cover in the northern hemisphere reaches the central parts of North America and Asia to 35° N latitude. New snow begins to deposit already in August in the north and reaches its maximum areal extent in January or February. According to the global grid-point tabulations with the snow depth accuracy of 1 mm by Schutz and Bregman (1988) I have interpolated 10 cm contour lines of snow depth of November and March (Figs. 2.14 & 2.15). The nodes of the grid are at intervals of 4° latitude from 2° N to the pole and of 5° longitude from the prime meridian east and west. This means that the grid is rather sparse in the south and, for example, in Iceland there is no observation point. The general trends according to these maps are that in the northern hemisphere there are three maximal areas of snow accumulation: Greenland, the West Siberian Plain and eastern Canada. With this accuracy small areas with heavy snow falls causing glaciers as in SE Alaska are not represented, as is also the case for west Norway. The Arctic Basin, covered by sea ice, and low latitudes in general have thin snow cover. Cold and dry snow provides good possibilities for wind transportation.

For example, in northern Alaska snowfall occurs throughout the year. During the summer months snowfall is relatively small. From October through May precipitation falls entirely by snowfall (Zhang *et al.* 1996: 514). Maximum monthly snowfall occurs in early winter, in October in the Arctic, and after that the cold air masses have lost most of their moisture (cf. Fig. 2.12), even the snow depth increases.

Snow cover tends to have some feedback effects on the conditions in the atmosphere. Because of its insulative and radiative properties snow cover increases the stability of the adjacent atmosphere. It has been noticed that in regions with seasonal snow cover in the autumn winds begin to decrease in strength approximately on the average date of occurence of first permanent snow cover, and in the spring, winds begin to increase about two months before the average date of disapperance of the snow cover (Berry 1981: 40–42).

The impact of snow cover on climate is mainly caused by its high albedo, small heat conductivity, heat expenditures on snow melting and the relatively small roughness at its surface. Together with the high radiating emissivity of snow this induces low surface temperatures and high temperature inversions above it (Kotliakov & Krenke 1982: 451).

2.4   Glaciers

Glaciers do not indicate only cold climatic conditions but a great amount of precipitation in snow form, permanent snow, and this means wind drift of

Fig. 2.14 Snow depth on the northern hemisphere in November. Compiled after the data from Schutz & Bregman (1988).

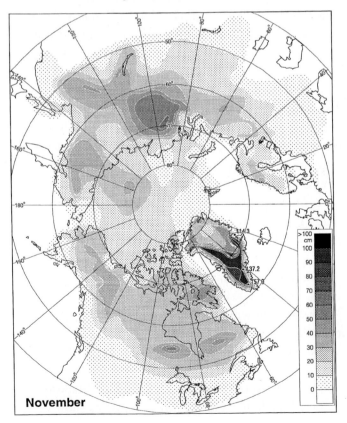

snow. Snow is a good indicator of a cold environment. Existence of a glacier indicates that in that certain part of the Earth's surface more snow has deposited than has melted during the previous years. The climatic reason for formation of glaciers can be a single one or a combination of several reasons supporting the ice formation. The thawing season might be so short that all the snow does not melt. Low temperatures are caused by high latitude and/or high altitude and the slope may be in shadow. The snow deposition may be locally so large that there is not enough energy to melt it. The snow drift by wind may be the primary reason for glacier formation. Local glaciation is often concentrated on one side of a mountain range (Evans 1977).

The areas covered by glaciers on the globe are well known and can be presented on maps (Figs. 2.9 & 2.16). The continuous Antarctic ice covers 13.9 million km^2 (grounded ice sheet, including ice rises, about 12.35 million km^2, and ice shelves 1.55 million km^2) and is by far the greatest ice mass

Fig. 2.15 Snow depth on the northern hemisphere in March. Compiled after the data from Schutz & Bregman (1988).

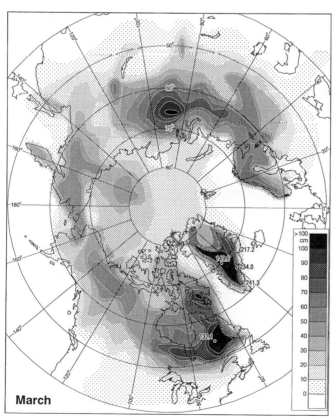

March

(91 volume%) on the Earth (Fox & Cooper 1994: 204). 95% of the Antarctic
continent is ice covered and only the remaining 0.62 million km^2 are ice-free cold
deserts (Miotke 1983: 22). From the southern hemisphere should be mentioned
other separate small glaciers in the Antarctic Peninsula, on Subantarctic islands,
on the Southern Alps of New Zealand and on the Andes in South America.

   When looking at the distribution of glaciers on the northern hemisphere
(Fig. 2.16) the dominant feature is the Greenland Ice Sheet (1.76 million km^2).
The great ice body covers Greenland except for a narrow belt along the coast.
The other large ice caps, ice fields and valley glaciers are mainly found in con-
nection with mountains in certain areas with high precipitation or short melting
season: eastern Canadian Arctic islands: Baffin Island, Axel Heiberg Island and
Ellesmere Island. Other large ice caps and glaciers exist on Iceland, Svalbard,
Novaya Zemlya, Franz Joszef Land and Severnaya Zemlya. Smaller, mainly val-
ley glaciers exist on several mountain regions: the Rocky Mountains in Alaska,
Alberta and British Columbia, Yukon Territory and the district of Mackenzie.

Fig. 2.16 Present glaciers and storm tracks in the northern hemisphere. Source: Flint
(1971: Figs. 4–8).

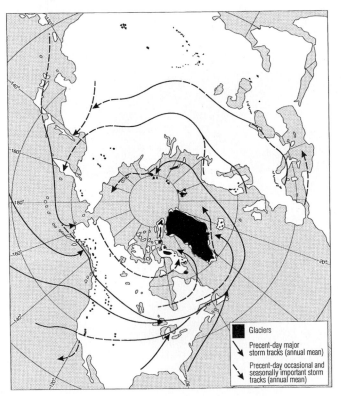

In continental Europe the major glacier occurences are in western Scandinavia, on the Alps and Caucasus (Fig. 2.16). In Asia can be found a great number of rather large glaciers on high mountains, eastern Siberia and Himalaya, which cover about as large an area as glaciers in continental North America, Europe and South America in total, but in this context we do not handle them in detail (see Flint 1971, Table 4-B).

During the Pleistocene great parts of the northern hemisphere were covered by ice sheets (Fig. 2.17). The deglaciation left vast amounts of sediment which were drifted by winds which were partly guided by the thawing ice sheets. But we have to keep in mind that during the Ice Age not all of the Arctic was ice-covered. For example, vast areas of central Siberia, the Arctic slope of Alaska, some areas of the Canadian Arctic Islands and Peary Land in north Greenland were largely uncovered by ice (Fig. 2.17). There the vegetation surely suffered from the cold and dryness, while deposits of glacial origin are missing.

Fig. 2.17 Maximum Pleistocene glacial regions and storm tracks in the northern hemisphere. Source: Flint (1971: Fig. 4–9).

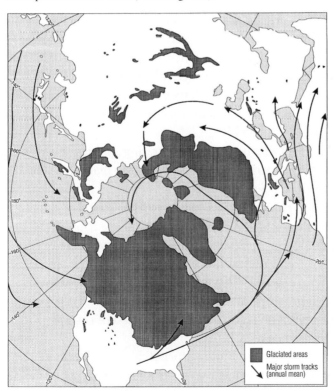

2.5   Soils

Surficial deposits in glaciated parts of the Arctic are young as they have only recently been exposed by the retreat of the ice-sheets. Around the present-day glaciers ablation is exposing new virgin soil surfaces which are not covered by vegetation. Typical cold climate deposits are frost shattered slope deposits, glacial drifts of different types (tills), glaciofluvially sorted gravels and sands, glaciolacustrine and marine silts and clays (e.g. Taber 1943). Very coarse tills and coarse stony gravels with pebbles are typical. The further from the centre of the glacial accumulation areas we move the finer are the sediments. Their origin is of course the weathering materials, eluvials of physical and chemical origin.

Clays are uncommon at the surface except for marine sediments which have become subaerial as a result of isostatic land upheaval. Tills contain some amounts of clays of preglacial origin and mixed and transported by glaciers.

Large amounts of fluvial flood deposits are characteristic for large river valleys. Arctic spring floods are violent. Large amounts of aeolian deposits, as loess, can be found in cold regions, for example in Alaska (e.g. Black 1951). In the river valleys some reworked fluvial deposits are common and in connection with them some aeolian loess can be found in favourable localities.

Unglaciated regions in cold environments, as parts of Yukon and Alaska as well as central or northern Siberia, are covered either by regolites due to chemical and physical weathering, and/or large amounts of fluvial deposits.

Physical weathering is intense because of temperature fluctuations around the freezing point. Arctic deserts are largely dominated by blocky weathering materials. Frost is also sorting mineral material in patterned ground which is often formed of mixed grained till.

The question has been whether the frost shattering could produce fine fractions, and in some laboratory experiments it has been shown that at least the silt fraction is produced (Lautridou & Seppälä 1986). Silt is rather common and because of its capacity to transport capillary water it is frost sensitive and supports ice segregation, frost heave and sorting.

Soil forming processes in cold environments are weak and very slow (e.g. Tedrow 1977) because of the short thawing season, limited amount of organic matter and low temperatures. That limits also the development of vegetation cover. However, large areas of the Subarctic and part of the Arctic also are covered by peat because of very slow decomposition of organic material on the surface.

The following factors are also limiting for deflation of soil: very short summers when the surface is uncovered and a very wet active layer above the permafrost table during the thawing season, and heterogeneous grain size composition. It takes a while before the surface is dried because the moisture cannot penetrate much downwards. It is often blocked by permafrost.

The composition of the soil microflora is specific in Arctic deserts: actinomycetes, non-sporeforming bacteria and blue-green algae predominate (Aleksandrova 1988: 17). They have their role also in stabilizing the deflation basins.

Chemical weathering and biological processes are restricted by cold and that means that the pedological processes are weak (Linell & Tedrow 1981: 53). Soil forming is activated especially by poor drainage usual in permafrost regions. This supports peat formation, and low temperature hinders the decomposition of organic matter, although bacteria of Arctic soils can grow in near-freezing conditions, and there is probably some activity even when the soils are in a frozen state (Tedrow 1977: 48–49). Some respiration from the soil microflora of northern Alaska has been recorded at temperatures as low as $-7\,^{\circ}\mathrm{C}$ (Benoit *et al.* 1972). Wet meadows are typical for the Arctic environments and they stabilize the landscape very well against deflation, for example. Frost has an important role in soil formation within Arctic and Subarctic environments. It forms also special uneven hummocky surfaces and patterned ground which increase the surface roughness, cause turbulence and limit aeolian processes in the summer and have their effects on drift snow distribution in the winter time (e.g. Sellmann *et al.* 1972). Soil formation on a tundra environment depends on the moisture conditions. Typical wet tundra soil is fine silty with an organic mat on a permafrost table. Details of tundra and Polar desert soils can be found in Linell & Tedrow (1981: 53–66). In tundra and the northern boreal zone we can find thick peat layers (e.g. Tarnocai 1978) which are also partly deflated by wind especially in the winter time. Soils at the boreal forest and low subarctic show strong morphological and chemical evidence of podzolisation (e.g. Jauhiainen 1969; Moore 1978).

According to Tedrow (1978: 413) typical polar desert soil consists of a deep, well-drained soil with A, B and C horizons, deep permafrost table, scattered vascular plants, and a desert pavement. Deposits of frost-shattered rock and recent landforms in well-drained sites show little or no soil profile development. The soils of Arctic regions have been presented in very considerable detail by Tedrow (1977). Our task is not to go deep into the pedological problems.

2.6   Permafrost

Permafrost means perennially frozen ground which is defined to be ground (soil or rock) that remains at or below $0\,^{\circ}\mathrm{C}$ for at least two years (*Glossary of Permafrost* 1988: 63). In a permafrost region the mean annual air temperature is at least 0 °C. The whole of Antarctica, 40–50% of Canada, 50% of Russia, 80% of Alaska and 99% of Greenland belong to permafrost regions (Washburn 1979: 22) as all the glacier covered areas are included, too. The southern border of sporadic permafrost (Heginbottom 1993) is a rather useful limit for the cold environment from the aeolian geomorphological point of view (Fig. 2.18). For

Fig. 2.18 Permafrost (Heginbottom 1993), forest limits (Hustich 1966; Tuhkanen 1980) and the geobotanical regions and subregions of the Arctic (Aleksandrova 1980).

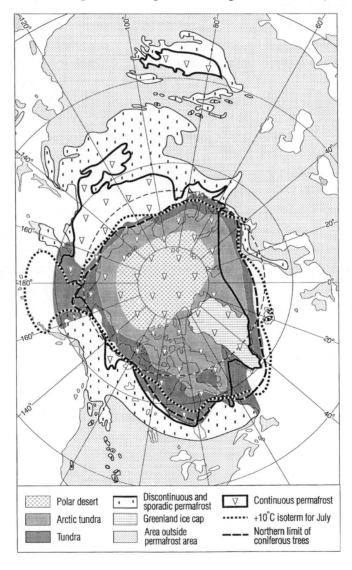

example, in Fennoscandia in the palsa region we can find also deflation of mineral material (Fig. 2.19). That limit is running from southern Alaska (ca 60° N) south of James Bay to southern Labrador just north of 50° N (Fig. 2.18). From northernmost Scandinavia (ca 67° N) the permafrost region spreads eastwards and in central Siberia its southern limit is close to 45° N turning somewhat northwards, and in Kamchatka it is about 55° N. In Mongolia and west China a separate discontinuous permafrost area exists between 40 and 27° N latitude.

Russian scientists, according to Jahn (1975: 5–6), speak about underground glaciation or the congelation, by which they mean an area commonly called the periglacial zone where temperatures below freezing prevail year round in the subsurface layers of the ground, in other words permafrost exists. We can roughly demarcate the boundary of the periglacial zone with the mean annual isotherm of −1 °C (Jahn 1975: 11). Frozen ground surface is as hard as rock and prevents sand drift almost totally.

We could limit this presentation to periglacial areas, neglecting the glacial regions, but then we forget an important characteristic, drifted snow which is an important part of the geomorphic story, because that has a key roll for mass

Fig. 2.19 Deflation area on Hietatievat, Finnish Lapland, (ca. 68° 26′ N) in the north boreal forest region in the zone of discontinuous permafrost with palsas.

balance of glaciers and for their location in mountainous regions. Permafrost does not mean unlimited wind action because of poor growing conditions of plants. Much vegetation is covering permafrost which also keeps the surface soil (active layer) moist and supports in this way the growth of the plants. The permafrost table does not let the soil moisture penetrate deep and form ground water, and the active layer often stays wet and immobile against deflation.

2.7   Seasonal frost

Seasonal frost which thaws away during the summer is a similar indicator of a cold environment to snow cover. According to my knowledge its distribution has not been mapped with any great accuracy. The seasonal frost can be very thin frozen layers of surface soil and last just overnight or it may be some metres thick and last most of the year depending on the weather conditions. Seasonal frost prevents the wind action for part of the year in the same way as permafrost. In middle latitudes it may also support the winter deflation when the surface layer of unvegetated sand may dry because of frost formation.

Long lasting seasonal frost is an effective limiting factor for plant growth and therefore it is sometimes important for the wind action, too. Both permafrost and seasonal frost may turn the soil surface and produce barren mineral surfaces uncovered by vegetation, and in this sense activate wind transportation.

2.8   Vegetation

Vegetation is the main limiting factor affecting the aeolian processes on Earth's surface. All kinds of vegetation effectively protect the ground from wind which is thus unable to pick up soil particles and drive them. Therefore a geomorphologist should pay reasonable attention to the different vegetation types and the possibilities of plants to colonize the sediment surfaces. The higher the vegetation cover the higher is the required threshold velocity of winds for effective eroding.

2.8.1   *Boreal forest zone*

The northern boreal tree line has often been considered as the southern limit of the Arctic (Figs. 2.1 & 2.18). In this cold environment the air temperature is so low that it limits the growth and development of the vegetation. As a result trees are growing slowly or they are missing totally (in so-called hemiarctic), the regeneration is limited and slow.

The aeolian geomorphic processes are sufficiently limited by the boreal forest zone (sometimes also called the Subarctic) and some external factors are needed to modify the conditions so as to be more favourable for aeolian landforms

(Fig. 2.19). The wind is one of the important factors responsible for the course of the forest-limits. Cold wind is drying up branches above the snow cover (e.g. Kihlman 1890). It has mechanical effects on the shape of trees (Fig. 2.20). Because of cold ground trees have flat roots. The wind fells considerably large numbers of trees (Hustich 1948: 10–11) and this can start deflation. Porsild (1929) reported from the Canadian Arctic: "The fact is everywhere apparent that in this country (Great Bear Lake) the cold and dry northeast wind determines the extent of tree growth. Where there is shelter from this wind a fair and sometimes even astonishing growth is found." (Citation from Hustich 1948: 11). Wind also deposits great amounts of rime ice on trees (Plate 1) and this may be one of the initial factors causing the orogenic tree line in Lapland (Fig. 2.21) and Labrador. In some recent studies about the effects of climatic parameters on the circumpolar forests the meaning of winds has been totally omitted (cf. Tuhkanen 1980, 1984).

Mean summer temperature is, of course, important for tree growth, but so is the length of growing season, which can be shortened by cold winds, for example at Great Whale River by Hudson Bay (Savile 1963: 95). Cold winds from the sea delay the spring warming, and though the effect is slightly reversed in late summer, the water is too cold to be of much value for warming.

Fig. 2.20 Wind formed juniper (*Juniperus communis*) forming a sand accumulation on a large deflation basin at Hietatievat, Finnish Lapland. Branches above the redeposited sand are only about 50 cm long. August 26, 1975.

Snow abrasion is a very important factor for the reforestation on the coastal areas and elsewhere with strong winds. Snow abrasion can prevent the growth of seedlings of spruce and the same thing can happen with sand drifting. All the seedlings close to blowouts are killed completely before there is any visible accumulation of sand at their bases (Savile 1963: 96). Savile concluded that snow abrasion increases with decreasing density of spruce at Great Whale River, and 'any factor that reduces the number of seedlings also endangers the survival of those that do occur'.

We can presume that trees hurt by sand and snow abrasion may be infected much easier by diseases such as rusts, which constitutes a serious drain of the vitality of seedlings and afterwards they are killed by abrasion or crowded out by other plants (Savile 1963: 97).

The northern limit of trees is located close to the southern limit of continuous permafrost (see Tarnocai 1978: Fig. 1) (Figs. 2.18 & 2.22). In this border region the annual snow cover lasts at least six months and the mean annual air temperature is somewhat below 0 °C. The forest limit should not be regarded as the limit of permafrost or aeolian activity. In the forest zone with a cold climate the wind action from a geomorphic point of view can be larger than in the forestless region

Fig. 2.21 Altitudinal forest limit formed by fell birch (*Betula pubescens*) on Muotkatunturit fells, Finnish Lapland (ca. 69°15′ N).

with a thick cover of moss. Vast coniferous forests grow on permafrost in North America and Siberia and these regions are often inactive with respect to wind action.

Typical boreal forest is a closed plant community (Fig. 2.23). In Siberia it is called taiga (a Yakutian term) which means a dense forest, difficult terrain to move in, often boggy. When going northwards we reach the forest limit. If deciduous forest forms the Arctic forest limit then it is less extensive than if coniferous forest does so. The further northwards we move the more scattered the trees get and lichen cover and shrubs increase and it is often called forest tundra (see Hustich 1951, 1953, 1966; Dansereau 1955: 85). Then we cross the tree-line and come to the tundra which does not bear trees and the vegetation consists of creeping shrubs, tufted grasses, lichens and mosses (Dansereau 1955: 86). Peat formation and deposition is characteristic for northern plant communities especially in moist habitats. A good cross-section of these habitats can be found around Hudson Bay (Moore 1978: Fig. 1) (Fig. 2.24). Lichen covered ground is very stable against aeolian processes, but if the cover is broken by feeding

Fig. 2.22 Major vegetation and permafrost zones in Canada, after Tarnocai (1978).

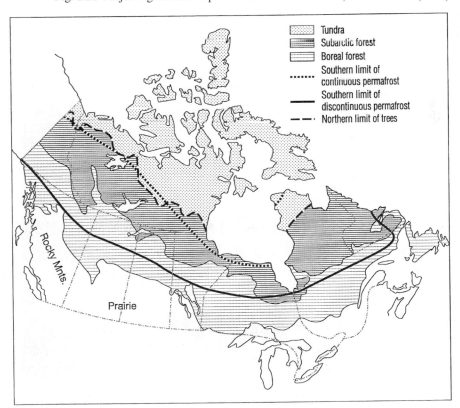

reindeer, for example, then it takes a long time to recover and deflation may start and keep the blowouts open for centuries (Fig. 2.19). Hustich (1953: 150) considered that the forest-tundra zone phytogeographically constitutes the Subarctic proper, and the polar tree-line represents the southern limit of the Arctic.

The Arctic tree-line in many regions is formed by larch (*Larix laricina*) together with dominating evergreen conifers such as spruce species (*Picea glauca* and *P. mariana*) in the North West Territories, south coast of Hudson Bay and Ungava-Labrador (on the Atlantic coast also fir *Abies balsamea*) (Hustich 1953: 151–156; Hare & Ritchie 1972). Close to the north coast of Russia and Siberia the tree-line species are *Larix sukatschewii*, *L. sibirica* and *L. dahurica* and in NE Siberia also pine, *Pinus pumila* (Hustich 1953: 155–157). Nowhere does the circumpolar pine forest (*Pinus silvestris*) reach so far north (70–71° N) as in the sheltered Norwegian fjords Alta and Porsanger and some river valleys in

Fig. 2.23 Typical vegetation of tundra, forest tundra and boreal forest. Wet areas hold aquatic and marsh vegetation, well drained areas bear climax mesophytic vegetation and dry areas such as dunes and cliffs support drought-resisting vegetation. Redrawn after Dansereau (1955: Fig. 22). © The American Geographical Society.

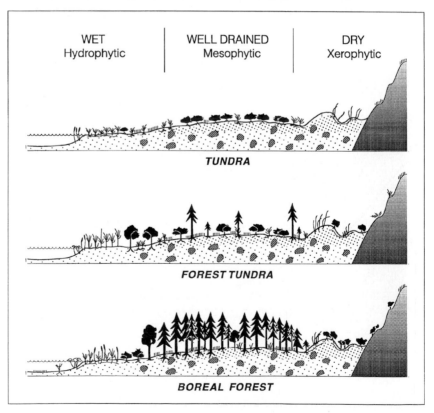

northernmost Finland (Hustich 1948: 15–16). In easternmost Siberia and northern Alaska broad-leaved trees *Betula*, *Populus* and *Chosenia* species are growing
at the Arctic tree-line (Hustich 1966: 19–25). In southern Greenland, Iceland,
northern Fennoscandia and the Kola Peninsula birch (*Betula*) forms the northern
forest tundra (Hustich 1966). All these regions are not sensitive for aeolian activity because many plants are prepared for cold seasons and are able to regenerate, too.

Forest fires belong to the natural regeneration of boreal forests (Fig. 2.25).
In the dry boreal forest in the Subarctic, as in northern prairies and southern

Fig. 2.24 Vegetation zones from the northern boreal to the Arctic lichen-heath of NE
Canada. Southern limits of continuous permafrost, widespread permafrost and
scattered permafrost indicated. Redrawn after Moore (1978).

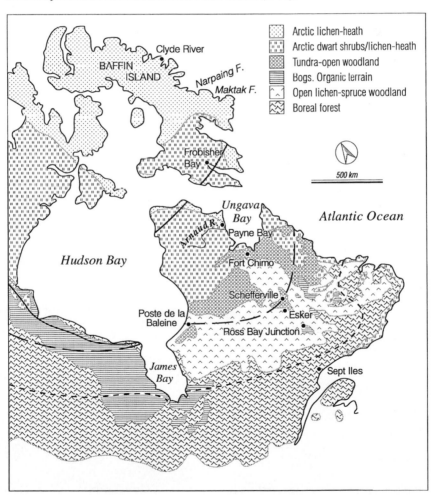

Northwest Territories, Canada, fire cycles of 80 to 150 years are common (Ritchie 1993: 110). In more maritime regions fire cycles are longer. In northern Lapland practically all the present pine forests have burnt at least three times in 800 years between 1160 and 1960 (Sirén, 1961: 35). Payette *et al.* (1989) demonstrated for Canada a south-to-north gradient of increasing fire-rotation period, reaching 1460 years in a shrubby forest-tundra zone.

As an example of the frequency of wild fires a remote sensing study of an area of 1000 × 1000 km across Saskatchewan and Manitoba in 1994 can be mentioned. The total burning area estimated is around 18 000 to 19 000 km^2. About 70 fires during the fire season were located mainly in the northern parts of boreal forests in so-called transitional forests bordering tundra (Li 1996).

If the forest fire takes place in a dry sandy area the regeneration of forest may take a longer time and the chances of deflation starting are obvious. Aeolian sand deposits reactivated by fire are characteristic for the northern parts of the boreal forest zone.

2.8.2   Arctic vegetation

Arctic lands even in the highest latitudes are not wholly devoid of plant life. Vegetation is subjected to severe conditions: dryness, short and cool

Fig. 2.25 Forest fire in the Mackenzie River valley close to Norman Wells in July, 1983.

summers, thin active layer of soil above the frozen ground, and thin snow cover which means exposure to frost. Plants get the greater part of their moisture from the snow melting, from the thawing frost, and from the coastal fogs. Most plants are perennial xerophytic forms with small leather-like leaves, succulent, hairlike coverings, tussock formation, low stature, and often very short stems. Bushes and even grasses often grow one-sided, as the windward side dries up (Samuelsson 1926; Nordenskjöld & Mecking 1928: 116) (Figs. 2.26 & 2.27).

Aleksandrova (1980: 9–14) gives a very comprehensive classification of the Arctic vegetation considering a complex of diagnostic characteristics of groups of species, the type of community under consideration, its structure and the life-form adapted to certain ecological conditions, so-called *ecobiomorphs*, e.g. *hekistothermal* dwarf shrubs or mosses, which means plants able to grow and propagate in average July temperatures ranging from 2 to 10 °C.

The main geobotanical regions and subregions of the Arctic are: (1) the tundra region; (2) the subregion of the Arctic tundras and (3) the region of the Polar deserts (Aleksandrova 1980: Fig. 1).

The tundra type of vegetation. Comprises various combinations of arctic shrubs, dwarf shrubs, herbaceous perennials, mosses and lichens. These

Fig. 2.26 *Empetrum hermaphroditum* growing on the lee side of a small aeolian sand accumulation on a deflation surface on a glaciofluvial delta at Peldojoki, Finnish Lapland. Effective wind from right to left. The scale is 1 m long.

plants belong to hekistotherms and microtherms and are mesophytes in their relation to moisture conditions. The tundra communities have cryogenic mosaic composition. Two subtypes are distinguished: Subarctic and Arctic tundra. Hyparctic shrubs *Betula nana* and *B. exilis* are present in the former but lacking from the Arctic tundra.

The polar deserts. Typical plant associations of lichen families (*Ochrolechia, Pertusaria, Collema, Cetraria* and *Stereocaulon rivulorum*), mosses (*Ditrichum flexicaule, Polytrichum alpinum*, etc.), and liverworts (*Cephaloziella arctica*, etc.) together with some hyperhekistothermal herbaceous plants (*Phippsia algida, Poa abbreviata, Papaver radiatum*).

2.8.3   *Tundra*

According to Passarge (1921: 44–45) the tundra zone (*Kältesteppen* in German) is covered by moss, lichens, shrubs, grasses, and dwarf and cushion plants (Fig. 2.27). To this zone, he considered, belong southern Greenland, Iceland, Orkney Island, the north coasts of Norway and Kola Peninsula, Kanin, the northern Petschora basin, the coastal areas of Siberia and of the Bering Sea together with Aleutes, west and north coasts of Alaska and Canada with some islands and the northeast coast of Labrador (Fig. 2.18). In Iceland and probably on Orkney Island the tundra is man-made.

Fig. 2.27 *Diapensia* bush on a deflated glaciofluvial delta surface close to Cape du Novelle France in the Ungava Peninsula, Arctic Canada. Effective wind from left to right.

In the southern hemisphere Falkland Islands, South Georgia, Kerguelen, Prince Edward and Crozet Island, Heard Island, St. Paul and Macquarie Islands belong to the tundra zone. Some of these regions are humid or dry having the mean air temperature of the coldest month above 0 °C (Kerguelen Island) and annual precipitation above 400 mm, but most of them have cold or very cold winters. The north Polar type has annual precipitation less than 400 mm, very cold winters ($< - 18$ °C) and short summers (2–4 months above 0 °C). Other subtypes with mild summers have winters that are mild (Falklands) or chilly (Iceland) or cold (Bering) and could be humid or dry (Siberia, Alaska, Canada) (Passarge 1921: 45–47).

Perrier (1982: 402) characterized the tundra as having a very cold climate (few months with a mean temperature above freezing), low annual precipitation (~200 mm) and strong winter winds, active layer (i.e. unfrozen soil) no more than 0.5 m which means excess surface waters creating mires. During most of the seasons (autumn, winter, spring) only sublimation may take place and during the short summer most of the surfaces remain wet and evapotranspiration is probably not far from potential evaporation. Perrier's definition of tundra is close to the Polar desert characteristics.

Aleksandrova (1980: 11–14) separates still 10 other vegetation types in non-zonal, non-mesic habitats: thickets of hyparctic shrubs, hekistotherm meadow-like communities, tundra-steppe and steppe types of vegetation, lichen and moss-lichen vegetation on boulder fields, homogeneous herb-moss tundra mires, hillocky and complex polygonal tundra mires, high-arctic 'mineral' mires without peat, meadows, open woodland, and *Krummholz* communities, but in this context we do not go into the details of these types.

On low Arctic tundra, for example Tuktoyaktuk, Mackenzie Delta, NWT, dominating plants are higher shrubs, dwarf willows and birches covering the mesic surfaces, but dry ridges and poorly drained sites will be different (Webber 1974).

On the Arctic tundra, for example on the north slope of Alaska, Ungava Peninsula, Spitsbergen, and large areas in Siberia, the plant cover is generally continuous, carpets of moss and lichen, and herbaceous prostrate shrubs. Tundra means a treeless area but not plantless. Wind action is possible only where the vegetation mantle is destroyed for some reason.

In the Arctic certain families of flora are circumpolar: dwarf willows (*Salix polaris, S. arctica, S. herbacea, S. reticulata*), cottonwool grasses (*Eriophorum*), sedges (*Carex*); the heath plants (*Rhododendron, Cassiope, Potentilla, Pyrola, Saxifraga, Draba, Vaccinium* and *Empetrum*). Of the mosses should be mentioned especially *Polytrichum* and *Sphagnum*. *Cladina* lichens are common.

Edlund (1993) published interesting conclusions from the High Arctic, Melville Island, NWT, Canada of the distribution of plant communities. He divided them into five bioclimatic zones which reflect the increasing impoverishment of vegetation with decreasing mean July temperatures. The zones are:

0: unvegetated, lichen free, no vascular plants;
1: entirely herbaceous, sparse, lowest diversity; mean July temperatures +3 °C and below;
2: herbaceous species dominant, local woody plants and sedges; temperatures from 3 to 4 °C;
3: woody species and one herbaceous species, or sedges, dominant, zone 2 herbs common; temperatures above 4 °C;
4: woody species and sedges dominant, greatest diversity, legumes, carices and compositae abundant; mean July temperatures near 5 °C or above.

According to Edlund (1993) the major factors controlling the distribution of vascular plants on Melville Island are chemistry, texture and moisture of surficial materials, and the intensity and duration of the thaw period.

Unvegetated areas are highly acidic or basic without nutrients or are physically unstable. Lichen-free surfaces suggest that the ice caps and snowfields expanded to more than six times their present size during the Little Ice Age. They are located at high elevations (>350–400 m) with extremely low temperatures, and too short snow-free periods for plants to establish themselves.

Barrens communities cover less than 20% of the surface and are common on well drained soils, such as gravels and coarse sands, on upper slopes and exposed knoll crests and ridges. These communities are the most common type of vascular plants on Melville Island. They are dominated by Dryas and Salix species. They do not give much protection against wind drift. The woody component of this community is confined to shallow depressions and runnels where snow accumulates.

Tundra communities nearly continuously cover the terrain with an extensive cryptogamic ground cover. They are generally confined to moderately or imperfectly drained soils on slopes which have have been under a persistent snow cover in winter. Typical species are: *Dryas integrifolia*, *Salix arctica*, *Saxifraga oppositifolia*, *Oxytropis arctica*, *Astragalus alpinus*, etc. *Cassiope tetragone* occurs on warm, sheltered south-facing sites with snow beds. Wind erosion has not much chance on tundra communities.

Wetland meadow communities on Melville Island are best developed on fine silt and sand with poor water penetration, often on the depressed centres of ice-wedge polygons. Dominating species are *Carex aquatilis* var. *stans* and to a lesser extent *Eriophorum triste* and *E. scheuchzeri*. On these surfaces wind is just transporting snow.

The Melville Island example is a very interesting one because this type of zonation can be found in the High Arctic in general and also in much wider areas. Vascular plant and cryptogamic communities are so covered and wet that wind is not able to erode or drift the soil. In connection with unvegetated areas Edlund (1993: 249) mentioned that deflation and deposition create local disruption of soils. He does not pay much attention to the protecting effect of snow cover and its meaning for the plants. Snow is mentioned as a source of moisture in soil which is important for the vegetation. The snowpack of the tundra is quite different from that of the other environments with trees and shrubs. The discontinuous shrub layer of the open woodland zone is a most efficient snow trap and the shrub species do not survive without snow cover (Hare & Ritchie 1972: 363). Microclimates are of great ecological importance in the Arctic, as pointed out by Corbet (1972), but it is not only the case with plants and animals but also the geomorphic processes: aeolian and frost action are highly connected with the very local physical conditions. For example, the subnival temperatures could be $0\,^{\circ}$C and only 100 cm above at the snow surface the temperature is $-40\,^{\circ}$C (Seppälä 1983a).

Around the northern subregion of the Antarctic Polar deserts is the region of the Subantarctic cushion plants which characterizes South Georgia, Prince Edward, Crozet, Kerguelen, Emerald, Heard and Macquarie Islands (Aleksandrova 1980: Fig. 23). The climate is always oceanic with cold summers and mild winters. A cryogenic mosaic of the vegetation is lacking. Herbaceous plants dominate forming tall cushions and thick mats, and together with covering mosses produce peat. Vegetation is different from the tundras of the northern hemisphere (Aleksandrova 1980: 171).

Tundra and boreal forests are very recent ecosystems and still under evolution in many places in glaciated regions, which can be seen from the evolution of pollen spectra of late glacial and postglacial time (see e.g. Huntley & Birks 1983). For example, spruce covered Finland from the east starting 5000 years ago and reaching the southwest coast about 3000 years ago (Aartolahti 1966). Most mires are less than 9000 years old and their present character as string bogs and palsas has formed during the last 2000–3000 years (see e.g. Seppälä & Koutaniemi 1985; Seppälä 1988). A minimum age of 4000 years for the development of complex tundra plant communities at Atkasook, Alaska is needed (Komárková & Webber 1980).

2.8.4 Cold deserts

The term cold deserts (*Kältewüsten* in German) was probably first used by Passarge (1921). He divided cold deserts into three main categories: (1) Polar deserts, (2) high deserts, and (3) Antarctic deserts. As Polar deserts he considered also ice deserts (glaciers, shelf and sea ice and glaciated land areas) (Passarge 1921:12–16). According to his definition (Passarge 1921: 1) they are regions where cold limits or completely annihilates the organic life. Vegetation cover does not play any role and scattered plants are mainly moss, lichens and some shrubs. In the northern hemisphere Passarge regarded as cold deserts only northeast Spitsbergen and Franz Josef Land and some islands in between them as well as the northeast corner of Greenland and all the glacier covered areas. This early limitation excluding the ice covered areas corresponds rather well with climatological analyses of north Polar deserts and the distribution figure evaluated according to the Turc evaporation formula (Bovis & Barry 1974: Fig. 2.4).

Characteristic features for Polar deserts are low air temperatures and very wide variability of temperatures day by day, month by month and even in the same months. As example Passarge (1921: 3) gives Snowhill, Grahams Land, Antarctica (64° 22′ S; 57° W) where on 6 August, 1902 the minimum temperature −41.4 °C was recorded, and the maximum of +9.2 °C was recorded on 5 August, 1903. Polar deserts can be defined with two climatological parameters: mean air temperature of the warmest month is less than 10 °C and mean annual precipitation is less than 250 mm (Péwé 1974: 33). In the northern hemisphere such areas (approximately 4.3 million km^2) (Fig. 2.3) include the northern slope of Alaska, northern coast of Canada and Siberia, most parts of the Canadian Arctic archipelago and northernmost Greenland (Péwé 1974: 33). In Antarctica the ice free areas about 600 000 km^2 (4% of the continental area) are also Polar deserts (Solopov 1967: 12). Ugolini (Ugolini *et al.* 1973: 247) searched for three days in a lateral valley of Beacon Valley, Victoria Land, Antarctica, but could find no signs of plants or insects. A real desert.

Passarge (1921: 44–45) separated the Polar deserts from the tundra zone. Compared with Passarge's (1921) definition Péwé (1974: 34) included in the Polar deserts rather wet treeless tundra regions of northern Alaska, Canada and Siberia. These areas are characterized by a thick mat of shrubs, grass, moss, and lichens and a great number of small lakes producing a swampy, surface-soaked desert. The northernmost islands of Canada, Peary Land of Greenland (Fig. 2.3) and the oases of Antarctica (Fig. 2.9) represent the vegetation-free arid areas characterized by a wind-swept, boulder-strewn landscape.

We have to keep in mind that the precipitation in Polar deserts is mainly in solid form and seasonally and perennially frozen ground affects geomorphic processes in the Polar deserts (Péwé 1974: 34). This means that in Polar deserts the ground could be saturated by water and permafrost is present.

In Siberia Polar deserts have thicker snow cover (>50 cm) which limits plant development and results in a more impoverished vegetation. Canadian Polar deserts are arid and there the vegetation benefits from snow cover, often less than 25 cm thick, since the snow may be the only source of moisture for the plants (Svoboda 1989: 316).

In Antarctica vegetation has practically no significance for aeolian processes. The whole continent except for the northern end of the Antarctic Peninsula belongs to the zone of the *southern Antarctic Polar deserts*. The vegetation covers just about 1% of the surface areas in the mountain oases and only a few per cent close to the coasts (Aleksandrova 1980: 182). The main limiting factor is dryness. The northern part of the peninsula and small islands to the east of it form the geobotanical region of *Antarctic Polar deserts*. That is much more humid, and the *Usnea-Andreaea* association forming cushions is the most often found in many places (Aleksandrova 1980: 177).

2.9 Some relevant ecological factors
2.9.1 Snow ecology

Snow plays a key role in the ecology of circumpolar and high-altitude regions (e.g. Aleksandrova 1988: 17). From the geomorphological point of view we should consider the factors affecting the vegetation which then either limits or supports the aeolian processes. Snow ecosystem functions at three critical levels that are defined by boundaries at the snow-air and snow-soil interfaces (Jones *et al.* 1994: 5):

(1) *Supranival* — above snow, including large plants such as trees, animals and the atmosphere;
(2) *Intranival* — within snow, including small plants such as shrubs, snow-pack properties, etc.
(3) *Subnival* — below snow, including small plants, mosses and lichens and the soil.

Snow itself is an ecosystem because the snow cover is a mediator among microorganisms, plants, animals, chemicals, atmosphere and soil. Snow functions as (Jones *et al.* 1994: 5–6):

(1) **Energy bank** — Snow stores and releases energy. It stores latent heat of fusion and sublimation and crystal bonding forces. The bonding forces are applied by atmospheric shear stress, and drifting snow-particle impact. The intake and release of energy at various times of the year thus make snow a variable habitat for intranivean organisms and are a cause of their migration within the snow environment.
(2) **Radiation shield** — Cold snow reflects most shortwave radiation and absorbs and re-emits most thermal infrared radiation. Its reflectance of

shortwave radiation is a critical characteristic of the global system. As snowmelt progresses, the snow cover reflects less shortwave radiation due to a change in its physical properties. This reflectance can be additionally reduced by sediment particles and/or microorganisms such as red snow algae.

(3) **Insulator** − As a porous medium with a large air content, snow has a high insulation capacity and plays an important role in protecting plants and other organisms from wind and severe winter temperatures, as well as in soil temperatures. Its insulation can result in strong temperature gradients, which fundamentally restructure the snow composition. In windswept areas specific organisms take advantage of enhanced snow cover insulation where vegetation is relatively dense.

(4) **Reservoir** − Snow is a reservoir for water, chemicals and organic debris, providing habitat and food sources for various life stages of microbes and small animals. The physical and chemical properties of snow, especially radiation penetration, gas content, temperature, wetness, porosity, pH, inorganic chemistry and organic debris content control intranivean biological activity and in turn are influenced by the behaviour of nivean organisms.

(5) **Transport medium** − Snow moves as a particulate flux as it is relocated by wind in open environments or intercepted by vegetation in forests. It moves as a vapour flux because of sublimation, resulting in transport to colder surfaces or to the atmosphere. During melt, snow moves as meltwater in preferential pathways within the snowpack to the soil or directly to streams and lakes or sea. These transport phenomena are taken advantage of by certain snow organisms.

Variations in snow conditions on the landscape scale, variation over tens of metres to kilometres, correlate strongly with the vegetation communities, elevation, slope, aspect, orography and exposure to the wind (Jones *et al.* 1994: 6). In Frans Josef Land, for example, in the Arctic–Atlantic desert areas close to sea level, the snow thickness ranges from 0 to 40 cm. Less snow falls on the coastal Polar deserts of Siberia where the thickness of the snow does not exceed 30–40 cm (Aleksandrova 1988: 18). Snow is blown off from raised sites and from exposed parts of slopes and accumulated in depressions and low-lying parts of slopes. The effect of snow cover on vegetation within the tundra and the Polar deserts is essentially different. In the Polar deserts the total amount of heat is much lower and the shortness of the vegetative period due to an only gradually disappearing snow cover can have catastrophic consequences for the vegetation. This does not occur in the tundra zone and the protective role of the snow, which is so distinctly evident there (especially in connection with shrubs), drops to practically nothing in the Polar deserts (Aleksandrova 1988: 18–19).

Thin snow cover is most satisfactory for the development of the vegetation within the Polar deserts, even if in part they are entirely free of any snow during the winter. The early disappearance of the snow from such localities is of primary importance. It extends the vegetative period and gives larger amount of total heat for the plants. Lichen covers do not like to be covered by snow (Aleksandrova 1988: 19). On Aleksandra Island (Frans Josef Land) the most favourable development of vegetation occurs where the thickness of snow is 25 cm and it disappears by 15–20 June. When snow cover is more than 1 m thick and stays until 25 July there is no sign of vegetation (25% of the total area is like this). Exceptions to this are the basins with thick snow cover which is absent for only two months, and they have rich mire vegetation and peat formation (Aleksandrova 1988: 19). Many places in polar areas have dry summers and there the snow drifts are important moisture sources for the vegetation.

Plant growth under thick snow cover is possible once snow is 50 to 80 cm deep. Soil heat is conducted upward, thawing the soil surface and raising the temperature to 0 °C. Especially blue wavelengths of light penetrate through deep snow, depending, of course, on the quality of snow (Salisbury 1985).

2.9.2 *Surface temperatures*

Of advantage to the vegetation in its short growth period are the steadiness of the summer temperature, the constancy of radiation, and the indirect effect of radiation from the ground. The temperature of the ground may be considerably higher than that of the air: in northeastern Greenland, with a July mean of 4.4 °C, the ground temperature was found to be 6 °C higher at a depth of 4 cm, and for the air between plant parts even 8.5 °C higher. For this reason tall trees are lacking, because they require a July mean air temperature of 10 °C, but their dwarf representatives are abundant, as they find the necessary warmth near the ground. The trunk of such a tree may be metres long, but it lies prone on the ground (Nordenskjöld & Mecking 1928: 116). This way they find also the protection of snow against frost drying, which causes wind-clipping of bushes of willows, birches and junipers (Fig. 2.20).

Rock and soil surfaces and the upper part of soil can reach high temperatures in bright sunshine in June and July in the northern hemisphere. We can give an example of temperatures from a deflation basin of a sand dune from Finnish Lapland (Fig. 2.28). Temperature on a barren sand surface rose to +40 °C and was even higher (to 45 °C) on the vegetated surface. In the same region the minimum winter temperatures range often from −35 to −45 °C and the mean monthly minimum can be close to −30 °C (van Vliet-Lanöe *et al.* 1993: Fig. 2). The drop of subsurface temperature in the blowout in the middle of a hot summer day can be $+39\,°C$ at the surface and decrease almost linearly from $+20$ to $+3\,°C$ from 20 cm below the surface to a depth of 130 cm in sand (Fig. 2.29). These measurements give us an impression of the conditions to which the plants growing on these sand dunes are exposed. Both in the summer and in the winter the plants have to be well prepared also against severe drying.

Fig. 2.28 Hourly maximum surface temperatures from 17 o'clock of 15th June to 9 o'clock of 17th June 1989 by a deflation basin at Hietatievat, Finnish Lapland. Air temperature measured in a shelter 2 m above the ground. Measured by B. Holmgren.

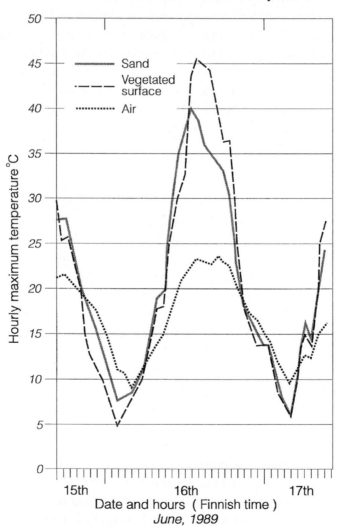

HIETATIEVAT Finnish Lapland

Fig. 2.29 Soil temperature of a deflation basin at Hietatievat, Finnish Lapland
measured from 20 cm above the surface to the depth of 130 cm on 18 June, 1974, at
noon. Air temperature 20 cm from the surface was +35 °C and in shadow at the same
level +30 °C.

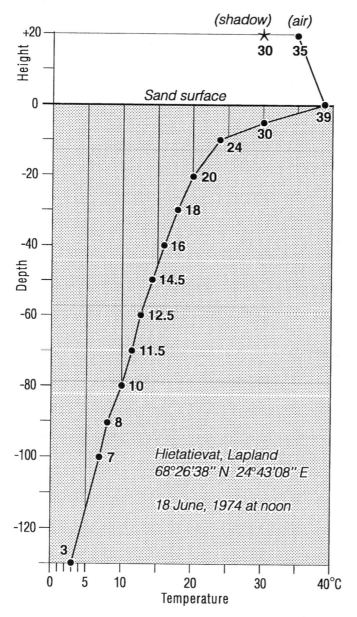

2.9.3   *Wind accumulating nutrients*

Wind drifted nutrients in the form of organic material are important for the small organisms and for the plants recolonizing the barren sand and silt surfaces in the Arctic and Alpine environments. Biologists speak of an aeolian zone and regions, and aeolian ecosystems (Swan 1963, 1967, 1992; Edwards 1987), which have a wide distribution in Alpine regions where wind is transporting nutrients. Wind-borne plant detritus is accumulated on the lee side of dunes and this supports the succession. Teeri & Barrett (1975) found windblown detritus in the High Arctic on Devon Island and concluded that it is an important source of nutrients in the terrestrial ecosystem. The same thing can be seen in almost every

Fig. 2.30 Profile of a snow patch, Green Harbour, Spitsbergen. Redrawn after Samuelsson (1926, Fig. 11).

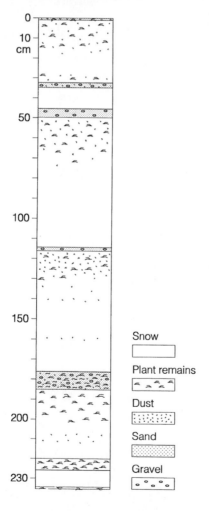

nivation hollow or basin in northern Fennoscandia and Spitsbergen (Samuelsson 1926) and Arctic Canada (Fig. 2.30). In the winter and on the thawing snow patches much organic material drifts in together with snow. Plant remains can form tens of cm thick layers in the nivation hollows. This is significant in environments which are short of nutrients, especially nitrogen.

3

General wind patterns in cold regions

The general circulation of the atmosphere is relatively independent of the surface. However, very large features such as mountains and oceans have their impact on the circulation. Winds close to the surface are greatly affected by surface roughness such as trees and other objects. That part of the atmosphere which is significantly affected by the direct influence of the surface is called the atmospheric boundary layer and it is about 1 km thick (Greeley & Iversen 1985: 40). Winds above this layer are independent of the surface and called geostrophic winds.

The direction and velocity of winds are recorded at the internationally recommended reference level for routine wind observations, 10 m above ground. From the geomorphic point of view the surface winds are the effective agents and the standard recordings do not reveal the importance of the surface roughness in this respect.

The characteristics of the wind near the Earth's surface are of major importance in determining its geomorphic impact. From the Earth's surface, the wind speed increases from zero upwards to its geostrophic value at the top of the boundary layer (Kind 1981: 339). The thickness of the boundary layer and the velocity profile of the wind depend on surface roughness. Over a rough terrain the boundary layer is thicker and the wind speed increases gradually with height. Winds in Polar areas are not much limited by friction at ground level because of quite even snow and ice surfaces and the lack of higher vegetation. In general in the open Arctic tundra, surface wind speeds in midwinter are about one-third greater than those of midsummer. Seasonal differences in wind speeds in the forested Subarctic are small because coniferous trees allow little seasonal difference in surface roughness. Wind speeds across the tundra are about one-half greater in midwinter than those of the Subarctic open forest, but only about one-fifth greater in midsummer (Rouse 1993: 80).

3.1 Arctic and Subarctic regions

In the winter time the sea level circulation in the Arctic is dominated by four large cells, two continental highs: the large, semipermanent Siberian high and the smaller Mackenzie Valley high, and two marine lows: the Icelandic low and the smaller Aleutian semipermanent low pressure, Fig. 3.1 (Dorsey 1951; Hare 1955). These vast cyclonic whirls produce resultant winds: SW flow over Europe, easterlies over Svalbard, and northerlies or NW winds over Canada. In North America the winter air pressure distribution leads to a contrast between the

Fig. 3.1 Mean sea-level air pressure in January and July with the positions of the Atlantic and Pacific Arctic frontal belts, redrawn after Dorsey (1951).

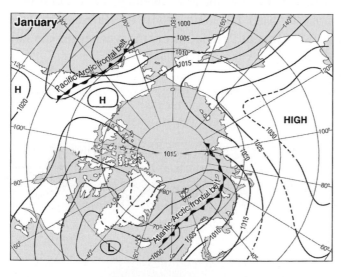

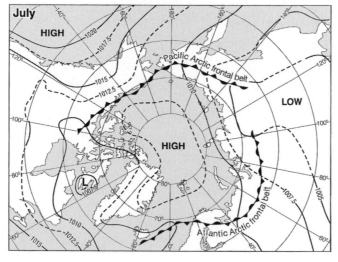

stormy east Arctic and the quieter climate of the Mackenzie and Yukon valleys (Hare 1955: 59–60). Southern Alaska and easternmost Siberia lie within the prevailing easterlies (Hare 1955: 59). In March there is pronounced weakening of the low cells and of the Siberian continental high, and the North American high shifts closer to the Pole. Then the strongest anticyclone activity is likely to occur over the central Arctic.

Storm tracks on the continents in the northern circumpolar region follow the latitudes quite well, moving from west to east, and over the oceans their routes turn to the left towards NE (Fig. 2.15). The direction of storm tracks does not indicate the direction of local wind which depends on the position of the cyclone. Winds circulate anticlockwise around the low pressure, which means that on the west side of the low pressure the north winds dominate and east of its centre the wind blows from the south. Cyclonic activity is greatest at the Arctic fronts (Fig. 3.1) where the Polar cold airmass meets the warmer middle latitude air.

Fig. 3.2 Winter cyclone frequency percentage expressed as the fraction of all occasions on which cyclonic centres occurred in the area. The shaded area indicates frozen sea. Redrawn after Petterssen (1950: Fig. 15).

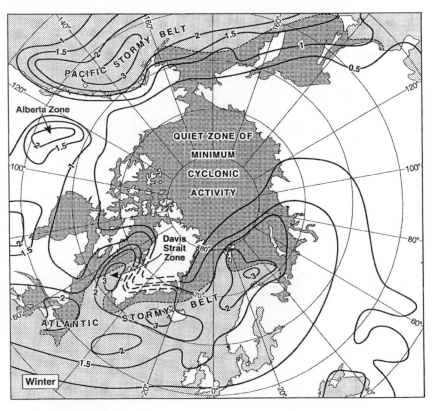

In the summer time these fronts are located close to the boreal forest limit (Figs. 2.17 & 3.1 and see Reed & Kunkel 1960; Hare 1968). The Arctic tree-line is located mainly somewhat north of the summer position of the Arctic front (Hare & Hay 1974: Fig. 3). Arctic frontal zones with maximum cyclonic activity (Fig. 3.1) cause cloudiness and a broad band of minimum net radiation which extends roughly parallel to the locus and 600 to 1000 km north of it (Hare 1968: 444). Distribution of cloud associated with the frontal belt is in turn reinforced by the outgoing surface radiation. The meaning of these climatic factors on the Arctic tree-line has not been explained in detail.

In the warm season with minimum thermal contrast between the polar and temperate zones (Fig. 2.5), the central Polar Ocean is under a feeble low pressure area (Dorsey 1951: 945). The Aleutian low fades, in summer, and the Icelandic low is also quite weak. The Siberian high disappears completely and is replaced by a thermal low pressure area over the heated Asiatic land mass (Fig. 3.1). The prevailing winds over the Eurasian coast are directed from the Polar Basin to the coast, with an easterly component, so that on the Siberian coast the winds are generally E-NE, while on the coast westwards of 90° E the wind is prevailing from the NE. By October the pressure patterns show a resemblance to the spring map. High pressure occurs over the central Polar Ocean, the sub-polar lows are back in their customary positions, the Siberian high attains moderate intensity and the pressure is building over the Canadian Archipelago, to the northeast of its winter-time position (Vowinckel & Orvig 1970: 210).

Stormy belts are located over the North Atlantic and Pacific (Figs. 3.2 & 3.3). The ice covered Arctic Ocean is a quiet zone with minimum cyclonic activity (Fig. 3.2). The Barents Sea between Svalbard and Novaya Zemlya is located at the Arctic front and is a very stormy area in winter time. In this region the frequency of the moderate winds (24 miles per hour, equal to about 10 m s^{-1}) is over 40% (Fig. 3.4).

In summer, when the northern continents are warm, the main temperature contrast is found also along the northern coasts. The main export of vorticity in the northern regions is then from sea to land, or from cold to warm ocean currents (Petterssen 1950: 134–135). Storms are less frequent on the cool oceans in the summer and their number increases on the warming continents (Fig. 3.3). Deep troughs of low pressure have great influence on mid-latitude cyclones. Atlantic cyclones follow the line of the oceanic North Atlantic Drift or move northward into Baffin Bay. The frequency of moderate winds during the summer months (June, July and August) is higher (10–30%) close to the coasts than in most places in the continents (Fig. 3.5).

In early autumn there is a strong thermal contrast between the Arctic, with already cold winter type air mass, and the more slowly cooling Subarctic continental mainland areas (Dorsey 1951: 949). Zonal westerlies dominate and cause strong storms at the coastal sections of Siberia and Arctic Canada.

Regions with dry climate and severe winters experience very strong winds at all seasons, due to depressions in which air masses of very different temperatures are superimposed in the lower levels of the atmosphere. In summer, these winds cause renewed attacks of frost and in winter the snow is swept away in blizzards (Tricart 1970: 19–20).

The most destructive storms in the far north are the fast-forming, quick-striking gales, so-called Arctic hurricanes (Businger 1991). They were not understood before scientists could see the cloud shapes revealed by satellites. According to Businger (1991: 18) 'these storms shared many of the characteristics of tropical hurricanes: highly symmetric, spiral cloud signatures, vigorous cumulonimbus clouds surrounding a clear "eye" and a band of strong winds reaching maximum strength at low altitudes close to the core'. These winds have a speed at least 30 metres per second. They are connected with a Polar low passing open

Fig. 3.3 Summer cyclone frequency. Redrawn after Petterssen (1950: Fig. 17). See Fig. 3.2 for explanation.

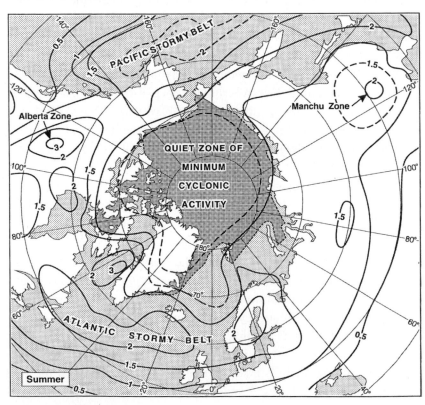

water uncovered by ice, and fluxes of heat at the sea surface play the dominant role in the structure and sustenance of the mature storm. Arctic hurricanes typical of the Polar winter occur from November through March. The reason for these storms is the very cold ($-40\ °C$) surface air which blows over ocean currents fed by relatively warm open water. This rapidly modifies its thermodynamic state by fluxes of heat and moisture from the sea surface (Businger 1991: 21). Energy stored in the air is transformed into the kinetic energy of the winds. Arctic hurricanes tend to make landfall quickly. However, while they are not very common features, they are important from the geomorphic point of view because they can move enormous amounts of material in a short while.

Fig. 3.4 Mean percentage of surface winds over 24 miles per hour ($10.7\ \mathrm{m\ s^{-1}}$) in the Arctic in December–February. Wind arrows fly with the prevailing 'all-speed' direction for the three month period. Redrawn after Hastings (1961: 21).

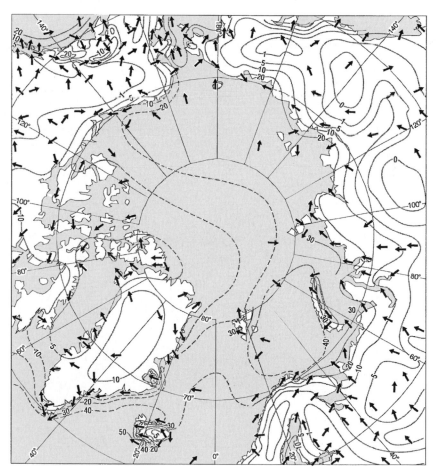

3.1.1   Russian and Siberian Arctic

A monsoon-like change in the direction of the wind characterizes the whole Asiatic Arctic zone: in winter winds from off the land predominate; in summer, winds from the sea (Nordenskjöld & Mecking 1928:167; Dewdney 1979 and Fig. 3.6). In the winter rapid cooling of the central Asian continent results in intense high pressure. The centre of the high pressure (above 1030 millibars) is located in the vicinity of Lake Baykal. A ridge of high pressure extends westward, roughly along 50° N latitude and from this ridge there is a pressure gradient northwestwards toward the Icelandic low pressure. The ridge acts as a wind

Fig. 3.5 Mean percentage of surface winds over 24 miles per hour (10.7 m s^{-1}) in the Arctic in June–August. Wind arrows indicate the prevailing direction for the three months. Redrawn after Hastings (1961: 22).

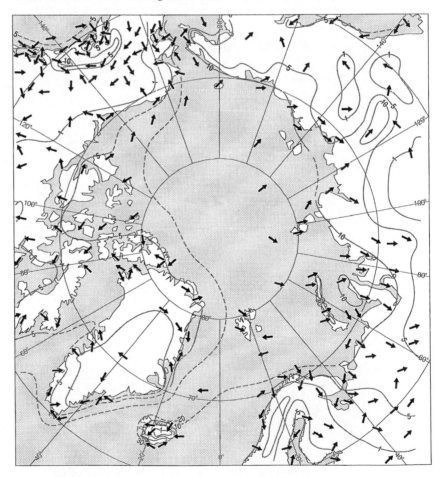

divide, winds to the north of the ridge being predominantly from the W or SW. Westerly winds are quite mild and winds from the east are cold. To the east of the Siberian high pressure there is a pressure gradient towards the low pressure of the Aleutian Islands and winds blow from N and NW (Dewdney 1979: 18). Pressure conditions in the summer are completely reversed. A low pressure is centred over Baluchistan and extends NE to the shores of the Arctic Ocean. Within Siberia the depression is shallow (minimum 1000 mb) and pressure gradients are gentle. Over the European plain and western Siberia, winds are from the W and NW, in Central Asia they blow from the N and NE. Along the shores of the Arctic easterly winds dominate (Dewdney 1979: 18; Fig. 3.6).

Atlantic warm water drift results in great variability in weather and higher mean winter temperatures in northern Europe. It means also cyclonic activity

Fig. 3.6 Air pressure conditions and prevailing winds in January and July in Russia. Redrawn after Dewdney (1979: Fig. 6).

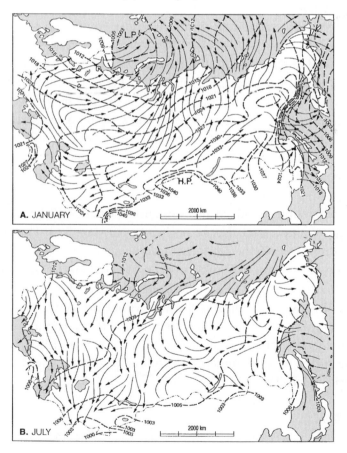

and storminess on the shores of the Barents and Kara Seas. The coasts of Novaya Zemlya suffer from the *bora* – strong winds of 100–120 km/h drifting great quantities of snow. On the western Barents Sea a strong easterly *bora* results when high pressure over the Kara Sea coincides with a vigorous depression in the southern Barents Sea (Baird 1964: 55). The cyclonic activity decreases during the summer and this means much cloudiness and fog in the Russian Arctic (Baird 1964: 59).

In winter the Laptev and Eastern Siberian Sea coast is really cold with little cyclonic activity. Winds are generally low but higher around the Lena delta. Further east the Siberian Arctic in the Chukhot Sea region gets effects of the Pacific and the climate becomes more comparable with neighbouring Alaska. The winters are windier as cyclonic activity increases. N and NE winds prevail carrying cold Arctic air and in the summer time SE winds keep the temperature cool (Baird 1964: 60).

3.1.2   North America

Baird (1964: 56) divided the North American Arctic into a stormy east and a 'Pacific' west. As a result of the NW Atlantic low the eastern parts of North America tend to have strong and persistent N and NW winds in winter. Winds are much lighter in the west, and blizzards in the west are infrequent. The Canadian definition for a blizzard requires the co-existence of 40 km/h winds, an air temperature equal to or below $-12\ °C$ and visibilities less than 0.8 km which last at least six hours (McKay & Gray 1981: 180). The frequency of strong surface winds (>24 miles per hour, equal to about $10.7\ \mathrm{m\ s^{-1}}$) in December–February is rather low ($<10\%$) in most parts of the Canadian Arctic and Alaska (Fig. 3.4).

The frequency of strong winds in June–August is lower ($<5\%$). Only on the western Canadian Arctic islands is it somewhat higher (Fig. 3.5). The direction of winds is then mainly from the sea to inland.

The early autumn thermal contrast between the cold Arctic air mass and still warm continental air causes cyclone deepening over the Mackenzie and Coppermine valleys, which completes the genesis of extreme types of autumn storms, which curve into the Arctic Islands or northern Baffin Bay and dominate the circulation probably for a week as they slowly fill (Dorsey 1951: 949).

In winter the Pacific Arctic frontal belt shifts southward lying over the interior of North America (Hare 1968: 444) (Fig. 3.1).

3.1.3   Greenland

The ice sheet of Greenland rising to more than 3000 m above sea level forms a continuous blocking high pressure which is a permanent hindrance to lows approaching from the west, which are therefore diverted northwards. The quite permanent outflow of cold air from the ice cap, the strong zonal temperature gradient, and the advection of warm air in the lower levels are conditions for this type of development (Putnins 1970: 5–6). Greenland is partly situated in the zone of west–east drift and therefore it has a great influence in the general atmospheric circulation (Figs. 3.2 & 3.3). Already Hobbs (1926: Fig. 29) showed that glacial anticyclones prevail on the continental ice sheets and slope winds similar to *Föhn* winds are blowing down from the ice plateau at the margins of Greenland (Figs. 3.4 & 3.5).

The Canadian Archipelago and Davis Strait are almost year round dominated by low pressure. The semi-permanent Icelandic low causes at the east coast of Greenland a nearly continuous outflow of Arctic air from the Polar Basin (Putnins 1970: 7). Cyclogenesis on the east coast takes place when the warm air crossing the ice cap meets cold Arctic air at the east coast (Putnins 1970: 6). When a real high is located over Greenland, lows pass Greenland far to the south. This is in contrast to the storm tracks, which are very close to the coast, during the presence of a shallow glacial anticyclone (Putnins 1970: 24).

Lows over NE Greenland and the eastern part of the Canadian Archipelago move with lower speeds than those over eastern Greenland and the Atlantic. The frequency of cyclones is at a maximum over the southern end of Greenland and also in Baffin Bay during January, April, July and October. The most frequently observed track for lows in the North Atlantic frontal zone is along Denmark Strait. In such cases a regeneration of the low usually takes place due to an influx of cold air from Greenland. The strongest cyclonic activity in the central Arctic is usually observed near the end of the summer season when the frontal zones are in this most northern position, and also in winter when the meridional exchange is quite intense (Putnins 1970: 22). The direction of the wind of north Greenland is invariably radial from the centre outward (Figs. 3.4 & 3.5), normally toward the nearest part of the coast-land ribbon (Peary 1898: 233).

3.2   Antarctica

Cyclonic storms form over the southern ocean around Antarctica as pearls in a chain and move clockwise around the continental coast (Fig. 3.7). This is explained by the temperature difference between cold land surrounded by appreciably warm water which causes a belt of low pressure with maximum cyclonic vorticity around Antarctica (Petterssen 1950: 133–134). Cyclones rarely penetrate deep inland except in the low-lying regions between the Weddell and Ross Seas. Because of the great strength of this circulation system, the ocean from about 40°S near the Antarctic Circle has the strongest sustained winds found anywhere on the Earth (*Polar Regions Atlas* 1978). On the Antarctic ice sheet, the continental anticyclone dominates with high pressure which makes the cold air drain down towards the coast. The centre of the anticyclone is located over the central part of East Antarctica near the Pole (about 85° S lat.) and the anticyclones are practically nonexistent over West Antarctica (Solopov 1967: 42–43).

The Antarctic 'oases' (Solopov 1967), as well as the whole coast of the continent, are characterized by cyclonic circulation. Cyclones observed in Antarctica form both on Polar and Antarctic fronts. Their formation is decisive for the meridional heat exchange between Polar and temperate latitudes (Solopov 1967: 43). Two of the cyclonic regions are located in the Bunger and Grearson Oases as well as at the Vestfold Oasis (Fig. 2.9) and when the cyclonic wind coincides with katabatic wind at the rear parts of the oases it causes fierce hurricanes (Solopov 1967: 44).

An overall picture of the flow of air near the surface was composed by Mather & Miller (1967) (Fig. 3.8). Flowlines come down radially from the east central highlands of the Antarctic continent towards the coasts following the topographic depressions. At the coasts winds circulate clockwise over the continent which means that very little moist air can penetrate the inner parts of Antarctica.

Fig. 3.7 Air pressure in Antarctica on 30 January, 1989 redrawn from the Russian meteorological satellite information for navigators. L means low pressure and H high pressure.

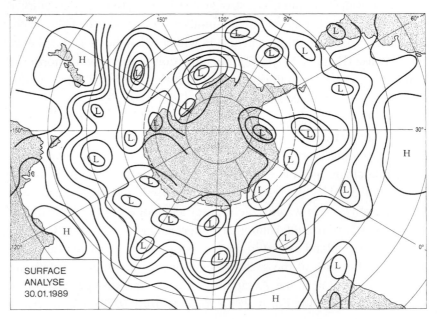

The profile of the Antarctic slope varies and this results in different intensities of the katabatic winds. For example, at the Burger Oasis with a slope of 0.012 no katabatic wind is observed (Solopov 1967: 14–15).

In Antarctica Cailleux (1963: 101) speaks about very violent winds, 180 km/h (50 m s^{-1}) but these are not surface winds drifting the material. They are high above the ground surface. His very general presentation is supported by many detailed observations.

At the Australian base Casey the average wind velocity is 20 km/h but blizzards with easterly winds 'can set in with very little warning and rapidly reach wind speeds well in excess of 150 km/h which can last for several days' (Potter 1987: 9–10). These winds occur at any time of the year and make travel extremely hazardous. From Mawson station Australians report the mean annual wind speed almost 40 km/h although gusts measure well above 180 km/h. 16 km inland from the coast the wind is much stronger. The winds usually increase after sunset and fall off around midday (Potter 1987: 19).

Fig. 3.8 General wind patterns in Antarctica. Redrawn after Mather & Miller (1967).

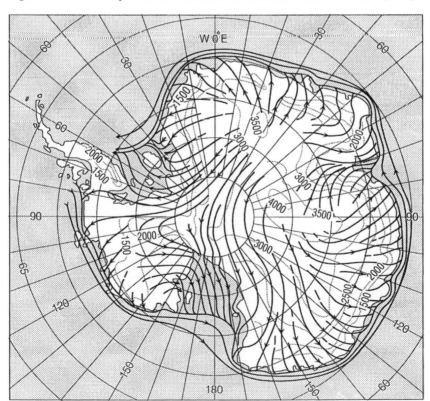

As an example of wind velocities we can present the measurements at McMurdo (77° 53′ S; 166° 44′ E, elevation 24 m), where the prevailing wind direction is year-round easterly and the monthly mean wind and maximum wind speeds have been recorded (Schwerdtfeger 1970: 337), as shown in Table 3.1.

At Halley Bay (75° 30′ S, 26° 39′ W, elevation 30 m) drifting or blow snow occurred every second day on average (Schwerdtfeger 1970: 338). At Hallett (72° 18′ S, 170° 19′ E, elevation 5 m) monthly means of wind speed are 2–3 m s^{-1} lower than in McMurdo (annual mean 3.6 m s^{-1}) but the peak gust measured was 51 m s^{-1}. There the relative frequency of occurrence of blowing snow was 7% of all weather observations (Schwerdtfeger 1970: 340). Respectively at Byrd (80° S, 120° W, elevation 1511 m) (annual mean wind speed 8.6 m s^{-1}) the relative frequency of occurrence of blowing snow was 29% of all weather observations (Schwerdtfeger 1970: 332).

The wind is the major climatic phenomenon in Adelie Land, East Antarctica (Fig. 3.9). Mawson (1915: 133) reported from Cape Denison, Adelie Land, Antarctica cyclonic gusts with momentary velocity of two hundred miles per hour (90 m s^{-1}) and a twenty-four hours' period with average velocity of 90.1 miles per hour (40 m s^{-1}) very constantly and from the same southerly direction (Mawson 1915: 117). The average velocity of the wind for autumn months was: March, 49 miles per hour (22 m s^{-1}), April, 51.5 miles per hour (23 m s^{-1}), and May, 60.7 miles per hour (27 m s^{-1}) (Mawson 1915: 134). At Cape Denison a

Table 3.1. *Mean and maximum wind speeds at McMurdo and Halley Bay.* Source: Schwerdt feger (1970: 337).

	McMurdo		Halley Bay	
	Mean speed m s^{-1}	Max. wind gust m s^{-1}	Mean speed m s^{-1}	Max. wind m s^{-1}
Jan.	5.3	24	4.5	31
Feb.	7.0	29	4.6	29
Mar.	7.3	27	4.9	38
April	6.1	28	5.0	35
May	6.9	43	4.8	35
June	7.2	43	4.6	40
July	6.5	36	5.1	39
Aug.	6.4	38	4.7	36
Sept.	6.9	41	5.4	35
Oct.	6.2	37	5.5	42
Nov.	5.4	35	4.4	34
Dec.	6.5	24	4.3	38
Annual	6.5	43	4.8	42

very high annual mean wind speed (>19 m s^{-1}) has been recorded (Meinardus 1938: Table 8).

At Port Martin, located 62 km west of Cape Denison (Fig. 3.9) the mean wind speed for March 1951 was 29.1 m s^{-1} (Périard & Pettré 1993: 313). Moreover, among the 32 years of data, a maximum wind, averaged over 2 minutes, of 90 m s^{-1} (corresponding to 324 km h^{-1}) has been recorded (Périard & Pettré 1993: 320).

From Adélie Land (67° 23′ S, 138° 43′ E, 1560 m above sea level) some 100 km from the coastline, 12.6 m s^{-1} annual wind speed with 0.94 wind direction constancy has been reported (Wendler 1987: 264).

3.3 Local winds

Winds with locally restricted influence and determined often by the topography are called local winds (Defant 1951: 655). Onshore and off-shore winds called land and sea breezes belong to the local winds, too, but they are not as typical for cold environments as mountain and valley winds based on air temperature differences and especially on density differences and gravity.

According to his observations in Spitsbergen and Greenland Samuelsson (1926: 86–87) grouped the local winds into the following types after their location:

(a) Winds of inland and cooling areas. Winds are irregular, no great differences in temperatures, air pressure or elevation, only weak airflows towards the outer limits of the region.

Fig. 3.9 Katabatic winds in Adelie Land, Antarctica. Contour interval 100 m. Redrawn after Mather & Miller (1966: Fig. 2).

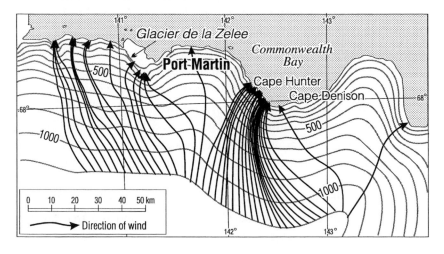

(b) Winds between cooling areas and valleys. Winds are often connected with snowing or more or less strong general flows and also with dry and warm winds with *Föhn* character.

(c) Winds of valleys and fjords. Wind directions are determined by the depressions, and flow direction is the long axis of the valleys. In the northern hemisphere wind erosion occurs especially on the right side and on the left in the southern hemisphere.

(d) Calming areas (*Auflösungsgebiet*) of the falling winds. These are often over the sea but the calming may take place also in the valleys. Strong winds never last far from the coast on the sea, because they are not canalized by land topography. Samuelsson (1926: 79) also noted that close to the edges of glaciers and on steep slopes the winds are strongest and transport a lot of snow.

Some local down-slope winds at some Antarctic coastal strips are of extraordinary strength and persistency. The average annual wind speed can reach as much as 20 m s^{-1}. More frequently such winds blow with surprising regularity during late morning and afternoon from one direction and then during the night and early morning from a nearly opposite direction. Such a diurnal cycle of winds indicates a thermal circulation produced by differential heating of cold and heat sources. In some cases orographical factors, such as narrowing of valleys, mountain gaps etc. are responsible for an intensification of winds from selected directions. In other cases the mere existence of large-scale relief causes a down-slope motion produced only by gravity (Flohn 1969: 139).

Mountainous areas with extreme relief, great thermal contrasts due to differential heating of slopes and the presence of glaciers often combine to produce complex local and regional wind regimes, which vary considerably from those anticipated from meteorological stations located adjacent to mountain barriers. On the continental interior margin of the Icefield Ranges near Kluane Lake, Yukon Territory, Canada, the summer winds show large velocity fluctuations and pulsations in the main glacial valley. The wind profiles display characteristics typical of the lower layers of a classic katabatic wind profile (Nickling & Brazel 1985). Diurnal cyclic motion in a down-valley direction has been observed. There was a distinct four hour period between lulls and peak winds down the valley at night. The daytime existence of these pulses and their long duration may be explained by the large glacier and cold air reservoir up-valley (Nickling & Brazel 1985: 130).

A variant of the thermal slope wind is the glacier wind. The continuous temperature difference between the glacier surface and the free air at the same altitude causes a glacier wind which blows downslope and reaches its maximum intensity and greatest vertical extent (50–400 m) in the early afternoon (Defant 1951: 663). Glacier wind, which has a strongly turbulent flow, fades out soon outside the glacier because its kinetic energy is dispersed by the ground friction (Defant 1951: 663).

Local wind directions affected by topography may differ very much from the prevailing wind directions. In Spitsbergen it has been often noticed that very local strong winds come down the valleys to the fjords. A common feature seems to be that some parts or a half of the fjord area is calm at the same time as other parts have a strong storm, depending on which valley the air is draining down to the coast. From Ice Fjord, West Spitsbergen we can find a typical example of the local winds guided by the valleys (Samuelsson 1926) (Fig. 3.10). Well known also are the intermittent gales on the fjords in Spitsbergen which were described already by Scoresby (1820: 402f) in the following way: 'The squalls continued from five minutes to half an hour at a time . . . a sudden calm

Fig. 3.10 Wind directions in Ice Fjord region, West Spitsbergen according to Samuelsson (1926: Fig. 2).

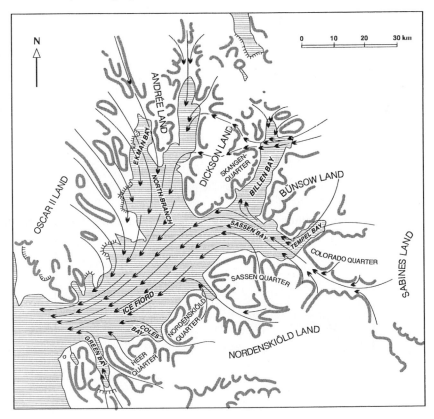

Fig. 3.11 Surface wind directions on icefields on the Southern Coast Mountains, some 120 km NW of Vancouver, British Columbia, Canada, interpreted with perennial snow drift features (Evans 1990: Fig. 3). Snow covered areas (white), mainly glaciers, are mapped with a satellite image NASA ERTS E-1385-18362-02, 12 AUG 73.

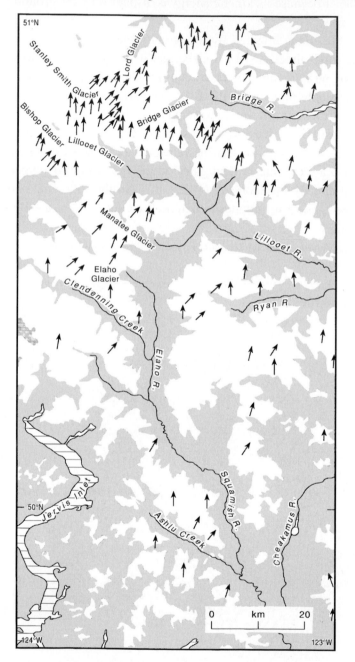

occurred and continued for an hour. The gale then suddenly recommenced with increased severity. . . . North, east, and south gales alternately prevailed, in rapid but irregular succession, during several hours.' The present author has seen how the southern side of Ice Fjord had a strong storm totally preventing boat traffic and at the same time the northern side was fully calm (cf. Nansen 1920: 139, 188; 1922: 190).

Icecaps and outlet glaciers have a strong influence on local wind patterns. Direct long-time measurements in these conditions are rare but aeolian drift features in snow reflect the direction of the locally effective surface winds (Fig. 3.11) which often turn off from the general wind pattern (e.g. Evans 1990).

Between Hofsjökull and Vatnajökull icecaps the gales have free play and that part of Iceland is very storm-harried and sand storms from this region have unhindered access to the lowlands through the large, flat-bottomed, barren valleys west of Hekla (Ashwell 1966) (Fig. 3.12).

In the Dry Valleys, Antarctica, wind directions are also greatly influenced by the local relief. For example, the ice-free valleys of McMurdo Oasis trend approximately east–west. According to Selby (1977: 949) 'up-valley and down-valley winds blow through the valleys, with easterly winds from the Ross Sea dominating their eastern ends and westerly winds from the Polar Plateau dominating their western ends.'

Fig. 3.12 Sketch of wave and lower circulations in Iceland with N-NE winds. Ice caps have a strong influence on wind directions. Redrawn after Ashwell (1966, Fig. 5). © Association of American Geographers, Blackwell Publishing.

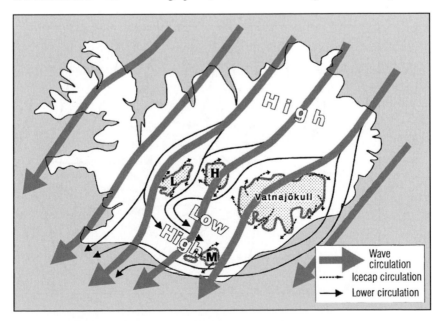

It is not only the valleys which cause very different wind patterns but also the mountains and hills. An example is the ground pattern of the prevailing NW wind on Kaimanawa Mountains, New Zealand (Fig. 3.13) which can turn more than 90 degrees when crossing the summit area (Zotov 1940). Similar observations have been presented from Mt. Pinneshiri (1100 m a.s.l.), Central Hokkaido, during winter monsoons by Nakamori (1994: Fig. 1), and Serra do Gerês, NW Portugal (Vieira 1999: Fig. 6).

Relief even with rather small relative heights (185–250 m) has strong impact on wind directions, as has been shown at Kevo, Finnish Lapland on an area about 2 by 3 km with deep fault valleys (Mansikkaniemi & Laitinen 1990). On the fell summit the wind directions differ rather significantly from the directions in the valleys which clearly guide the winds (Fig. 3.14).

With modern instruments we are able to record wind speeds and directions continuously or with short intervals. This means that we can collect much data. To handle them a technique has been developed to present the frequency distribution of winds (Seppälä 1977) without classifying the observations as has been necessary when using wind roses. With the frequency isopleth diagram (Fig. 3.15) we can find out the relations between winds and aeolian landforms much better and present the critical wind speeds and directions more precisely and compare different sites with each other (e.g. Seppälä 2002).

Fig. 3.13 Diagrammatic contour map of Mount Patutu (1731 m), Kaimanawa Mountains, New Zealand, showing ground pattern of the prevailing north-westerly winds (Zotov 1940: Fig. 6).

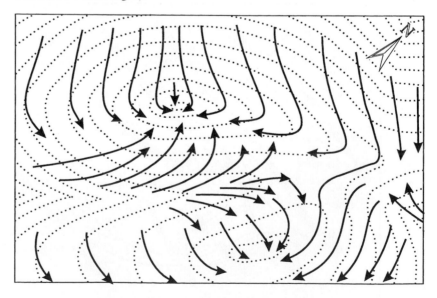

3.3.1 Föhn winds

A special wind begins with forced ascent of moist air against a mountain range. In the troposphere, ascending cloud air condenses water-vapour; due to the liberation of condensation heat, the cooling rate is only about 5–7°/km. In contrast to this, subsiding air warms at the usual dry-adiabatic lapse-rate about 10°/km. Since much of the water-vapour has been precipitated out during the ascending part of the motion, the descending air is desiccated and its relative humidity drops to desert-like values. Along many valleys of the northern side of the Alps in Switzerland and Tirol, this unusually warm and dry wind is known as *Föhn* (foehn) (Flohn 1969: 168). Similar descending warm winds are frequently observed at the lee-side of mountain ridges in the cold environments, too. At the eastern flank of the Rocky Mountains in Canada and the United States it is called the *Chinook* and in other regions it has many local names (Barry 1992: 145), e.g. *zonda* in Argentina (Buk & Trombotto 1995). At the windward side of

Fig. 3.14 Distribution of wind directions (%) at Kevo region, Finnish Lapland, from 6 June to 8 September, 1983. Percentage of the calm periods is indicated with a figure at the centre of the wind roses. Kevo is an official meteorological station. Jesnalvaara fell is the highest place in the vicinity and it rises above the tree-line. Data from Mansikkaniemi & Laitinen (1990).

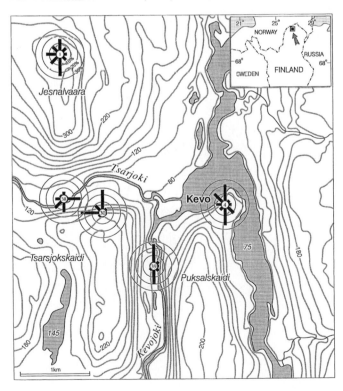

Fig. 3.15 Frequency isopleth diagrams indicating the direction and speed (see Seppälä 1977, 2002) of the official wind observations from Kevo meteorological station (see Fig. 3.14) and from Vaisjeäggi palsa mire some 10 km NE of Kevo and at 200 m higher elevations measured in September, 1975 on a palsa 2 m in height and 2 m above its surface. Percentage of calms is given in the centre of the figure.

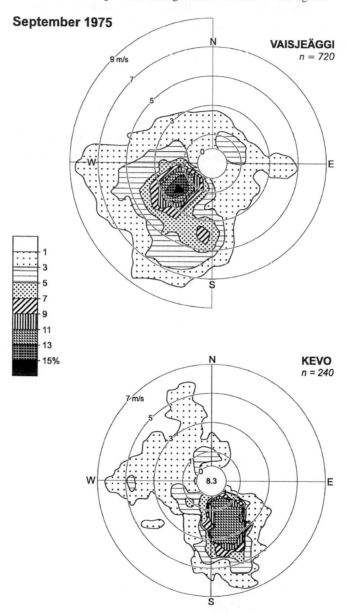

the mountain the weather is very rainy and the leeward side is characterized by dry conditions, e.g. Scandinavia, Japan and South Island in New Zealand. *Föhn* winds occur occasionally in Spitsbergen (Migala & Sobik 1984).

McGowan (1994; McGowan *et al.* 1995) measured near surface wind speeds of 30 to 50 m s^{-1} during *Föhn* windstorms at Lake Tekapo, New Zealand. A mean wind speed at 2 m height of 7.5 m s^{-1} was observed during *Föhn* events to entrain dust grains.

Nielsen (1933: 245) presented an idea: 'In arctic and sub-arctic regions the erosive force of the wind is not only governed by its strength, but just as much by its drying power, for moist winds are only slightly erosive, whereas dry, steady winds are capable of carrying even very considerable masses of earth away.' He based his conclusion on the observations of *Föhn* winds in Iceland where erosion is especially active in periods of descending winds of a *Föhn* character and as a consequence, the northwest–southeast axis of elevation divides Iceland into two regions which differ as regards aeolian landscape form: a northern part with south wind erosion and a southern part with north wind erosion (Nielsen 1933: 248).

In West Greenland the *Föhn* is most effective in winter, when the temperature at times rises from -20 °C to considerably above freezing. The absolute temperature maximum of the winter months has been recorded in December in Godthaab, $+15$ °C, and the temperature fluctuations and changes are then very large, while in summer they are small (Nordenskjöld & Mecking 1928: 240). On the coastal areas of Greenland are found also sea and land wind variations. In many hilly sites in the Arctic the coastal gales are notorious (Hare & Hay 1974: 65). Often they are the canalized type of winds and reinforced by katabatic effects (Section 3.3.2).

Bora is a cold, dry and gusty wind that blows in winter over the Dalmatian Mountains of Yugoslavia towards the Adriatic Sea. The same term is used also elsewhere as on the Caucasian shore of the Black Sea and on Novaya Zemlya. Favourable conditions for *bora* winds are a short topographic drop and a sharp climatic division between a cold plateau and a warm plain. Under suitable pressure conditions cascading of the chilled continental air can take place (Defant 1951: 670). Wind speed in gusts can be 40 m s^{-1}. Temperatures are around freezing on the coast and relative humidity may fall below 40% during anticyclonic *bora* (Barry 1992: 149).

3.3.2 Katabatic winds

Some of the strongest polar winds are katabatic or downslope winds blowing off icecaps. Gravity or katabatic winds are common in Antarctica and Greenland where the ice sheets dip towards the coast. Cold, heavy air tends to drain downslope and forms katabatic winds which affect only the lower layer about 100 m thick. Katabatic winds are especially strong near the foot of the plateau where the ground surface is steepest around the margins of the ice sheet, and are most persistent where a surface depression or valley favours channelling of the air drainage (Mather & Miller 1967). The wind on the ice cap in Greenland is generally a katabatic one (Putnins 1970: 33). The wind and the temperature gradient should be related. The strongest wind is observed when the temperature differences are the largest.

Katabatic wind is distinguished from the nocturnal downslope drainage of a heavier, cold air mass (Barry 1992: 156). During the night, the air near the surface is cooled by the long-wave radiation of the soil. Cold air flows downslope, driven partly by the density gradient directed towards the relatively warm air off the slope, but mainly by its own weight, due to gravity. These can be as more or less regular avalanches of air moving in 5 to 30 minutes pulsations down the slope (Flohn 1969: 157).

Aeolian processes are locally important where katabatic and regional winds are funnelled by the long narrow fjords as on Baffin Island. They may reach velocities in excess of 150 km h^{-1} (40 m s^{-1}) especially during the winter (Gilbert 1983: 164).

Katabatic winds may be observed with remote sensing by using the Advanced Very High Resolution Radiometer (AVHRR) carried on board NOAA satellites. Observations can be made by two different methods: the detection of thermal signatures, and the detection of snow lineation features on the snow surface (Cotton & Michael 1994: 537). Katabatic winds caused warm thermal signatures in the imagery, probably being due to a combination of adiabatic warming and the destruction of an inversion layer. On the snow surface the winds produce detectable 'flow lines' which are believed to be either lines of blowing snow or linear snow dunes (Cotton & Michael 1994: 537–538).

Ohata (1989a; 1989b) has described and studied special types of katabatic winds due to the snow and ice masses. He calls them downslope katabatic winds which occur when the ambient air temperature over the snow and ice is higher than 0 °C, and the air above the surface cools because the sensible heat is transported to the surface for melting of snow and ice. This type of wind is also called glacier wind or snow patch wind. These winds have rather limited geomorphic effects because they blow mainly on ablating glaciers and do not enter far (1.5 to 2 km only) from them and their maximum speed is about 5 m s^{-1}. There is a positive correlation between the wind speed and the air temperature difference between sites on the glacier and the nearby ground surface (Ohata 1989a: 111).

3.4   Winds and temperature

The connection between wind and temperature is interesting. 'As a rule temperatures in the Arctic regions are lowest during the calm, clear weather; and generally the stronger the winds the warmer it is, even though storm winds from the polar side are sometimes followed by lower temperatures.' (Nordenskjöld & Mecking 1928: 18). This author has noticed the same feature, that in Fennoscandia the extreme cold weather is windless but often in southern Canada cold weather is combined with strong winds from the Arctic and that is why the 'windchill' effect is regularly reported in the official weather forecasts.

Eastern edges of cyclones at the Arctic front cause southerly winds which can be mild and moist. When they have passed eastwards the winds turn to the north and get cold and drier.

When an Arctic hurricane passes, the sea-level pressure drops drastically accompanied by a sharp spike in temperature (Businger 1991: Fig. 2).

In Antarctica it was different: 'during the storms the mean temperature was more than 2° below normal, while during a calm it was above normal. The stronger the wind, the lower the temperature sank.' (Nordenskjöld & Mecking 1928: 18). The reason for the strong storm from inland to the coast might be the temperature difference between the very cold inland ice and the adjacent sea with its low air pressure. The relief might also be a reason (Nordenskjöld & Mecking 1928: 18).

4

Wind drift of mineral material

Among the studies of fluid dynamics concerning gas-solid mixtures the pioneer work was Bagnold's (1941) book on blown sand. Since then several authors have considered sediment transport by wind (e.g. Chepil 1945–46; Zingg 1953; Kuhlman 1958; Belly 1962; Kawamura 1964; Anderson & Hallet 1986). Comprehensive reviews of this topic can be found in many recently published books (Greeley & Iversen 1985; Pye & Tsoar 1990; Easterbrook 1993; McEwan & Willetts 1993; Nickling 1994). In this text we shall repeat some basic principals which are also basic facts for the understanding of snow drift.

Geomorphologically effective wind in nature is always a turbulent air flow, not a laminar one, which means in fluid dynamics that it has shearing force or drag and eddies and the velocity at any point varies as the logarithm of the height (Bagnold 1941: 44). The forces affecting the Earth's surface change all the time and the effects for winds are difficult to treat from the theoretical or practical point of view.

When sand starts to move, the air above the sand surface behaves as a heavy and non-homogeneous fluid, so that the wind velocity distribution is altered by the sand movement (Belly 1962: 1). There is a maximum at some height (some 5 mm) above the surface in the wind velocity profile which is caused by the blown sand (McEwan 1993).

4.1   Sand drift mechanisms

The type of grain movement depends on the wind velocity and grain size. The particles are moving either by creeping, in saltation or in suspension (Fig. 4.1). This is why the grain size of deposited sediment does not reveal the wind speed unambiguously. The characteristics of a sand grain population have been used for palaeoenvironmental interpretations of aeolian sediments, which is difficult because we do not know which transport mode the deposited sediment followed.

When sand size particles with diameter ranging from 0.06 to 2 mm are set in motion by pressure of a turbulent wind they initially roll along the sand surface but suddenly some of them jump up from the ground surface and move by bouncing. This mode of transportation is termed *saltation*. Grains are carried forward a small distance in the airstream in flat trajectories 5 to 10 cm above the surface. Immediately when the sand grains start to move it is not only the wind energy affecting their transport but also the impact of high-speed saltating grains which can impel a surface grain six times greater in diameter and more than 200 times greater in weight (Bagnold 1941: 35). This makes the measurement of the velocity of turbulent wind rather questionable. The energy load at a certain point is not only a function of the wind at that point but also what has taken place upstream when the grains have started to move.

Saltation is sustained through dislodgement of further particles by the impact of descending grains. The rate of travel of sand grains is about half the wind velocity measured at 1 m. Saltation normally accounts for about 95% of the bulk transport of sand in dune formation and involves mainly the fractions between 0.15 and 0.25 mm, with an upper limit of 0.5 mm (Mabbutt 1977: 218).

Experiments show that the downward velocity of the grain is very near to the terminal velocity when it hits the ground. By this time the forward velocity of the grain is nearly that of the wind (Raudkivi 1990: 51). Hence the angle at which the grain hits the ground is given by tan β = terminal velocity of grain/wind velocity.

Fig. 4.1 Threshold velocities of winds for various aeolian transport modes in relation to the grain-size for homogeneous groups of ball-shaped quartz grains. Wind velocities are at a height of 10 cm above the surface V_{10}, and the friction velocity V^*. Compiled by Kuhlman (1960).

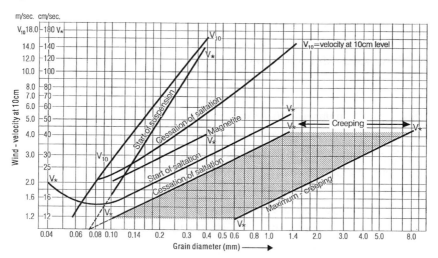

Observations show that the angle of impact remains remarkably constant over a wide range of conditions at $10° < \beta < 16°$ (Bagnold 1941: 16).

The grain in saltation strikes the surface at a flat angle. A portion of its energy is passed on to grains which are ejected upwards as it hits the loose surface. The rest of the energy is transmitted to the loose surface in disturbing many grains. This continued bombardment results in a slow forward creep of the grains composing the surface. Grains creeping on the surface are not affected directly by wind and receive their momentum by impact from saltating grains. The surface creep amounts to about one-quarter to one-fifth of the total transport, but it plays a very important part of sand movement by wind. Saltating grains and those in surface creep differ in speed. Surface creep is responsible for the changes in the size grading of sand deposits (Raudkivi 1990: 52).

Wind has to reach a threshold velocity before transport occurs. The threshold wind speed at which particle motion just begins is a critical parameter of any accurate formulation that expresses the transport rate of heavy particles such as sand, soil or snow as a function of horizontal wind speed. Wind near the threshold speed for particle movement is turbulent from over a rough surface. Threshold wind speeds are usually reported as average values measured at a specified height for a certain period, but the maxima or gusts actually initiate particle motion. These horizontal velocity maxima and the vertical component of the turbulent wind transfer the momentum to the surface (Schmidt 1980: 453).

Bagnold (1941) differentiated between the 'fluid' and 'impact' thresholds. When the fluid forces become large enough, small grains are removed from the surface, after that the surface is stable at the wind velocity. Increasing wind velocity will move the next larger grains and so on, until only the largest grains remain exposed on the surface. Above this velocity the motion of the sand bed is continuous.

When wind reaches a critical velocity over loose, dry sand, grains begin to roll and accelerate and after a few centimetres may jump almost vertically into the air, travel many times their diameter, and finally return to the surface in a long parabolic path. Successive jumps of a grain are called saltation. The saltating sand of sand storms, in contrast to the silt and clay, has a clearly defined upper limit, usually about a metre. As a rough approximation 540 cm s^{-1} has been suggested as the minimum velocity needed to entrain dry sand. When a grain returns to the surface it may bounce back up, its impact may entrain another grain, or the grain may simply roll forward.

Saltation of particles creates additional forces when moving particles impact those at rest on the surface. This disturbs the sand surface and then the wind-speed needed for sand drift is 20% lower than the fluid threshold and Bagnold (1941: 94) called this the 'impact threshold'.

Loose particles on a surface experience a horizontal drag force and a vertical lift force from wind. When these forces exceed the gravitational force and cohesion holding the particle on the surface, then the particle starts to move. The threshold velocity or the threshold shear velocity (u_t^*) for any size of particle is the velocity at which it starts to move. Sokolów (1894) gave some threshold velocities required to transport quartz grains of different sizes (Table 4.1).

Pernarowski's threshold velocities (Table 4.1) are systematically somewhat lower than Sokolów's ones. This is not unusual since many investigators have given divergent values and the reason might be, for example, the grain shape and angularity of the quartz grains.

Particles smaller than 0.15 mm are carried in *suspension* close to the surface. The coarse silt fraction 0.05–0.15 mm moves as a suspension a few metres above the ground but is deposited with the saltation load immediately the wind slackens. Particles smaller than 0.05 mm form dust clouds which can extend to a great height and stay as a suspension for hours after a sand storm (Mabbutt 1977: 218). Suspended fine grains may remain aloft for long periods of time. The finest particles in dust storms can be drifted thousands of km, for example crossing the Atlantic Ocean from the Sahara to the SE United States (Schütz *et al.* 1981) (Table 4.2).

Table 4.1. *Threshold velocities required to transport quartz grains of different sizes according to Sokolów (1894: 12) and Pernarowski (1959: 53)*

Grain size mm	Wind velocity 10 cm above the surface m s^{-1}	
	Sokolów	Pernarowski
0.25	4.5–6.7	3.9–5.2
0.5	6.7–8.4	5.5–7.5
1.0	9.8–11.4	8.0–10.4
1.5	11.4–13.0	9.7–12.8

Table 4.2. *Calculated and corrected flight time and height of particles for a wind of 15 m s^{-1} (Committee on Sedimentation 1965: 276, table 2–1.3)*

Diameter mm	Fall velocity cm/s	Flight time	Distance	Maximum height
0.001	0.00824	9–90 yr.	4–40 × 10^6 km	6.1–61 km
0.01	0.824	8–80 h	4–40 × 10^2 km	61–610 m
0.1	82.4	0.3–3 s	46–460 m	0.61–6.1 m

Silt and clay particles are transported easily and lifted to great heights by winds once they are entrained in the wind. Clay particles are generally aggregated or cling to larger particles. Silt seems to be the most mobile soil constituent at the wind velocity of 15 m s^{-1}, but at lower wind speeds the cohesion of particles and small surface areas compared with sand grains keeps the silt fraction from becoming mobile (Figs. 4.1 & 4.2). An increasing threshold velocity is needed for both fine and coarse particles. Field data (Chepil 1953) show that the 0.01–0.005 mm fraction is the least wind erodible, which is in agreement with the curves (Fig. 4.2).

Large particles such as coarse sand, gravel and pebbles are in normal wind conditions transported by a *surface creep*, rolling along the ground. Sand grains between 0.25 and 2 mm are rolled and pushed along the surface in advances of a few mm by the impact of saltating grains. This much coarser load is moving less than 1 cm min^{-1} and the saltating finer grains are moving with a rate of a few metres per second (Mabbutt 1977: 218).

Surface creep refers to the bedload transport of sand and has been estimated to account for about 25 per cent of total sand transport. It is surface creep that

Fig. 4.2 Threshold shear velocity of wind versus median grain size of quartz grains. Sources: Savat 1982; Iversen & White 1982. Chepil's and Logie's curves present the fluid threshold (Savat 1982: Fig. 1).

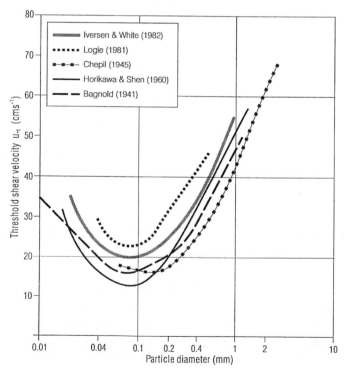

permits grains too large to move by saltation to move downwind. Surface creep and saltation, like many other sedimentary processes, are gradational to one another.

In the present stage of sand transport studies, one speaks not about sand but about single sand grains, and investigators like to measure the collision forces, shear stress and wind velocities on the surface of the grain itself to understand the real physical background of the mechanism of the process (McEwan *et al.* 1992). If we can reach this level we are still far away from the reality of the aeolian landforms which are not formed of single-sized spheric sand grains but of a variety of grain sizes, of different mineral grains with different densities, with impurities, moisture and transported by whirlwinds with changing speed and directions.

4.2 Basic elements affecting sand drift
4.2.1 Wind velocity and turbulence above the surface

At meteorological stations wind speed is normally measured 10 m above the surface with hemispherical cup anemometers. It is easy to record the wind speed 2 m above the surface, and it can be done with small cup anemometers even 10 cm above the surface (e.g. Stengel 1992: 37) but then the moving and changing surface can affect the measurements.

With the so-called hot wire technique the wind velocity can be measured very close to the surface. The cooling by the air stream of a very small electrically heated probe is measured just at the site of observation. The trouble with this technique is that the saltating sand grains damage the probe easily.

The third technique to measure the wind velocity is based on the pressure which the wind exerts on an obstacle placed in its path. With the so-called pitot tube, it is possible to make very accurate measurements within a millimetre above the boundary surface (Bagnold 1941: 38–39) but when it is placed very close to the surface it will disturb the sand transport. When the wind speed is low and the pressures are small then the pitot tube is not accurate enough.

Butterfield (1971) designed a simple, inexpensive, mechanical wind monitoring system to measure gusts. It consists of a system of brass rods suspended from stainless-steel needle-bearings and mounted in a Cassella thermograph body. A thin aluminium drag-plate attached to the rods is compressed against a coil return-spring by the wind pressure on it and activates the rod system to move a fine siphon-pen across the chart. The small friction of the needle bearings, the counter-balanced construction of the rods and pen arm, and the lightness of the whole system ensure a rapid response to very small pressure fluctuations on the plate. A detachable, battery-operated chart mechanism producing a chart speed of 6 cm/min is fitted for short-period sampling of winds.

Wind velocities high above the ground surface are measured with a meteorological balloon whose position is recorded with radio signals, and a computer
records its position at different levels and the horizontal drift speed (e.g. CORA
equipment by Vaisala). From the geomorphic point of view the wind speeds
several kilometres above the surface are only of secondary interest. Volcanic
microparticles are transported in the upper atmosphere. Normally the dust clouds
are in the atmospheric boundary layer at a height of less than one kilometre. Sand
and snow drifting phenomena depend directly on wind characteristics close to
the Earth's surface (height <25 m). The wind near the Earth's surface is always
turbulent or gusty, fluctuating with time and location, and therefore it is normally
spoken about as mean speed.

A typical wind speed profile for the atmospheric boundary layer is logarithmic.
The speed increases with the height. The tangential stress and turbulence exerted
by wind blowing over a surface is called *shear stress*. For a laminar sublayer
the surface friction velocity (or shear velocity), u^*, is the square root of the
ratio of shear stress to fluid density, $(u^* = (\tau/\rho)^{1/2})$. High wind velocities near
the surface provide high shear forces, too. The kinematic viscosity, v, is the
ratio of absolute viscosity (μ) to fluid density, $v = \mu/\rho$ (Greeley & Iversen
1985: 42).

For an aerodynamically rough surface the velocity profile is independent of
viscosity, such that

$$U/u^* = 1/0.4 \cdot \ln(z/z_0),$$

where z_0 is the equivalent roughness height, z is the distance above the surface
and 0.4 is the Kármán constant (κ).

For a sand surface, z_0 is about 1/30 the particle diameter. For rougher surfaces
with pebbles, blocks, shrubs or trees the equivalent roughness height must be
determined by measuring the wind speed for at least two heights. An increase
in roughness also results in an increase in surface shear stress and a significant effect on wind flow and surface friction speed (u_*) (Greeley & Iversen
1985: 42).

Sherman & Hotta (1990:19) must be quoted at length:

> Air flow across a surface will impart a shear stress to the surface as a result of
> a net downward flow of momentum. The shear stress (τ) is the force the wind
> exerts on the sand surface:
>
> $$\tau = \rho u_*^2 \tag{1}$$
>
> where τ is air density (approximately 1.22 kg m^{-3}) and u_* is shear velocity.
> For turbulent flow, momentum transfer depends on the aerodynamic roughness
> of the surface, and is proportional to the change in velocity with elevation.

Assuming that velocity increases logarithmically with height, this relationship is described by the general wind profile equation (von Kármán 1934):

$$u_z = (u_*/\kappa)\ln[(z - h)z_0], \qquad (2)$$

where u_z is wind velocity at an elevation z above the sand surface, κ is von Kármán's constant (approximately 0.40), z_0 is the roughness length (corresponding to the elevation at which wind speed goes to zero), and h is the displacement height (the aerodynamic boundary, usually assumed to be at the sand surface, $h = 0$ m). For unvegetated sand surfaces, the displacement height is small relative to the roughness length and can be disregarded. Equation (2) predicts shear velocity based on wind measurements and a knowledge of the roughness length and thereby links wind velocity with shear stress. This is one of the most fundamental relationships in understanding sediment transport.

Wind near the ground surface directly influences the movement of particles. The surface roughness causes the wind to be turbulent and this supports the occurrence of wind erosion. The minimal fluid threshold velocity, necessary to initiate movement of the surface grains, is likely only to be exceeded during infrequent and irregular velocity fluctuations of less than one second duration (Butterfield 1971: 128). Velocity fluctuations superimposed upon the mean flow and having a periodicity of one second or less are generally defined as turbulence (Kalinske 1943). The higher the frequency and the shorter the periodicity of such turbulent fluctuations from the mean flow, the greater will be the erosive power of the wind (Butterfield 1971: 128).

Conventional, inexpensive wind-monitoring equipment, such as cup anemometers, are inadequate to record gusts within the turbulent spectrum. One peak gust of 1 to 5 seconds duration may be sufficient to initiate sand drift on a dry surface but cannot be recorded with a slow-response cup anemometer (Butterfield 1971: 128).

4.2.2 *Air temperature, pressure and humidity*

Very little is known about the effect of the air temperature on the drift ability. Smith (1966), when describing the very coarse pebble ripples from Antarctica, launched the idea as follows: 'Movement of this coarse material is attributed to the effect of very low temperatures on the force of the wind. Drag is directly proportional to density of the air, and this is inversely proportional to absolute temperature. At winter temperatures, this results in a substantial increase in transporting power of winds of a given velocity'. Of course, the air density also varies with altitude: the higher the location the lower the air density. At an altitude of 5500 m the air pressure is about half of the pressure at sea level at the same temperature.

Temperature affects the density of air and its fluctuation affects the motion energy of wind (Table 4.3).

The relationship of air pressure (P in hPa) and temperature (T in °C) to the density of the dry air (ρ in kg m^{-3}) have a simple equation known as the ideal gas law ($P = \rho \cdot R_d \cdot T$, where $R_d = 0.287\,053$ kPa $\cdot$ K^{-1} $\cdot$ m^3 $\cdot$ kg^{-1} and T in °K) (Stull 2000: 8). With this equation can be compiled a practical nomogram (Fig. 4.3) in which we can read, for example, that if the air pressure is 1013 hPa and temperature is −40 °C the density is then 1.515 kg m^{-3}, and in a temperature of +15 °C the density is only 1.225 kg m^{-3}. The higher the density of air the more effective is the aeolian process with the same wind velocity.

Using the Bagnold entrainment equation (threshold friction velocity) it has been calculated that a mineral grain of 2 mm diameter can be carried to a height of 2 m by a wind of velocity about 36 m s^{-1} when the air temperature is −70 °C but it requires a velocity of some 65.4 m s^{-1} at a temperature of +50 °C (Selby *et al.* 1974) (Fig. 4.4).

Humidity increases the density of air but at the same time it decreases evaporation and sublimation ability of the soil surface.

4.2.3   *Roughness of the surface and air-ground surface contact*

When thinking of the wind drift of solid particles we have to take into account the effects of friction in the thin boundary layer which exists in the immediate neighbourhood of the solid body. Boundary layer theory constitutes one of the most important pillars of fluid mechanics (Schlichting 1979: 8). Air motion is always turbulent flow which means irregular and eddying fluctuations on the main stream. In turbulent motion the velocity and pressure at a fixed point do not remain constant with time but perform very irregular fluctuations of high frequency. In natural winds these fluctuations manifest themselves very clearly

Table 4.3. *Density and kinematic viscosity of air in terms of temperature at a pressure of 0.099 MPa (Schlichting 1979: 8).*

Temperature °C	Density [ρ] kg/m^3	Kinematic viscosity $v \times 10^6$ m^2/s
−20	1.39	11.2
−10	1.34	12.1
0	1.29	13.0
+10	1.25	13.9
+20	1.21	14.8

in the form of a tendency for squalls and often attain a magnitude of 50% of the mean wind speed (Schlichting 1979: 557). Energy is dissipated preponderantly by the small eddies, and the process occurs in a narrow strip inside the boundary layer (Schlichting 1979: 555).

The part of the atmosphere which is significantly affected by the direct influence of the surface is called the atmospheric boundary layer which is typically

Fig. 4.3 Nomogram for the density of dry air (ρ) in kilograms per cubic meter. P is air pressure in hecto Pascals and T is the air temperature in centigrade. Two examples are drawn on the nomogram. See the text on page 88.

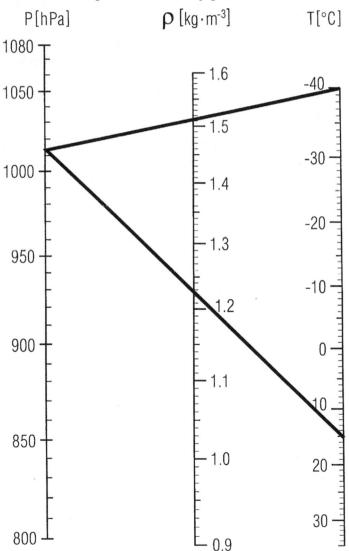

about 1000 m thick. The particle transport takes place mainly in the atmospheric boundary layer. An exception is the volcanic dust blown to very great heights by eruptions and then deposited, for example, on ice sheets years after the eruption.

Because a rougher surface eventually causes a decrease in wind speed gradient near the surface, the temporarily high gradient from the upwind surface causes a temporary increase in surface shear stress. The ability of the wind to move loose particles is therefore greater at the leading edge of a rough surface than it is further downstream and we would expect erosion to occur in this area. For the opposite change from rough to smooth, the stress level at the leading edge is less than the downwind equilibrium value. This creates a shelter effect or a wind shadow and a net deposit in this area (Greeley & Iversen 1985: 44–46, Fig. 2.9; Cooke *et al.* 1993: Fig. 17.2) (Fig. 4.5).

Wind velocities and turbulence close to the threshold of motion are not very important for the transport of major particle sizes. For dust transport the turbulence is more significant. Nickling (1978) found that the rate of dust transport is better correlated with levels of turbulence than with friction velocity (u^*).

Fig. 4.4 Threshold wind velocity required to lift a 2 mm granule to a height of 2 m at various temperatures (Selby *et al.* 1974).

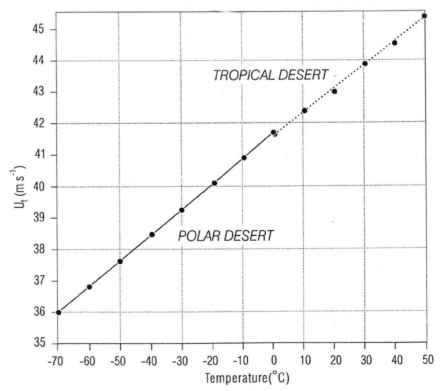

Every single sand grain causes tiny turbulences and pressure differences in moving air at the boundary layer. Turbulence causes erosion on the bed and creates an uneven surface, which then makes more turbulence in flow. This feedback produces sorting on the sand surface and ripple formation starts, as shown with wind tunnel studies by Seppälä & Lindé (1978). The coarser the particles the rougher the surface and the lower the threshold velocity needed to produce drifting eddies.

Roughness of the surface is related to any kind of change of surface geometry. When the wind speed is strong enough to induce drift the flow near the surface is dominated by the shear forces. Under these conditions the mean wind speed U depends on the height z, the shear stress at the surface τ_0, the density of the

Fig. 4.5 Mutual effects between sand dune accumulation and separation of the boundary layer. B-C and E-F are rough surfaces causing disturbances in boundary layer flow and sand dune accumulation C-D. A-B is a smooth surface on which is the basic air flow. Source: Kádár (1966: Fig. 12) based on the diagrams of Schlichting.

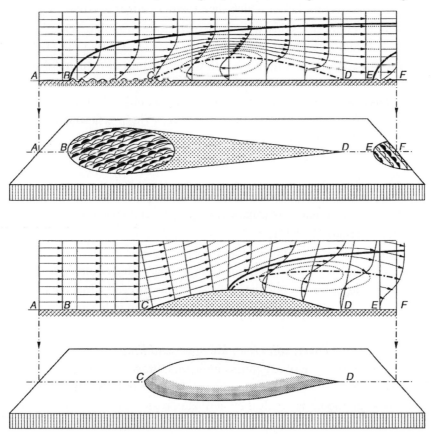

air ρ, the mean height of the roughness elements κ_0, and their non-dimensional spacing λ, according to the following relationship:

$$U/U^* = 2.5 \ln(z/k_0) + 5.5 - C(\lambda),$$

where $U^* = \sqrt{\tau_0}/\rho$, and $C(\lambda) =$ a constant depending on λ (Kind 1981: 341). The parameter U^* has dimensions of velocity and is known as the *friction velocity* or *shear velocity*.

4.2.4   Obstruction and topographic features

Roughness elements of the surface, bushes, stones and other obstacles such as snow fences cause a very complex flow pattern and affect the velocity profile close to the obstacle. This can give a rapid spatial variation of the shear stress τ_0 or the shear velocity U^*. This results in the formation of drifts in some areas and in erosion from others (Kind 1981: 341). If the flow has a smaller cross-section the flow must accelerate and if the streamlines diverge the flow must decelerate. A good example of this is the rough surface (Fig. 4.5) causing sand accumulation which then further causes vortices. An obstacle can cause a closed leeside eddy or so-called separation bubble behind it or a closed windward eddy, as shown by Robertson-Rintoll (1990: Fig. 7) (Fig. 4.6). A common occurrence found in wind flow over a three-dimensional surface is the formation of a horseshoe vortex around its base (Fig. 4.6). The surface shear stress under the vortex tends to be fairly high and usually erodes the loose material around the base (Kind 1981: 343) as happens often when using vertical sand traps (see section 6.11). This process works in mesoscale around the nunataks rising above glaciers and forms moat lakes and snow accumulations on the lee side (see section 14.2).

During wind drift the surface roughness changes because of erosion, accumulation and sorting and this causes changes in wind speed and vorticity close to the surface (Bagnold 1941: 53–54).

Bagnold (1941: 67) showed that transport rates are increased across surfaces with 'sand with a very wide range of grain sizes'. This could be true at the initial phase of drift, but very soon the surface will be covered with coarse grains forming a lag deposit on the surface and the transportation will decrease to a minimum.

4.2.5   Surface cover, salt crust, hardpan and lime

The presence of roughness elements: gravel, stones and blocks on a dry sand surface, affects the erosional capacity of wind. When the density of the roughness elements is small they activate erosion and they protect the ground when their number is large (Logie 1982). According to the wind tunnel studies by Logie (1982) the minimum cover density needed to shelter the sand surface from the wind depends on the size of the obstacles. The larger they are, the higher is the minimum cover density.

Wind tunnel tests prove that small amounts of soluble salts such as NaCl and KCl can significantly increase the threshold velocity of the sand because cement-like bonds formed between grains tend to hold individual particles in place (Nickling & Ecclestone 1981). The relationships between surface salt concentration and threshold shear velocity for the monovalent chlorides were quite similar to those with $MgCl_2$ and $CaCl_2$ (Nickling 1984). A significant exponential relationship exists between threshold shear velocity and surface salt concentration (Nickling & Ecclestone 1981; Nickling 1984). With increasing concentration the salt crystal growth causes the initially smooth sand surface to develop an irregular frothy texture which also directly affects the threshold shear velocity (Nickling & Ecclestone 1981: 505).

Different salts, gypsum and lime may concentrate at the surface layer also in the cold, arid environments and form a surface layer as a hardpan, plaster the

Fig. 4.6 Diagrammatic flow structure evolution within the near-surface wind flow on a parabolic sand dune. Source: Robertson-Rintoll (1990: Fig. 7). © John Wiley & Sons Limited. Reproduced with permission.

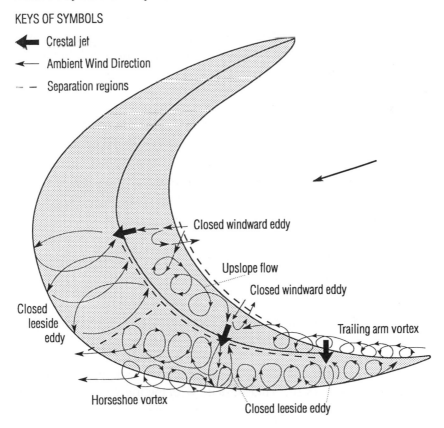

KEYS OF SYMBOLS

sand grains together and prevent the wind drift. Fristrup (1952a: 96) reported salt encrustations over wide areas in Peary Land, north Greenland: 'The salts are composed of chlorides and sulphates of magnesium and calcium. The salt crusts are partly deposited by the drying of ponds formed by (ice) melting during the summer, and partly by salts being extracted from soil during strong evaporation (with upward water movement) in summer. In the month of August most of the watercourses are dry, and there is a violent evaporation from the small lakes and ponds. Apart from this, and perhaps of greater importance, is the winter development of thick ice layers in the valleys that were completely dry in August.'

Saline soil conditions are quite common in Antarctica. In the basins of dry valleys of Victoria Land so-called evaporite soils are found. These sites may at times impound saline waters. The soils are characterized by the veneer of salt (calcium chloride) up to 2.5 cm thick (Tedrow 1977: 235–236, 536–539). Famous salt lakes are Lake Bonney in Taylor Valley and Lake Vanda in Wright Valley (Tedrow 1977: 536).

Free calcium carbonate or lime and organic matter in soils may cause severe erosion by wind. This was observed in Canada by Hopkins, but on loamy sand with high $CaCO_3$ content it decreases erodibility because of increasing soil cloddiness and mechanical stability of clods (see Chepil & Woodruff 1963: 267). In fine soils $CaCO_3$ seems to produce clods which have a suitable size for wind drift and probably this is why loess deposits are so rich in lime.

4.2.6   Grain size and shape of surface material

The friction of the ground surface retards the wind velocity. It reaches zero in the region of 1/30 of the particle diameter of the grains or stones on the granular surface. This means that if the boulders are 30 cm in diameter, the air around them is still to a height of 1 cm, above which the wind movement begins (Bagnold 1941: 50). The velocity as a function of the logarithm of the height is expressed by Bagnold (1941: 52):

$$V_h = 5.75 \, V_d \log h/k, \tag{1}$$

where V_h = velocity at a given height h

$\quad$ V_d = drag velocity

$\quad$ k = height at which the velocity is zero or $d/30$, where d is the diameter of particles.

Over flat surfaces and steady wind conditions V_d is directly proportional to the rate of increase of the wind velocity with the log-height. This means that in order to find the surface drag it is only necessary to measure the wind velocity at any two known heights, to plot these velocities on a graph against the log-height and to draw a straight line through the points (Bagnold 1941: Fig. 18).

The shear stress produced at the sand surface by wind is one of the most important factors for the sand movement by wind action. Shear stress is a maximum at the Earth's surface and decreases gradually with height, becoming zero in the geostrophic wind above the boundary layer (Kind 1981: 339). When the shear stress exceeds a critical value, the sand and snow particles start to move. Several investigators have given somewhat different threshold friction velocity (critical shear velocity) values for sand grains of different diameter (Savat 1982; Iversen & White 1982; Cooke *et al.* 1993: Fig. 17.5) (Fig. 4.2).

A puzzling feature in sand transport by wind compared with fluvial transport is that according to several authors the threshold velocity of aeolian erosion is lowest for grains about 0.1 mm in diameter but the median of dune sand is normally close to 0.2 mm. This is the size most easily drifted by *fluvial* processes according to Hjulström (1935) (see Sundborg 1955) (Fig. 4.7). Chepil (1945) believed the lowest *aeolian* threshold velocity was for particles about 0.2 mm in diameter (Fig. 4.2) – but later he changed his mind because he wrote 'the most erodible particles of 2.65 density are about 0.1 mm' (Chepil & Woodruff 1963: 237).

The distribution curve of the aeolian sand is very sharp and many grains are around 0.3 mm, and grains less than 0.1 mm or more than 0.6 mm in diameter are relatively few according to Bagnold (1941) (who studied the cause for this distribution), and Kawamura (1964: 3), although some up to 2 mm diameter will be moved by high wind velocities. Kawamura (1964: 3) used in his studies sand of 0.2–0.3 mm with specific gravity of individual grains on average about 2.5. Natural sediments consist of varying grain sizes and therefore the fluid threshold for any sediment cannot be defined by a finite value but should be viewed as a threshold range which is a function of the size, shape, sorting and packing of the surface sediment (Nickling 1988).

On coasts the wave action will wash the coarse grains and uncover new unsorted sediments for further drift by wind. Nielsen (1933: 246–247) observed in Iceland that the gales roll small stones up the slopes and throw them into the air from the ridge.

The original shape of sand grains depends on the mineralogy and the shape affects the drift. Wind action causes shape-sorting which means that angular grains are transported less easily than well rounded grains (MacCarthy & Huddle 1938). Thin flat grains are light and they have a large surface area compared with round grains with the same diameter. The critical factor is the ratio of the surface area and absolute weight of the grain. Wind is effective in sorting flat quartz grains, which have poor rolling qualities, and concentrating them on the dunes in Lapland (Seppälä 1971a: 59). Biotite and muscovite grains drift easily because their form can be very large but thin.

4.2.7   Density of the particles

Wind moves the finer and lighter particles faster than the coarser and denser ones (Chepil 1965: 127). The common mineral materials are quartz, the feldspars, calcite and aragonite. Their density is roughly 2×10^3 times the density of air at normal temperature and pressure (Allen 1985: 21). The so-called accessory minerals normally total no more than a small fraction of 1% by weight. These minerals are significantly more dense than the common ones (Table 4.4).

Fig. 4.7 Threshold value of wind and water velocities for erosion within a uniform material. Source: Sundborg (1955: Fig. 1) compiled after Bagnold and Hjulström (see also Bagnold 1941: Figs. 28 & 29).

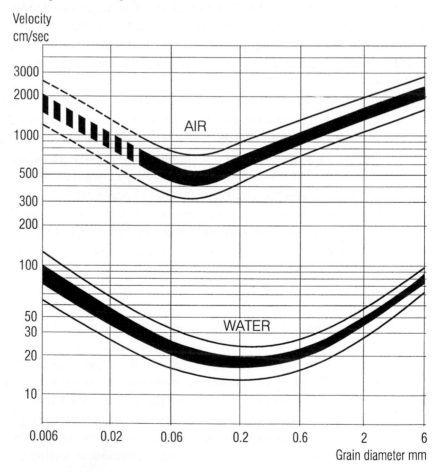

4.2.8　Texture of the surface and sorting of the material

Texture of the surface and sorting of the material have their influence on the further transport of material. Entrainment of grains from the bed is influenced not only by the characteristics of individual grains, but also by bulk sediment properties which include the grain size, sorting, orientation, packing arrangement, porosity and cohesion (Pye & Tsoar 1990: 44). Particles at the surface may be bound together by silt and clay to produce erosion-resistant surface crusts. The effectiveness of silt and clay as bonding agents depends on their relative proportion in relation to the quantity of sand fraction (Nickling 1994: 454). Soils with 20–30% clay, 40–50% silt, and 20–40% sand are least affected by abrasion (Chepil & Woodruff 1963: 262).

During the transportation grains are sorted according to size, shape and mass. This process starts immediately when the sand starts to move. The surface is pitted because of the collisions of saltating grains on the sand bed (Bagnold 1941: 57; McEwan et al. 1992: Fig. 2). Sorting and differences in velocity of grain movement seem to be the main reasons for ripple formation (Sharp 1963; Seppälä & Lindé 1978; Anderson 1987). Savat (1982) speaks about selective transportation, which means that either finer particles are removed or coarser grains shelter and protect the finer grains and the result is that the coarse fraction is in motion. Both denudation and accumulation processes take place at the same time in many places. Sorting results and lag sands and gravels can cover the surface and the finer particles. Wind erosion occurs only when soil grains capable of being moved in saltation are present in the soil (Chepil & Woodruff 1963: 215). The presence of coarse grains by sheltering and fine dust particles

Table 4.4. *Density (g cm^{-3}) of common minerals and rocks. Source: Allen (1985: Table 2.1).*

Common minerals		Accessory minerals		Rock types	
Name	Density	Name	Density	Name	Density
aragonite	2.93	corundum	3.99	diorite	2.72–2.96
biotite	2.80–3.40	epidote	3.25–3.50	gabbro	2.85–3.12
calcite	2.71	garnets	3.56–4.32	gneiss	2.66–2.73
dolomite	2.87	haematite	5.27	granite	2.52–2.81
microcline	2.56	hornblende	3.00–3.47	granodiorite	2.67–2.79
muscovite	2.80–2.90	olivines	3.21–4.39	sandstone	2.17–2.70
orthoclase	2.55	pyroxenes	3.20–3.55	schist	2.70–3.00
plagioclase	2.62–2.76	rutile	4.25		
quartz	2.65	tourmaline	3.03–3.10		
		zircon	4.67		

in the soil by cohering hinder the movement in saltation (Chepil & Woodruff 1963: 244).

4.2.9   Content of organic material

Soil formation and the content of organic matter as humus and roots fix the mineral material mechanically and aid soils in absorbing moisture effectively, and thus decrease the influence of wind. Organic matter may form aggregates of mineral particles that are less susceptible to entrainment by wind than the individual grains, but very high organic contents in dry soils increase erosion potential because of the very loose soil structure (Nickling 1994: 455). Organic soil cover can be broken by fire, mass movements, frost, animals or wind felling trees. These events are needed to reactivate wind action if mineral material is anchored by organic material. Peat is fibrous material and not easily eroded.

Organic matter, soil microorganisms and various products of organic matter decomposition as well as soil colloids are limited in amount in cold environments but still their effect on stabilizing the wind erosion is considerable.

4.2.10   Moisture of surface material

Humidity is a rather critical factor for sand drift by wind in all conditions. Water among sand grains fixes them together. Rain will immobilize a sand surface with cohesive forces between sand grains and form larger grain aggregates (Fig. 4.8) and the threshold shear velocity needed for transport is much higher already when the water content is a few percent (Fig. 4.9). Water content in sand is related to air humidity (Fig. 4.10). Experimental studies by Logie (1982) show the impact of soil moisture on the wind erodibility of the sand. Small quantities of soil moisture have a great influence on the susceptibility of the sand to deflation. The resistance of wet sand surface to deflation depends not only on the wind speed but also on the air temperature and relative humidity of the air.

Fig. 4.8 Wetting of particle aggregates. The capillary forces decrease from A to C. Source: Stengel (1992: Fig. 21 after Schumacher 1988).

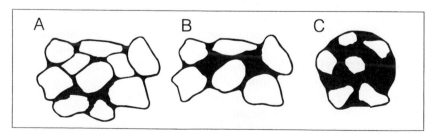

There is a good relation between the soil moisture of the surface and the period of time during which the surface could resist the wind (Fig. 4.11). Samples with water content below 1.2% could not resist the force of the wind. Deflation started immediately in the wind tunnel tests of Logie (1982: 170).

Cohesion among soil particles and aggregates decreases the erodibility of soil. Rain followed by drying produces a certain degree of cementation (consolidation) among the individual particles, aggregates them and tends to reduce erodibility of the soil surface. The cementing materials are composed mainly of water-dispersible particles smaller than 0.02 mm in diameter (Chepil & Woodruff 1963: 255–257). Conversely, increasing erodibility results from frost action and

Fig. 4.9 Variation of the threshold shear velocity with moisture content. Source: Belly (1962: Fig. 12).

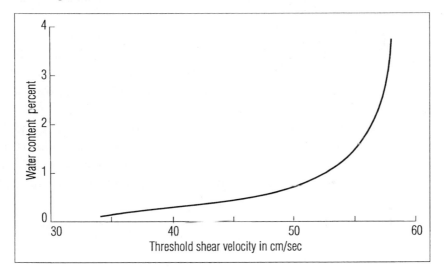

Fig. 4.10 Relationship between water content of sand surface and air humidity and variation of rate of transport with different air humidity percentages and their relationship to threshold shear velocities of wind (U_*). Source: Belly (1962: Figs. 7 and 8).

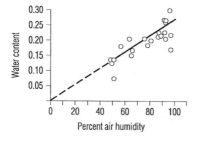

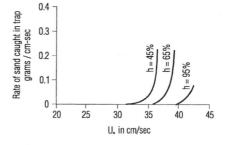

the abrasive force of wind drifted sand particles, which all tend to loosen the bond between soil particles and aggregates (Chepil & Woodruff 1963: 239). Silt particles are dispersible by water much more than clay particles, which form a compact, massive soil structure and clods.

Melting of surface material releases moisture which cannot penetrate downwards because of the frozen layer underneath and this keeps the surface moist and unfavourable for wind entrainment until it dries.

4.2.11 Soil temperature and frost

Surface temperature affects evaporation and dryness of the sand surface. Especially important for sand drift are temperatures below freezing – the freezing water cements the sand particles together. When the soil surface has a temperature well below freezing the moisture from humid air can condense on the surface and form rime ice.

Melting and evaporation take energy and this keeps the surface temperature low. Strong evaporation can cool the subsurface layer so much that the frost table can even rise temporarily during the thawing period, as was observed on a palsa (Seppälä 1976b: Fig. 4). Strong radiation together with strong wind increased evaporation and affected thawing of frost.

Repeated freezing and thawing of the surface soil tend to soften and disintegrate the surface crust and aggregates, and to enhance wind erosion (Chepil & Woodruff 1963: 235). Very strong frost activity causes bare soil surfaces as in mudboils and patterned ground and these can be the starting points for wind drift.

Fig. 4.11 The time a wet sand surface can resist a wind force of 50 km/h (c. 14 m s^{-1}) measured at 5 cm height. Source: Logie (1982: Fig. 6).

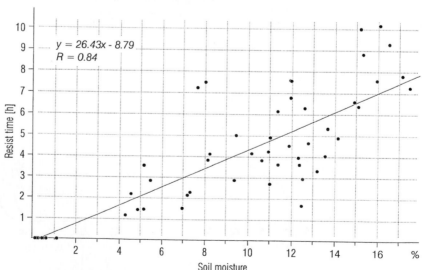

4.2.12 Sublimation

Water possesses the special property that it can pass from the solid phase into the gaseous and vice versa without passing through the usual liquid phase (Fig. 4.12). When ice changes to water vapour this evaporation is called sublimation. The term may also be used in the broader sense of the whole sequence of change, from solid to vapour, movement of vapour, and change from vapour to solid (Paterson 1981). For example, snow turns into ice in a dry snow zone through sublimation. Ice always has a thin surface equilibrium layer of water molecules. The vapour pressure is expressed in millimetres of mercury (Table 4.5).

Table 4.5. *The rise of the vapour pressure of ice with its temperature (Seligman 1980: 27)*

T, °C	Pressure, mm Hg
0	4.579
−10	1.947
−20	0.770
−30	0.280
−40	0.094
−50	0.029

Fig. 4.12 Low-pressure region of the H_2O phase diagram. Source: Maeno 1981.

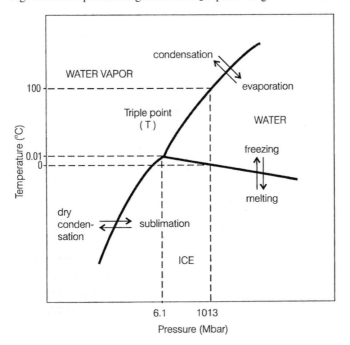

Thus as the temperature rises from $-40\,°C$ to $0\,°C$ the water vapour pressure becomes 50 times higher. When the air temperature changes the temperature in the upper snow layer changes too, and this means a temperature gradient in snow and a vapour pressure gradient, as well. The increase of vapour pressure gradient with rising temperature causes hoar formation, firnification and ice and snow evaporation (Seligman 1980: 27). When the temperature rises ice will tend to give off increasing amounts of water molecules in the vapour, and as the temperature sinks, more and more of the molecules in the vapour will re-attach to the solid ice. Sublimation of ice, snow or frozen ground surface depends on the water content of the air and wind blowing over the surface. If the wind is dry (its relative humidity is low) then it can pick up water molecules from the surface. If the wind is saturated (relative humidity is 100%) no sublimation will take place. If the air is supersaturated in relation to the surface it may deposit some molecules on the surface forming some more ice. Water vapour condenses upon the surface. If the surface is cold and the wind is relatively warm the condensation takes place easily because the cold surface cools the air at the boundary layer and it reaches dew point. Warm surface and dry wind support sublimation, and cold surface and wet wind blowing over the surface will encourage condensation (Seligman 1980: 28).

In dry frost climates with severe winters the snow cover is very thin, easily blown away by the wind and it offers the soil no protection against frost or wind deflation. In this dry climate, the soil dries out easily and is attacked by wind, even when frozen (Tricart 1970: 19). This is why sublimation has recently received much attention (e.g. Huang & Aughenbaugh 1987; McKenna-Neuman 1990a, 1990b, 1993; Law & van Dijk 1994). Mainly the studies considered sublimation from frozen sediments and were done under field conditions, and therefore there is some disagreement over the importance of sublimation for material entrainment.

The sublimation process is influenced by air temperature, relative humidity, wind speed, net radiation and particle surface area (Law & van Dijk 1994). These variables together with sediment moisture content influence the amount and rate of pore ice sublimation. However, the sublimation of frozen moisture from sediments can produce dry dust and loose sand and makes possible the aeolian movement of sediments in cold environments in winter time. The conditions in Polar regions in the middle of winter, however, are not very favourable for sublimation (see 2.1.4).

In cold environments in winter time mineral particles even on the surface layer are stable because they are cemented by ice. If the sand surface is free of snow or nearly so, then sand grains can be drifted in certain conditions. Cold air contains very little moisture. Moving air may cause sublimation, however, the temperature is well below freezing. Sublimation of the cementing ice would liberate the frozen mineral particles. Porosity has its effect on sublimation, which is in inverse relation to the porosity (Paterson 1981).

Hobbs (1974: 362) defined the physical proportions of sublimation in the following way:

> The latent heat of sublimation L_s of ice is the change in enthalpy when a unit mass of ice is converted isothermally and reversibly into water vapour. At standard atmospheric pressure and at the ice-liquid-vapour triple point (273.16 K), the latent heat of sublimation of ice is 2838 kJ kg^{-1} (Rossini *et al.* 1952) or 0.53 eV per molecule. About 0.506 eV of this amount is caused by the change in internal energy and the remainder is contributed by the pdV term in eqn. (5.10). It is interesting to note that the latent heat of fusion of ice is only about 12 per cent of the latent heat of sublimation, so that only about 12 per cent of the hydrogen bonds must break when ice melts. The latent heat of sublimation of ice at 0 K is 0.491 eV per molecule (Eisenberg and Kauzmann 1969) and this is twice the energy of the hydrogen bond. (Fig. 4.11).

According to Huang and Aughenbaugh (1987: 172) 'the rate of sublimation of pore ice in the natural environment depends mainly on the balance between the force driving the water molecules away from the ice and the bonding force resisting this movement. Sublimation can occur at any temperature below the melting point of ice. The vapour produced dissipates through the processes of diffusion and convection. Thus, sublimation of ice in frozen soil is a heat-mass transfer process that is controlled by the moisture and temperature gradients as well as by the thermophysical properties of the soil.' Laboratory measurements of sublimation of frozen silt show that sublimation of pore ice is primarily temperature dependent (Huang & Aughenbaugh 1987: 179). It is most pronounced at a temperature of $-1\,°C$ and it drops off dramatically at the colder temperatures (Huang & Aughenbaugh 1987: 176). Other factors influencing sublimation besides the ambient temperature include relative humidity of air and water content of silt (Huang & Aughenbaugh 1987: 179–180).

These physical facts explain the conclusions by McKenna-Neuman (1990b: 331) that particle loss with sublimation appears to increase with increasing temperature, and with decreasing volumetric ice content, which she could prove with laboratory experiments. The experiment also provided evidence that particle loosening through sublimation is sufficient to support autumn aeolian transportation and to build small dunes (McKenna-Neuman 1990b: 334).

The capacity of the atmosphere to retain moisture is a function of temperature. The lower the temperature the less moisture the air is able to contain. The concept of precipitable water explains the depth of liquid water produced if all water vapour were condensed from a vertical atmospheric column of uniform cross-section (Schemenauer *et al.* 1981: 131). Hay in his unpublished thesis (Hay 1970; after Schemenauer *et al.* 1981 and after Langham 1981; Hare & Hay 1974) computed the distribution of mean precipitable water for January in Canada. The minimum values of 2 mm or less occur over Keewatin, N.W.T. and the Arctic Islands, since the air temperatures during January are very low and these regions are distant from the major moisture sources (Fig. 2.12).

Mean relative humidity of air over ice in the Arctic in January is above 100%, characterized by supersaturation (Zav'yalova 1987: Fig. 2b) (Fig. 2.10) and this means that the sublimation cannot be very active in the winter time when the sun is below the horizon. Solar radiation in early winter (October) is still effective and later (April, May) gets stronger and that means also possibilities for sublimation. Probably the early winter is more important in this respect because the snow cover is still thin and partly scattered because of strong winds.

Relative humidity in Antarctica is lower than in the Arctic. The humidity values in the Arctic are comparable with those of Graham Land. In Antarctica the decrease of humidity is linked with the effect of wind regime of katabatic winds (*Föhn* effect) (Zav'yalova 1987: 94). The most humid place in Antarctica is Graham Land, where, throughout the year, relative humidity exceeds about 80% and the driest regions are the oases where humidity oscillates around 50% most of the year. The range of relative humidity in the oases is high. During the *Föhn* winds it may decrease 30–40% in a short period. When the winds come from the sea the humidity could increase more than 40% in three hours (Zav'yalova 1987: 94–95). Low humidity of air supports also snow drift by wind.

Loewe (1962: 5171) reviewed some evaporation observations made on a lake at Cape Evans in McMurdo Sound, Antarctica and they can be given as an example from a cold climate: at $-13\,°C$ the daily evaporation was 0.5 mm, at $-19\,°C$ it was 0.17 mm, and at $-25\,°C$ it was 0.023 mm of ice. Evaporation should not be considered only as a function of temperature alone, but it will increase with the wind velocity and with the gradient of water vapour pressure between the ice and the overlying air (Loewe 1962: 5172). At Cape Evans the mean wind velocity is rather high, 9.5 m s^{-1} during the winter half-year, and there *Föhn* conditions occur sometimes, which means low relative humidity, and high evaporation. On the ice cap the mean temperature is much lower and the mean relative humidity is often over 100% which means a prevalence of rime formation over evaporation (Loewe 1962: 5172). This is often the situation in the Arctic, too, especially during the winter and then sublimation does not exist.

We should not overestimate the meaning of sublimation for the wind drift of sand. We should get more field observations of the conditions, especially from the early and late winter periods when the sun shines. Cooling of the sand surface will form hoarfrost in the upper pore spaces and even thin ice lenses in the deeper layers, cementing the sand very efficiently.

Sublimation of a frozen sand surface leaves a dry layer of sand on the surface and before this is drifted away the sublimation cannot continue. Drifting of dry sand means that it is deposited somewhere else and it will protect the frozen sand layers there against sublimation.

A small amount of drifted mineral material in snow will support the evaporation and melting of snow, and therefore the deflation areas in cold environments are often free of snow already in late winter, some months before the total snow melt. Thick layers of sand (3–5 cm) protect the underlying snow from thawing. The same feature can be seen on glaciers where a thin layer of moraine causes moraine-covered ice ridges, and glaciofluvial material protects the dirt cones from melting on the ablation areas.

The Thornthwaite method was applied in the study of 'potential evapotranspiration' of all the Canadian weather stations to obtain the map of the climatic regions (Sanderson 1950: Fig. 1) (Fig. 2.11). This study indicated that the water need or potential evapotranspiration in the northwest is high and a subhumid climate results.

In low latitudes with strong solar radiation sublimation is very significant. For example, in the White Mountains of California and Nevada 50–80% of total springtime snowpack sublimates (Beaty 1975) and is therefore geomorphologically inert (Thorn 1978: 415).

The strong winter storms (up to 300 km/h) drive the dry snow so that extensive areas in Dry Valleys, Antarctica, remain almost free of snow. The very low air humidity ensures that the snow almost exclusively disappears through sublimation, leaving the soil and rock dry (Miotke 1983: 23). The upper layers of soil are mostly extremely dry, moisture content below 0.1%, and this means that the active layer above the permafrost is only 20 to 40 cm (Miotke 1983: 23–24) and moisture of melting seasonal frost does not prevent the wind drift of mineral grains.

According to Sekyra (1969: 280) in oases of Antarctica there occur: '(3) Sublimation ice (in skeletal development found at depths at a maximum of 0.1 m on deflation platforms and detritus, mostly at the peripheries of firn fields);' and '(4) Ice filled cracks (only in isolated parts of the oases; short-term occurrence). In mountain areas this type of ice in crevasses, passes – due to intensive sublimation – into a homogeneous cementation fissure ice up to non-homogeneous ice (permafrost).'

Sublimation takes place in strong solar radiation in spring time even if it is cold. For example, on the fells in Lapland (c. 70° N lat.) an ice layer can be observed hanging fixed on the shrubs and grasses 10 cm above the surface of thin snow cover. Underneath is an open empty space. The thin ice layer was formed by sublimation and refreezing when the air temperature was about −15 °C.

On Baffin Island, Arctic Canada, autumn and winter transport of coarse mineral particles is observed which is much more significant than is usually assumed. This has been explained as a result of sublimation and abrasion (McKenna Neuman 1990a).

4.3   Aeolian sand

Sand in general is loose, non-cohesive granular material with grain size 0.0625–2.0 mm in diameter. According to the Wentworth size classes very fine sand is from 0.0625 mm (4.0 Phi) to 0.125 mm (3.0 Phi) and very coarse sand has a grain size from 1.0 mm (0 Phi) to 2.0 mm (−1 Phi). In between are the fine, medium and coarse sands. Sand consists of mineral grains which are generally in tangential contact only and thus form an open, three-dimensional network. As a consequence, sand has a high porosity (35–40% – Pettijohn *et al.* 1987: 1 – where 'Phi' is also explained).

Definitions of 'sand' as a deposit – as distinct from a size term – are diverse. For example, definitions include if the average, median, or modal size of material falls in the sand range, or if more than 50% of it is within these limits (Pettijohn *et al.* 1987: 1–2). Aeolian sand is not a problem because normally it is very homogenous, well sorted and the median grain size (0.1–0.35 mm) lies predominantly in the fine sand range (e.g. Seppälä 1969: 169; 1971a: 42–43) see Figs. 4.13 & 4.14.

Sand, unlike the finer materials, is largely transported by rolling and sliding along the bottom or by saltation and only to a smaller extent by turbulent suspension (Pettijohn *et al.* 1987: 2). Bagnold (1941: 6) places the lower limit of 'sand' as that size where the terminal fall velocity is less than the upward eddy currents and the upper limit as the size such that a grain resting on the surface ceases to be movable either by direct pressure of the fluid or by the impact of other moving grains. This kind of behavioral definition of sand depends on the nature of the flowing medium (air or water) and could only be valid for 'average' conditions of flow. The size limits thus defined approximate those set by tradition (Pettijohn *et al.* 1987: 3). Bagnold notes further that materials designated 'sand' have one peculiar characteristic which is not shared by coarser or finer materials – namely, the power of self-accumulation – of utilizing the energy of the transporting medium to collect their scattered components together in definite heaps, leaving the intervening surface free of grains. The common mode of transport of sand is by the migration of such heaps or dunes, be they subaerial or subaqueous (Pettijohn *et al.* 1987: 3).

Sedimentary structures are used to identify how the sand has moved while deposited. Textural studies should receive attention in such analyses. Texture of sand grains includes the shape, roundness, surface features, grain size, and fabric of the components (Pettijohn *et al.* 1987: 69). The physical processes at the site of deposition impart a distinctive texture by the arrangement of sand grains.

4.3.1   *Grain size and sorting of aeolian sands*

The grain size of aeolian sand is usually analysed by dry sieving. For aeolian sands it is recommended to use 10 test sieves with mesh sizes from 0.074 to 2.0 mm to get a characteristic grain size curve and to find out the fine differences. The sieving results are presented in histograms of weight of different grain size classes. In a histogram typical to aeolian sand there is one

Fig. 4.13 Cumulative frequency grain size curves of sand dune material and redeposited aeolian sands from Finnish Lapland. Grain size parameters on Table 4.6. (Seppälä 1995: Table 1).

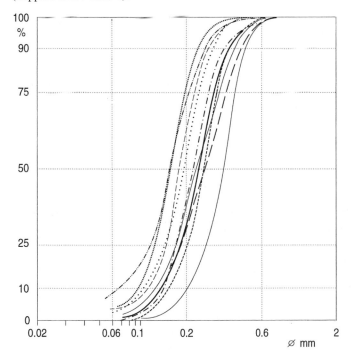

dominating bar and two or three smaller ones, and the rest are very limited in size (depending of course on the size classification). Aeolian sands resemble a normal or Gaussian distribution since the size scale is actually logarithmic, the distributions approximate log-normal (Pye & Tsoar 1990: 49).

Fig. 4.14 Cumulative grain size curves of sand dune material from Finnish Lapland on logarithmic and probability scales. Source: Seppälä (1971: Fig. 29).

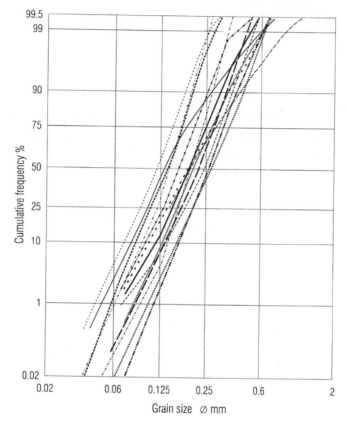

Table 4.6. *Grain size in mm and sorting parameters of some aeolian sands from Finnish Lapland (Seppälä 1971a; 1995a). For explanation see the text.*

Sample	P_{90}	Q_3	Md	Q_1	P_{10}	S_o	Sk	K
Kuttanen	0.36	0.29	0.26	0.185	0.14	1.25	0.79	0.24
Hietatievat	0.262	0.195	0.14	0.11	0.07	1.33	1.09	0.22
Kiellajoki	0.34	0.26	0.19	0.14	0.11	1.36	1.01	0.26
Kiellajoki	0.46	0.36	0.28	0.21	0.17	1.31	0.96	0.26
Kiellajoki	0.20	0.17	0.14	0.12	0.095	1.19	1.04	0.24

The best source material for aeolian sand lies in the grain size fraction 0.2 to 0.1 mm and for loess 0.01 to 0.05 mm (Troll 1948). Typical sand in sand dunes has a median grain size between 0.1 and 0.3 mm. Grains with diameter above 1.5 mm and below 0.02 are rare. Large grains are usually some light minerals and the smallest some heavy minerals.

Grain size data is conveniently presented in the form of a cumulative frequency distribution curve with grain diameter plotted on the abscissa in logarithmic scale and weight per cent of different fractions on either a linear scale (Fig. 4.13) or a probability scale (Fig. 4.14). Aeolian sand is typically unimodal, well sorted, symmetrically skewed and highly peaked, in the distribution curves.

There are several ways to compute the graphical statistical parameters and to make quantitative comparisons about the grain size of mineral material from the cumulative frequency curves. Folk and Ward (1957) proposed parameters Φ_5, Φ_{16}, Φ_{25}, Φ_{50}, Φ_{75}, Φ_{84}, and Φ_{95} corresponding to the Phi size values at the fifth, 16th, 25th, 50th, 75th, 84th and 95th percentiles read from the cumulative frequency curve. Dispersion around determines sorting. Thus an estimate of sorting based on the 84th and 16th percentiles should be superior to one based on the 75th and 25th percentiles, that is, one closer to the central tendency (Pettijohn *et al.* 1987: 75). Folk and Ward (1957: 13) combined each of these measures in their inclusive graphic standard deviation, which is the average of sorting based on both the central and exterior parts of the size distribution. With these values are then calculated the *graphic mean* $M_z = (\Phi_{16} + \Phi_{50} + \Phi_{84})/3$ and the *inclusive graphic standard deviation* $\sigma_1 = (\Phi_{84} + \Phi_{16})/4 + (\Phi_{95} - \Phi_5)/6.6$ showing the sorting, i.e. the spread around the mean; low values of σ_1 characterize the well sorted sediments.

The *inclusive graphic skewness* Sk_1 represents the asymmetry of skewness of the distribution.

$$Sk_1 = (\Phi_{16} + \Phi_{84} - 2\Phi_{50})/[2(\Phi_{84} - \Phi_{16})]$$
$$+ (\Phi_5 + \Phi_{95} - 2\Phi_{50})/[2(\Phi_{95} - \Phi_5)]$$

Positive Sk_1 values indicate a dominance of coarse material and negative values of Sk_1 indicate a deficiency of fine particles compared with the log-normal distribution.

The *inclusive graphic kurtosis* K_G indicates the peakedness of the distribution of grains.

$$K_G = (\Phi_{95} - \Phi_5) : [2.44(\Phi_{75} - \Phi_{25})]$$

Close to log-normal the values of K_G are around 1. Much below 1 means a wide range of grain sizes and well above 1 a very narrow range of grain size distribution.

Another widely used method of computing the graphical statistical parameters uses the grain size diameters P_{90}, Q_3, Md, Q_1, and P_{10} in millimetre scale corresponding to the values of 90, 75, 50, 25 and 10 percentiles read from the cumulative grain size frequency curve (Köster 1960: 141; Krumbein & Pettijohn 1938: 238).

$$\text{Sorting } S_0 = \sqrt{Q_3/Q_1}$$
$$\text{Skewness } Sk = (Q_1 Q_3)/Md^2$$
$$\text{Kurtosis } K = (Q_3 - Q_1)/2(P_{90} - P_{10})$$

As an example can be given the characteristics of some dune sands and redeposited aeolian sands from Finnish Lapland (Fig. 4.14; Table 4.6).

Aeolian processes often leave a lag deposit on the ground surface, for example on interdune areas and blowouts. The lag consists of poorly sorted coarse grained sediment remaining from the source material from which the aeolian sand fraction has been removed by wind. The question now is should we call the lag deposit aeolian because large stones and blocks are also left on the surface and they may contain signs of abrasion as ventifacts (for example, in northern Iceland). They are surely wind modified (abraded) but not wind drifted sediments.

In the Polar regions coarse aeolian deposits are commonly found forming large ripples and small dunes which consist of gravel and even stone size grains (e.g. Selby *et al.* 1974). They have the characteristics of aeolian deposits but are deposited by extremely strong winds and finer fractions have been drifted further in suspension. Katabatic winds in Greenland and Antarctica can remove stones in suspension. Sometimes very coarse gravel layers have been found in sand dunes, for example in Spitsbergen (Samuelsson 1926: Fig. 36).

4.3.2   *Shape and roundness of aeolian sand grains*

Grain shape and roundness are results of the modification of angular grains of many shapes by abrasion, solution, and current sorting. It has been even presented that much of the superior roundness of aeolian sands is a product of selective sorting rather than of intensive wearing (MacCarthy & Huddle 1938: 64). Long distance transportation surely rounds the quartz grains. This has been experimentally demonstrated by Kuenen (1960) who suggested that rounding of grains by wind abrasion is 100–1000 times faster than rounding by fluvial abrasion. Roundness reflects the abrasion history. Shape has hydraulic importance because it determines the ratio of surface area to volume of a grain – the greater the ratio, the more easily will the grain be entrained by hydraulic forces acting on the surface and the slower will it settle (Pettijohn *et al.* 1987: 77).

Often the grain shape has been determined only by visual comparison to certain roundness and sphericity tables (e.g. Cailleux 1942, 1952). This is a rather subjective method and the results of different investigators may vary. A more exact method is to measure different parameters as axes or circles drawn inside and around the particle projection or to imagine spheres round or inside the grains.

Shape is defined also by various ratios of a particle's long (*L*), intermediate (*I*), and short (*S*) axes. They can be determined using a binocular microscope and graduated eyepiece. The dimensions of the small axis can be determined using a horizontal Perspex plate attached to a mirror inclined at 45° (Willetts & Rice 1983). To measure the axes of sand grains can be quite a wearisome task, and therefore for the estimates of grain sphericity only two-dimensional data has often been used, observed from the thin sections or photographs using the so-called Szadeczky-Kardóss method (Köster 1964: 180–181).

The most exact methods are those where the particles are mechanically handled as three-dimensional objects. Krygowski (1965) constructed an instrument to measure the rolling ability of mineral grains which characterizes the roundness or sphericity of the particles. The instrument consists of a matt glass plate which can be inclined and an electrically driven 'bulldozer' which pushes the grains on the glass plate up and down. The smaller the dip of the glass plate when the grains roll down the plate the more rounded are the grains.

Winkelmolen (1969, 1971) constructed another type of instrument and made studies of the same aspect of particle morphology as Krygowski and called it rollability – the minimum angle of slope which will cause a grain to roll or fall. Rollability is a property derived from roundness and shape. It is in essence a functional shape description – in contrast to a purely geometrical one – and is determined for loose sand in the rollability machine. Winkelmolen concluded that plots of rollability against grain size were helpful in discriminating between some beach and dune sands. Winkelmolen's rollability machine is better because all grain sizes are suitable for analyses, while in Krygowski's one could use only grains bigger than 0.5 mm in diameter.

Well-rounded grains are the result of either many cycles of transport, each contributing its small share of rounding, or of intensive abrasion in a special environment where rounding was accomplished very rapidly (Pettijohn *et al.* 1987: 79). Beach and dune areas are environments producing well-rounded grains. Often the well-rounded grains are reworked older sand deposits (Fig. 4.15) or from sandstones.

In cold environments the mineral material is often a very fresh weathering product of crystalline rocks, or ground by glaciers and after that transported only short distances by water before aeolian drifting. This means that the sand grains are still very angular in aeolian deposits as in sand dunes in Lapland (Seppälä 1969, 1971a) (Fig. 4.16).

Flat sand grains have large surface area compared with their mass and therefore they are easily drifted by wind. That being so, in mechanical roundness analyses (Krygowski 1965; Winkelmolen 1969), coarse grains get lower values because flat particles do not roll fast but on the contrary they slide on the instrument (Seppälä 1969). Naturally flat mica and biotite grains in dune sands are often very big in diameter compared with quartz and felspar grains. Wind is effective

Fig. 4.15 Well-rounded quartz grains from sand dune deposit in Poland (left) and angular quartz grains from dune sand, Finnish Lapland (right). Scale in mm. (Seppälä 1969: Figs. 7 and 8).

Fig. 4.16 Roundness histograms of quartz grains of dune sand from Lapland and Poland. Source: Seppälä (1969: Figs. 3 and 4).

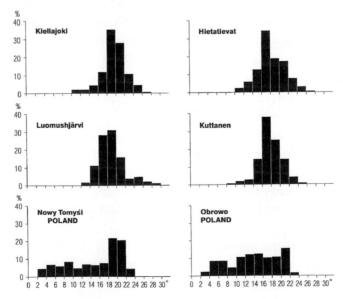

in separating flat grains with poor rolling qualities, and concentrating them in the sand dunes (Seppälä 1971a: 59).

4.3.3   Surface features of aeolian sand grains

Wearing and abrasion by geomorphic processes leave characteristic features in the microrelief of sand grain surfaces. Deposited grains can provide a tool to study the source, weathering and transport history of the grains. Many different kinds of microstructures: striae, cleavage plates, chattermarks, solution pits etc. can be identified. They may be polished, weathered or fresh. The features occur relative to surface relief, grain surface summits, depressions or even parts in between. The study of surface texture of sand grains is mostly applicable to environmental discrimination of modern sands (Pettijohn *et al.* 1987: 82).

Surface textures of sand grains can be studied by binocular and polarizing microscope and is done nowadays more often by scanning electron microscope (SEM) – Margolis & Krinsley (1971); Krinsley & Doornkamp (1973); Le Ribault (1978). Quartz grains are especially studied because they are ubiquitous and both mechanically and chemically stable.

Aeolian drift causes especially frosting on the quartz grain surfaces. The matt surfaces of well rounded and polished grains can be seen by the naked eye (Fig. 4.15). The originally clear quartz grain surface is covered by closely spaced tiny pits caused by impacts of saltating grains during the transport. Frosting is related to the scattering of light due to the small pits (Kuenen & Perdok 1962).

In cold climates etching has been found on clear quartz grains outside of aeolian deposits. Etching pits can be also due to frost in other sediments such as particles from frost shattered grains, glaciofluvial deposits and till (Seppälä 1969, 1971). This has been found also in laboratory experiments (Kuenen & Perdok 1962; Lautridou & Seppälä 1986). However, frost does not produce such totally frosted quartz grains as does long lasting aeolian transport.

4.3.4   Mineralogy of aeolian sands

The mineral composition of aeolian sands depends on the mineralogy of the source material which may be original regional rocks or reworked till, glaciofluvial deposits or beach material which are formed of a mixture of several rock sources.

Typical for cold environments is a multimineral composition of aeolian sands because the source material is often very fresh and has gone through only a few geomorphic processes and drifted only short distances, compared with old aeolian sands in middle latitudes as in Central Europe in former periglacial environments where quartz content is often more than 90%.

Recycling of sand, repeated weathering, erosion and deposition of the material enriches the quartz content because the other common minerals disappear easier than quartz. If the bedrock has undergone just mechanical dis-aggregation and there has been no chemical weathering, then aeolian sands can be composed of many different minerals besides quartz, as is the case in Victoria Valley, Antarctica, where there is a large proportion of feldspar (10 to 60 per cent) (Cailleux 1968: 63).

The specific gravity of individual sand grains (quartz, feldspar) is about 2.5–2.7. Values less than 2.4 or more than 2.9 are rare (Table 4.4). Only very fresh, recently weathered and locally transported dune sands have reasonable amounts of accessory or heavy minerals, for example in Finnish Lapland (Seppälä 1971a). On average heavy minerals formed 34.6 ± 5.0% by weight of the dune sands in Lapland, with granulite bedrock (a metamorphic rock with garnets), and the maximum content 50.7% of heavy fractions was found from a particularly dark stratum of dune sand (Seppälä 1971a: 49). The most common heavy minerals are hornblende, garnet, augite and other pyroxenes, and magnetite. The percentage of heavy minerals increases significantly towards the fine fractions (Seppälä 1971a: 50–51). The maximum heavy mineral content found was 70.2% by weight, in the 0.125–0.074 mm diam. fraction (Fig. 4.17). This fraction formed only <8% of the total weight of the sample.

In regions of recent volcanic activity as in Iceland aeolian sand is black, containing volcanic material: basaltic fragments, mafic minerals, volcanic glass and reworked volcanic ash.

Fig. 4.17 Heavy mineral content by grain size fractions of aeolian sands and their source materials from Finnish Lapland. Source: Seppälä (1971a: Figs. 32 and 33).

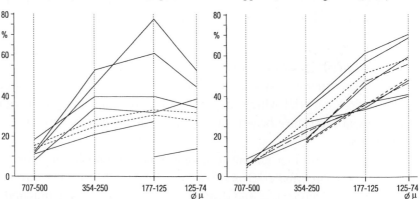

4.4 Aeolian dust and loess

Aeolian dust is solid particles transported in suspension in air. A very comprehensive review of aeolian dust is presented by Pye (1984, 1987) and the task here is not to repeat all of what he has written. He has not widely touched upon the present-day cold environments (Fig. 4.18). Most of the loess deposits in the middle latitudes are relict – formed during the Pleistocene glaciation.

Airborne dust is in origin deflated soil dust and volcanic dust of eruptions. Mineral material transported in suspension in the atmosphere has a grain size smaller than 0.1 mm. In very strong storms even stones can be drifted in suspension but grains larger than 0.02 mm settle back on the ground quite quickly after the wind speed decreases. Dust which stays in the atmosphere for a long time and is transported long distances, for example across the seas (Pye 1992), is mostly smaller than 0.01 mm in diameter (Pye 1987).

Loess deposits are formed of the airborne particles between 0.01 and 0.06 mm in size. In the particle size distribution grains smaller than 0.06 mm form 70–95 per cent and 97–99.5 per cent are below 0.2 mm (Butzer 1965: 194). 'Loess is an unconsolidated porous silt, commonly buff in colour (locally gray, yellow, brown or red), characterized by its lack of stratification and remarkable ability to stand in a vertical slope. It commonly shows a crude columnar structure. It is generally highly calcareous and effervesces in weak acid. Loess is essentially a silt.' This is a sedimentological definition of loess by Pettijohn (1975: 290–291) quoted from Smalley & Smalley (1983: 56).

Fig. 4.18 Distribution of major loess occurrences. Redrawn after Pye (1987: Fig. 9.1).

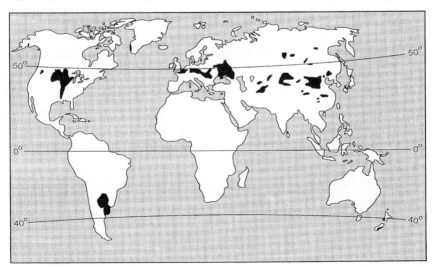

The mineralogy of loess is more complicated than that of aeolian sands. Mineral composition of loess is related to the source rocks. Quartz grains form often half or even three quarters of loess but can be up to 92%, as in Cypress Hills, Alberta, Canada (Catto 1983: 1163). Feldspar and calcite content in loess varies very much. Typical loess consists of carbonates (10–30 per cent) and clay minerals (10–20 per cent) (Butzer 1965: 194). In loess deposits signs of soil formation and weathering can be seen. Carbonate concretions ('loessdoll', Lösskindel) (Seppälä 1969: Fig. 3) are typical examples of diagenetic features and indicate secondary mineral movements after the essential deposition.

Some heavy minerals (hornblende, garnet, fluorite, magnetite and apatite) are found in loess especially in its finer fractions (Catto 1983: 1163). Alaskan iron-rich loess contains much biotite and chlorite derived from schists and gneissic rocks (Péwé 1955). Loess also often contains clay particles which adhere to the larger quartz and other grains.

Quoting mainly Taber (1943) and Péwé (1955) the loess of Alaska can be briefly characterized as follows. The colours of the dry silts (loess is aeolian silt) range from light grey to brown and from dark grey to dark brown and almost black when wet (Taber 1943: 1473–1475). The silt is fine-grained (0.03–0.07 mm) with a clay content of less than 5%. All silts contain some portion of coarse material which consists of angular fragments of quartz and micaceous quartzite, plant remains, small concretions or aggregates of particles slightly cemented by limonite, and fragments of weathered schist. The analyses of mineral composition of silts show a great variety. Because the presence of lime is often considered a defining characteristic of loess, and calcite is rather common in Alaskan silts, it is well-founded to call them loess, even though Taber (1943) and Péwé (1955) prefer the term silt. The fresh rock flour with light grey colour can be identified from most of the mineral grains which show staining by iron oxides. Volcanic glass is present in practically all the silts but in most less than 1% (Taber 1943: 1481). In silt deposits can be found plant remains showing that the flora during the accumulation of the silts was essentially the same as that of today (Taber 1943: 1481). That is also good evidence that the accumulation of silt took place on the dry land and among vegetation. In some regions the climate was probably somewhat milder during the silt accumulation than at present. For example, today Seward Peninsula is treeless, but large spruce logs are found there in the silts in many places (Taber 1943: 1483).

Wind drifted volcanic ash can be found in loess layers as well as on ice sheets in Greenland and Antarctica and especially on Icelandic and Alaskan glaciers. These so-called tephra layers can be used as trace layers when dating loess deposits (e.g. Begét & Hawkins 1989; Begét *et al.* 1991) and ice layers of glaciers.

Often the quartz grains of loess are sub-rounded but also angular grains with chattermarks can be found (Pye 1987: 216) probably indicating frost shattering (cf. Lautridou & Seppälä 1986) or mechanical breakage by glacier movement. Frost weathering is able to produce silt size material and can be considered as a major mechanism for supplying loess material (e.g. Brockie 1973; Lautridou & Ozouf 1982).

Butzer (1965) defined loess according to its origin as two types (1) periglacial loess, deflated from outwash deposits, from freshly exposed till, and from barren rock and tundra surfaces, and (2) desert or continental loess originated from desert areas. At present we are interested in the former type.

Aeolian silt without calcite is called loess-like sediment. Airborne silts are rather common in glacial environments where they are drifted from outwash plains by wind. Also subaquatic fluvial and lacustrine silts can be loess-like, as well as some colluvial deposits (e.g. Konishchev 1987). They are sometimes difficult to separate from aeolian loesses. Subaquatic silts often have some stratification (Péwé 1955) which is normally absent from airborne loess.

4.5   Salt drift and weathering

There are many studies of salt weathering, which means crystallization pressure caused by salt crystal growth, hydration and thermal expansion of salt (e.g. Yatsu 1988; Goudie 1997). Our task is not to go into the details of weathering processes but to mention that wind plays a certain role also in salt weathering processes in cold environments.

Wind drifted salt spray is a common phenomenon everywhere in coastal areas but very little studied. Samuelsson (1926: 155–156) mentioned that windows some 5 km from the west coast of Sweden became salt coated after a strong storm.

In Antarctica 100 km inland from the open ocean we have found potassium and sodium in snow, indicating airborne distribution of small salt water drops. Soil salts of calcium carbonate, mirabilite (hydrous calcium sulfate), chlorides, iodates, and others are common and widespread in all ice-free areas of the Antarctic (Black & Berg 1963: 127). Preglacial accumulations from weathering, marine salts concentrated in different ways and local hydrothermal emanations all seem likely sources for the initial salts.

Clear water lakes with high salt content are common in the arid parts of Greenland and the Canadian Arctic islands (Davies 1974: 57). What role the sea water spray plays in this salt accumulation is unknown.

Honeycomb structures called tafoni are best developed in coarse grained crystalline bedrock and morainic boulders, Antarctica (Cailleux & Calkin 1963; Selby 1977). Péwé (1974: 46) wrote that one of the least understood processes in arid Polar deserts is cavernous weathering or formation of the tafoni. Cavernous weathering forms recesses, hollows, pits, niches or tafoni in rock faces (Fig. 5.3). Tafoni may occur in small pits up to 10 metres in height and 20 metres long especially in granitic rocks (Péwé 1974: 46–47). Sekyra (1969: 282) uses the term *aeolian corrasion* and strongly supports the aeolian hypothesis for honeycomb relief of various shape. Wind is surely involved in tafoni formation and probably in at least two ways: by drifting salt water drops which activate the weathering and then by removing the debris (Blackwelder 1929a).

Evaporation of salt-bearing solution causes mechanical disintegration through the growth and expansion of salt crystals. In Dry Valleys, Antarctica, there is very little liquid water flowing from melting glaciers. Precipitation is limited and mainly in the form of snow. This means that rubble and salts ensuing from weathering and the marine salts brought by the snow and wind are not washed away. They percolate into the soil and rock and play a dominant role in the disintegration of rocks (Miotke 1983: 23).

Salt accumulation in cold environments can also decrease the effects of wind. Deflation is partly limited on the vegetation-free polar deserts by the presence of an evaporation-caused salt crust or hardpan at the ground surface (French 1976: 203).

5

Abrasion

Abrasion (sometimes called corrasion) means the polishing effect of solid particles carried by the wind. In geomorphology abrasion is defined as erosion of solid rocks by any agent: glacial, fluvial, currents, surf or wind. The amount of abrasion depends on sediment supply, wind speed and turbulence, and the heterogeneity of target materials (Malin 1992: 27).

Wind abrasion is a common feature in regions with deflation, when particles are moving close to the ground surface and bombarding the immovable rock fragments or rock surfaces (Plates 2 and 3). We should not underestimate abrasion caused by snow and ice crystals (cf. McKenna-Neuman 1990a; Schlyter 1995a): very intense abrasion takes place on tiny nunataks without large sand or silt sources.

Mawson (1915: 123–124) described abrasion in Antarctica : 'Of the effect of wind and drift upon rock, there was ample evidence around the Winter Quarters. The northern aspect of the rocks was quite different from the southern. The southern, windward faces were on the whole smooth and rounded, but there was no definite polish, because the surface was partly attacked by the chipping and splitting action of frost. The leeward faces were rougher and more disintegrated. More remarkable still were the etchings of the nonhomogeneous banded rocks. The harder portions of these were raised in relief, producing quite an artistic pattern.'

Abrasion does not only concern rock surfaces but the snow surface and hard, blue ice may be channelled and pitted by wind drifted snow and ice crystals (Plate 14) (Mawson 1915: 124).

5.1 Signs of abrasion on rocks

On the west coast of Greenland, Jensen (1889: 68) observed smooth polished grey gneiss blocks which had the form of tables and mushrooms with thin stalks or legs. In south Jameson Land, east Greenland, Schunke (1986: 26) found table rocks of sandstone origin on a largely unvegetated surface. From interior Jameson Land wind polished crystalline boulders are reported, some-times perched on top of the tors. The most prominent of these so-called ventifacts are the very resistant Scolithos-quartzite blocks (Möller *et al.* 1994: 342–343). Similar forms are found in Greenland where ventifacts and rock bosses have been etched by sand blast (Flint in Boyd 1948) and in dry valleys in Antarctica (e.g. Nichols 1966; Sekyra 1969: 282; Péwé 1974) (Fig. 5.1); simi-lar effects occur by snow blast on many nunataks in Antarctica (Seppälä 1991) (Fig. 5.2, Basen, basaltic blocks). In the inland oases and on old Pleistocene moraines in the mountains, Antarctica, were found wind troughs, mushroom-like tombs and honeycomb relief remodelled by wind (Sekyra 1969: 282). On the west coast of Ross Island, McMurdo Sound, Antarctica, Debenham (1921: 65–69) found polished rock surfaces and potholes worn by wind abrasion.

In Victoria Dry Valley, Antarctica, the wind polished moraine consists of dolerite boulders up to 90 cm in diameter with a small proportion of gneiss and granite erratics which have a surface veneer of ventifacts (Webb & McKelvey 1959: 126–127). The divides are formed of dolerite boulders sandblasted to honeycomb patterns. The resistant cap of each mushroom granitic rock is iron

Fig. 5.1 Ventifacts in Wright Valley, McMurdo Sound, Antarctica. Photo by K. Moriwaki, 15 December, 1970.

stained and little abraded, while the lower cavernous sides have been actively sandblasted (Webb & McKelvey 1959: 132).

From the Rockefeller Mountains, King Edward VII Land, Antarctica, pitted surfaces, wind-formed potholes, 'pock-marked' surfaces and wind-eroded rock surfaces on very coarse grained granites are described. The igneous rocks seem to have been more susceptible to wind erosion than have the metamorphics (Wade 1945: 72–75). Wind together with frost action is playing a key role in the formation of these features, but we should not forget the melt-water on rock surfaces subject to direct sunlight (Wade 1945: 72). This author has seen, on nunataks in Queen Maud Land, stones and rocks with cuplike holes on them in which meltwater collected and produced weathering gravel. Wade (1945: 72) found large quantities of fine rock material on the snow surface sometimes a mile away from the nearest exposure.

On southwestern Greenland near the margin of the ice-cap (Nordenskjöld 1910: 25–26, Fig. 5) and at the edge of a small nunatak in east Greenland (Teichert 1939: 147) wind-corraded rocks were found. (See Fig. 5.3).

Near the mouth of the Kolyma River on Four Pillar Island, northeast Siberia, Sverdrup (1938: 370–372) found up to 20 m high granite pillars that had been drastically reduced in diameter about 3.5 m (12 feet) above the base, and polished by wind erosion. His explanation was the abrasion of drifting snow which may

Fig. 5.2 Wind polished basaltic rocks with rock varnish on Basen nunatak, Queen Maud Land, Antarctica, January 1989. (Seppälä 1991: Fig. 2).

reach the maximum speed and erosional impact at that height. Rather similar rock pillars resulting from abrasion of sandstone and siltstone are depicted from Western Sverdrup and adjacent islands, Arctic Canada (Hodgson 1982: 15, 30).

The height of the abrasion signs on big blocks does not always indicate the effective height of the moving particles during the whole abrasion process. If the block has been buried in sand and deflation has uncovered it then abrasion has taken place during deflation close to the surface. It means that the upper parts of the block were first abraded and the most recent processes affected the lower parts. The other alternative factor in cold environments could be the changing snow depth which could explain the marks of abrasion found very high above the ground surface. Accumulating snow protects the lower parts of rocks and the last abrasion takes place at the upper levels. The melting snow is not easily drifted by wind. In the most active abrasion environments snow cover seems to be rather thin but the light dry snow crystals move much higher than the sand grains.

From Iceland Cailleux (1939: PL. III, Fig. 2 and VII, Fig. 1) described wind worn 'palagonitic' (stratified tuff) blocks on which the softer strata were strongly scoured by wind and harder elements were embossed. In northern Iceland wind has faceted basaltic boulders on large areas with pavements. Also fluting and smoothing on volcanic breccias results from wind action (Samuelsson 1926; Cailleux 1939; Bout 1953). (Fig. 5.4).

Fig. 5.3 Wind-corraded coarse-grained granitic stone from the silt terrace west of the Søndre Strømfjord, West Greenland. Photo by W. Tobiasson.

In Ungava peninsula, Arctic Canada, this author found wind abraded blocks on vegetated surfaces on tundra (Seppälä *et al.* 1991). The lee side of them had lichen indicating that the stoss side was still actively abraded but in this case by wind drifted snow and ice crystals. Those diorite blocks had some quartz veins which were more resistant and raised up from the surface. The lower parts of the about 1 m high blocks were less abraded than the higher parts. The snow protects the lower parts although wind usually blows a wind furrow in snow around the blocks (Hellemaa 1991: 314). The prospecting road close by was marked with wooden posts and they showed signs of abrasion (Fig. 5.5) from 20–40 cm above the surface upwards. This has been the common depth of snow in northern Ungava.

Large faceted boulders and windblown sand deposits occur at the head of Pangnirtung Fiord, Baffin Island (Gilbert 1983: 164). The rocks in the desert pavement, for example in northern Iceland, are subsequently faceted by the wind, and as ventifaction proceeds, new faces are cut on the rocks.

5.1.1 Ventifacts

For the wind worn rock features the term *eologlyptolite* was proposed by Jan Dylik (Karlov 1969: 221). It means rock fragments bearing well-marked

Fig. 5.4 Wind abraded erratics in northern Iceland. 1973.

traces of eolization, i.e. smoothing and polishing work of mineral particles blown by the wind. We shall use the term ventifacts (*Dreikanter* in German) for these wind abraded stones caused not only by sand grains but also by moving snow and ice crystals.

Fig. 5.5 Pieces of initially painted wooden track markers from Ungava Peninsula, Québec. 1984.

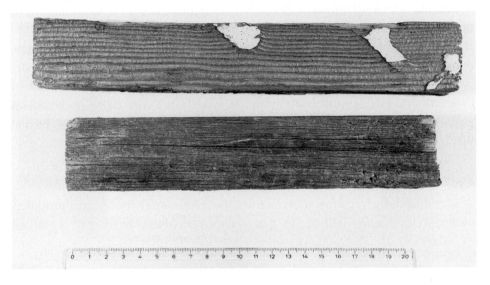

Fig. 5.6 Ventifacts from Delta Junction area, Central Alaska. 1994.

Several types of wind-worn rocks have been described by Karlov (1969): facetted stones on which facets are separated by reasonably well-defined edges. The stones can be multi-faced and he has also found non-facetted eologlyptolites.

Sharp (1949: 177–178) described for Big Horn, Wyoming ventifacts 20–30 cm in diameter having up to twenty wind-cut faces exhibiting cutting all over the surface. Their shape can described as irregular or polygonal. Ventifacts of intermediate size, 15–35 cm in diameter, are mostly ridge-shaped, with two major opposed faces, or pyramidal, with three to six faces converging toward an apex. Pyramidal ventifacts are confined to the upper part of the stone, probably because it was partly embedded in the ground when the cutting took place. Bigger boulders rising several tens of centimetres above the ground usually show a curved, wind-carved surface.

All wind-worn surfaces of ventifacts show some degree of pitting, fluting, or grooving (Fig. 5.6) which are primarily lithologically controlled. Pits are irregularly shaped and can be up to 3 cm in diameter. In the Big Horn Region, Wyoming they are best developed in medium- to coarse-grained crystalline rocks, although elsewhere common in lava (Sharp 1949: 178). Hard minerals such as quartz cap the pitted surfaces. Pits form on faces making angles between 55° and 90° with the wind. Flutes are scoop-shaped in plan and broadly U-shaped in cross-section. They may be nearly indiscernible to the naked eye or up to 15 cm long, 4 cm wide, and 2 cm deep. Usually one end of a flute is open downwind and the other closed (Sharp 1949: 179). Flutes form best on surfaces inclining less than 40° to the wind. Flutes are independent of mineral hardness and rock structure. Grooves are longer than flutes and open at both ends. The largest reported by Sharp (1949: 179) were 45 cm long, 5 cm wide, and 2.5 cm deep. They are best developed on surfaces gently inclining or parallel to wind.

Transverse erosional marks on rock surfaces were found in Greenland (Fristrup 1952–53). Very good examples of 'transverse' erosional marks and scoured rock surfaces are described by Selby (1977) from McMurdo Oasis, Antarctica. In Antarctica the transverse erosional marks, flutes, may be up to 2 m long, 1 m wide, and 0.2 m deep (Selby 1977: 950). He found these exceptionally large marks to be attributable to the high lifting capacity of cold and dense air, and large particles available for abrasion. The most common type of erosional mark is the parabolic mark with mean dimensions of about 15 mm length, 9 mm width, and 1.7 mm depth (Selby 1977: 952). Selby expected that wind directions can be identified from the erosional marks. The orientation of hollows in large, cavernously weathered boulders and ventifacts on the valley floor in the lower part of Victoria Valley, South Victoria Land, Antarctica (Cailleux & Calkin 1963) can probably be used for the same purpose.

A *facet* is one particular type of face. The term *facet* appears to mean a relatively plane surface cut at right angles to the wind, regardless of the original shape of the stone (Sharp 1949: 180). King (1936: 203) used the term 'face' of the original surfaces of the rock fragment. Facet he applied only to those surfaces which have been cut under wind action.

Most wind-cut surfaces facing into the wind are inclined 30–60° from the horizontal. Schoewe (1932: 128–130) experimentally indicated that reduction of faces at angles below 30° proceeds slowly, but cutting does not cease entirely at that angle and he concluded that sand corrasion does not occur on horizontal surfaces. Sharp (1949: 193) concluded his observations of ancient ventifacts that 'cutting has occurred on horizontal surfaces, but it cannot be shown that faces are reduced to horizontal attitude by wind action, although there is no apparent reason why they should not be'.

5.1.2   Examples of ventifact occurrence

Several students have reported ventifact observations from Greenland. Ventifacts are found everywhere with sand blasting, in deserts, on beaches and close to glaciers with unvegetated sand surfaces. A considerable amount of ventifacts, mostly *Einkanter* (one worn facet surface only), occurred on Sand Island and close to Sandodden (about 74° 20′ N, 20° W) and NW of Renbugten (about 73° 25′ N, 26° 30′ W), northeast Greenland (Flint 1948: 208–209).

In Peary Land at Brønlunds Fjord, north Greenland the ventifacts are very abundant (Troelsen 1949: 22–23; 1952: 221). An experimental abrasion study was made there, with the placing of a number of brick blocks of various shapes among the natural ventifacts. The study lasted only five months from midwinter to 1st July. Violent gales were frequent throughout the period. Because a great deal of sand of varying grain size was carried along by the wind, it was not possible to determine which proportion of the wear of the blocks was due to the effect of blown sand and how much to abrasion by snow and ice crystals (Troelsen 1952: 222). Soft silica-bricks had been strongly eroded on the side of the prevailing winds, and a little on other sides due to infrequent winds, too. The lowermost part of the blocks showed little erosion, but already 1 cm above the ground the scars were significant. Narrow practically horizontal furrows and ridges extend to the leeward from small pits and knobs on the sides of the blocks. The flat upper surface found on some of the blocks was absolutely untouched by erosion (Troelsen 1952: 222, Fig. 2).

Nichols (1969: 82–83) found numerous well-developed ventifacts in Inglefield Land, north Greenland close to an ice-dammed lake. Wind-cuts more than 1 m above ground level are located on the sides of the blocks towards the lake or glacier. The leeward surfaces of fragments not submerged by the high stage of the lake are stained by limonite; those surfaces which were submerged are veneered with yellow lacustrine silt. The windward sides are grey, as the yellow lake sediments and any limonite that is formed are annually removed by sand-blasting. This difference in colour is easily recognized.

Nichols (1969: 82) considered that wind-cutting takes place during the summer months when coarse sand is moving. He did not visit the area in the winter. The location of sand dunes and the distribution of the wind-cut surfaces prove that winds blow off the glacier. Nichols does not give any role to drifting snow as an abrading factor even where there are numerous less spectacularly developed ventifacts. Aeolian sand deposits are not large compared with the widespread block fields, considering the aridity of the climate in north Greenland (Nichols 1969: 90–93). We may conclude that the work of wind is of greater importance than Nichols believed but much happens in the winter.

The effects of wind polishing are evident in the Queen Elizabeth Islands of Canada (Pissart 1966; Tedrow *et al.* 1968). Particularly the cliffs around the bays on the eastern margin on Prince Patrick Island are subject to wind sculpturing (Tedrow 1977: 74–75). The example from Paulatuk, N.W.T. on the Canadian Arctic coast proves the winter abrasion by snow because the ventifacts there are surrounded by stone pavement which prevents sand blasting (Fig. 5.7).

Fig. 5.7 Block with one facet abraded by wind drifted snow from Paulatuk at Darnley Bay, N.W.T., Canada. The block is surrounded by coarse stone pavement. Photo by J. Ross Mackay, 1968.

Cailleux (1939) gave many examples of volcanic boulders, pebbles and blocks worn by wind. They cover large areas in Iceland. Upper parts of boulders are shiny and lower parts of greater blocks are eroded and not covered by vegetation. Boulders have flat faces forming sharp edges which are often at right angles to the prevailing NE wind direction. The lengthened spatula-shaped cupules are mostly directed parallel to the same wind direction, and their deeper end is always windward (Cailleux 1939: Pl. VII, Fig. 2). Only small pebbles, beneath 5–10 cm, have been turned upside down during the process by the wind. Wind scour causes undermining on the sand surface on the sides of stones (Evans 1911: 334–335; King 1936: 208; Sharp 1949: 184). Evans (1911) concluded that after the turning over toward the wind on its abraded surface 'a plane of abrasion [would] then be formed on the stone making an angle of about 60 degrees with the first'. King (1936: 208) considered it far more probable that the pebble would merely sag toward the forming depression into an intermediate position. Therefore the wind directions could not be determined reliably at all places where only small pebbles exist. There might have been changes in position of the stone between the periods of cutting. Stirring of ventifacts by frost action was thought possible by Cailleux (1942: 45). Big blocks only partly exposed on the surface are especially sensitive to shifting by frost heave if the substratum is silty till (Seppälä 1987b: 48). Other factors shifting ventifacts are thought to be heavy rains, melt waters, very strong gusts of wind, animals, solifluction, settling, and solution (Schoewe 1932: 132; Sharp 1949: 184–185). Each facet on a ventifact does not mean a different wind direction.

Ventifacts on West Spitsbergen have lichen cover 1–4 cm above the ground surface also on their windward side. This has been protected by a small amount of snow packed against the rock surface. The snow accumulation also causes abrasion of small hollows a few cm above the ground surface (Åkerman 1980: 250–251, Fig. 4.19). This is also a clear evidence of the abrasion effect of drift snow. The lee sides of wind facetted blocks in Spitsbergen are all covered by lichens, as they are in Paulatuk, Arctic Canada (Fig. 5.7).

Ventifacts have been reported from several piedmont areas in the Rocky Mountains (e.g. Wentworth & Dickey 1935; Sharp 1949). Ventifacts are lacking on many of the highest erosion surfaces presumably because the stones were protected from the wind by vegetation or topography, and another explanation could be the scarcity of dense siliceous stones resistant to weathering (Sharp 1949: 176). Stone type is important for ventifact formation.

East of Big Horn Mountain in Wyoming ventifacts were cut by wind blowing consistently from the NW (Sharp 1949: 177). In South Park, Colorado the wind grooved surfaces on undisturbed boulders were found to face N 60° W, indicating abrading winds from that direction (Powers 1936: 218).

Wind polished rock surfaces are rather common features in Antarctica (Fig. 5.1). Lindsay (1973b) reported from Wright Valley that wind initially removed the fine sand grains from the soil surface and produced a lag gravel. Rocks on the desert pavement are then subsequently facetted.

5.1.3   Rock types in ventifacts

From the United States ventifacts are reported with different origins: sandstone, basal conglomerate, kame gravel, glacial boulders and Pleistocene moraine materials (Wentworth & Dickey 1935: 102). Powers (1936: 214) added to the former list three regions where abundant evidences of former wind abrasion are to be found: the sandy plain of the glacial Lake Wisconsin, the upper valley of the Arkansas River in the Rocky Mountains, central Colorado, and the broad semi-arid intermontane basin called South Park, Colorado. In Wisconsin wind-carved pebbles have accumulated at the surface as a true 'desert pavement' of early postglacial origin (Powers 1936: 215). Smith (1949a: 211) pointed out also from southern Wisconsin ventifacts of hard, siliceous rock which may have originated under periglacial conditions.

On Big Horn, Wyoming, ventifacts are composed of chert, quartz, quartzite, quartzitic sand stone, gneiss, hornfels, pegmatite, diabase, and different granites (Sharp 1949: 177). From other regions have been found wind-cutting on limestone and dolorite, but presumably the lack of this type of ventifact can be attributed to destruction of wind worn surfaces by solution (Sharp 1949).

On dense resistant rocks, such as quartz and chert, sandblast action polishes especially smooth faces. On fine-grained diabase more conspicuous markings can be seen, but its markings can not be compared with those on granitic and gneissic rocks (Sharp 1949: 179).

In northern Iceland and on Basen nunatak in Queen Maud Land, Antarctica, this author has seen well formed ventifacts of basaltic rocks (Fig. 5.2). In Finnish Lapland some diorite, gabro and hornblende stones have wind-worn facet surfaces; however, they are rather rare.

In Central Alaska (Shaw Creek Road) the ventifacts are vein quartz from the bedrock. They are up to 15 cm in diameter, well cut, polished, and form almost a continuous horizontal sheet which is covered by aeolian sand and loess (Péwé 1965: 48–49). This author has seen north of Delta Junction by the Alaskan Highway a similar layer of ventifacts of micaceous gneiss, hornblende gabro, greenstone, metagabro, basalt and diabase (identified by Dr. Martti Lehtinen, Geological Museum, University of Helsinki) (Fig. 5.6).

In Poland the old sediment-covered ventifacts from the Pleistocene are often formed of sandstone, granodiorite, quartzite and phyllite. These ventifacts are not always smooth surfaces but they have pits and other signs of abrasion formed by wind. Schlyter (1995: Fig. 4) described this type of relict pitted ventifact from southern Sweden.

5.1.4   *Formation*

The formation of ventifacts has been studied in the field and experimentally in the laboratory. King (1936) discussed the meaning of the original size and shape of rock on the face forming. He made his observations on beaches in New Zealand. According to him, cutting is most rapid when the face is nearly vertical, and the facets develop quickly attaining an angle of 50° or 60° from the horizontal. Often two faces are directed obliquely toward the wind and both sides are simultaneously trimmed into facets by the sand blast. In the case of a rectangular stone so placed that it presents two faces at right angles to the directions of the two prevailing winds, two facets are cut facing in opposite directions (King 1936: 206–207). Angular fragments give more possibilities of formation for multi-facetted ventifacts, and rounded boulders may be expected to offer greater opportunity for the facet positions to be governed by wind direction (King 1936: 210–211).

Cailleux (1942: 54) thinks that fine material transported by wind may help cut ventifacts. This material may be fine enough to follow wind currents in vortices and thereby aid in cutting flutes and grooves. Schoewe (1932: 127) discovered that sand grains impinging at low angles on hard, smooth surfaces skid instead of rebounding directly into the air. This could occur on wind-cut rock surfaces and may also have something to do with the development of flutes and grooves (Sharp 1949: 179).

Powerful wind can also support the fragmenting of rocks in certain conditions by packing sand particles in rock fractures, as reported from Taylor Valley, Antarctica (Hall 1989).

5.1.5   *Snow as an abrading material*

Sharp (1949:185–186) does not regard the wind drifted snow as responsible for ventifact formation on Big Horn. He considers questionable ventifact formation by snow drift: 'It is yet to be demonstrated that wind-blown ice particles at any temperature can produce ventifacts of the type found. Big Horn ventifacts should be more widely distributed if cut by ice particles, for snow was certainly more widely available than sand. Snow is often so abundant, when present, that it buries small stones and protects them from cutting.' This is an odd conclusion in Wyoming where winters are very severe and winds on the open prairie are really strong and even more effective on the mountains.

It has been somewhat unclear whether blowing sand or blowing snow is the more effective polishing agent in the ventifaction process (Tedrow 1977: 81). A somewhat questionable observation from Canada was that abrasion by snow is not significant (McKenna-Neuman & Gilbert 1986: 218).

Many ventifacts found in the Arctic and Antarctica on pebble paved surfaces (Figs. 5.7 and 5.2) without fine sediments which wind could drift, large ventifact fields in northern Iceland (Fig. 5.4), erratic blocks which have signs of abrasion well above the vegetated earth surface in Ungava Peninsula and wooden track markers (Fig. 5.5) eroded above the snow surface also in Ungava, are all evidence of the abrasion ability of wind drifted snow.

In West Spitsbergen most of the wind abrasion is carried out during the winter when soils are drier, weathering produces more fine material and wind is drifting snow and ice crystals (Åkerman 1980: 245).

Schlyter (1995: 20) and Schlyter *et al.* (1995) conclude that snow abrasion proved to be less efficient than abrasion by quartz grain impact: given the same kinetic impact energy, snow is 200–300 times less efficient than quartz. If compared on a particle size basis, abrasion by relatively coarse snow (0.4 mm) is only about as effective as abrasion by coarse silt. In this kind of comparison we have to think also of the time. In the Arctic conditions which prevailed in southern Scandinavia during the late glacial time, the summer was short and the snowy season probably over 10 months, and the winds stronger in winter and dry snow available everywhere. To produce a sand storm or a dust storm we need strong winds and much material. If silt drift was the reason for widely distributed ventifact formation in late-glacial time, then the silt should be deposited somewhere. Even thin vegetation very effectively accumulates the drifted material. We cannot expect that all the drifted silt was blown to the sea.

Low density particles have large surface compared with the high density particles of the same mass (Table 5.1). Therefore the low density particles can be transported easier with high velocity, and the impact force will be more effective. Snow density in the calculations is low compared with pure ice density 0.91.

Physical characteristics of ice and snow form the basis for cold environment abrasion of rocks. Heim (1885: 286) wrote in his textbook that very cold ice is

Table 5.1. *Particle diameter (ϕ), radius (r), sphere volume ($V = 4/3 \cdot \pi \cdot r^3$), density ($\rho$) and mass (M) of drifted particles.*

	ϕ	r	r^3	V	ρ	M
snow	0.5	0.25	0.015625	0.0654	0.3	0.01962
sand	0.2	0.1	0.001	0.004187	2.6	0.0189
snow	0.2	0.1	0.001	0.004187	0.4	0.001675
silt	0.06	0.03	0.000027	0.00011305	2.6	0.000294

hard as glass, shattering at −50° to a fine powder ('Eis ist bei sehr grosser Kälte hart wie Glas. Bei −50° wird es kaum noch von der besten Feile angegriffen, es zerspringt, gepresst, unter heftigem Knall in schmeidend pulverige Splitter.'). Steenstrup (1893) told an interesting episode. He had asked Heim where he got that knowledge and Heim replied that he could not remember, 'probably from some North Pole explorer'. Steenstrup (1893) had noticed in Greenland that ice was not so hard but broke as easily as glass when it was −30 °C. He had also made some experiments on ice hardness and concluded that it is between 2.5 and 3 at −70 °C.

Different investigators have come to different results for the hardness of ice. All agree that the hardness of drifting snow and ice crystals increases as the temperature decreases. The hardness of the ice stated in the geological handbooks is 1.5 to 2.0. This is the case at freezing point. Koch and Wegener (1930) found that the Mohs hardness of drifting ice was exactly 4 at −44 °C and 2–3 at −15 °C. Teichert (1939; 1948) took the observations by Koch & Wegener (1930: 302) and in the Alps and Greenland and from these data he presumed that probably at −50° the hardness of ice may be 6 and concluded that: 'There is no doubt that snow at low temperatures possesses the physical properties required for corrasion even on hard rocks.'

Blackwelder (1940: 61) supported Teichert's conclusion with data showing that ice at very low temperature (−78.5 °C) has a hardness approximately 6, or that of orthoclase feldspar. Then he concluded that 'ice is hard enough to abrade limestone, shale, and many other common rocks, even including some igneous masses' (Blackwelder 1940: 62).

These experiments are regarded as having proved that the ice crystals in drifting snow often will be of a hardness of about 6 or perhaps even more in the high Arctic and Antarctica (Fristrup 1952–53).

Laboratory studies show that abrasion by snow occurs already at relatively high temperatures (Dietrich 1977a, 1977b; Schlyter *et al.* 1995) and some field experiments also indicate the same (Åkerman 1980). In west Greenland snow abrasion was observed taking place at a temperature as high as −2.5 °C (Frich 1988).

5.1.6   *Experimental abrasion measurements in natural conditions*

Most geomorphologists in the High Arctic and in Antarctica have drawn attention to the abraded rock surfaces, but very little information is found in literature of the quantitative measurements of the rate of abrasion in these regions. 'Evteev (1962), working on the oases in the vicinity of Mirnyy, found honeycomb and cell-like cavities developing in the rocks at a rate of 1 to 1.5 mm per year as a result of wind action' (citation from Tedrow 1977: 81).

Åkerman (1980: 266) started his experimental measurements by turning some lichen covered blocks in West Spitsbergen and found that after one winter the lichen was heavily affected and after the third winter almost all lichen was destroyed. Then he placed wine bottles upside down in the field to register the local wind directions from the blast-peening effect on the surface of the glass, which was in one winter so small that bottles were changed for softer plexiglass tubes with a diameter of 5 and 10 cm and 50 cm long. 40 cm of the tubes was left above the ground. The abrasion of the plexiglass tubes very well reflected the prevailing wind directions of the winter months. No abrasion was noticed in July–August or September–October which means that snow and ice crystals were the abrading agents (Åkerman 1980: 267). From the surfaces of the tubes Åkerman (1980: 286–287) could determine the relative intensity of the abrasion which was at its maximum 0.5 to 1.0 m above the ground surface (Fig. 5.8). This means that the wearing agent was drifted snow and ice, while Sharp's (1964: 797) measurements in Coachella Valley, California showed the maximum abrasion

Fig. 5.8 Wind abrasion with height above the ground upon plexiglass and clay pigeons from December to April 1975 at Isfjord Radio, Svalbard. 0 = no wear, 5 = maximum wear. Source: Åkerman (1980: Fig. 4.54).

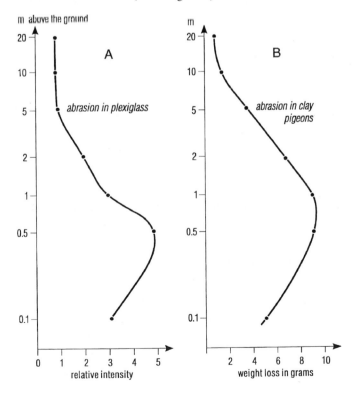

at the height of 20 cm at which grain size, number and velocity combine to give maximum impact energy and wear by mineral particles. Åkerman (1980: 283) also used for abrasion measurements artificial clay pigeon plates (hardness around 2.5 on the Mohs' scale) exposed to wind in West Spitsbergen. Weight losses were recorded up to the height of 20 m above the ground and the maximum was again from 0.5 to 1 m above the ground (Åkerman 1980: Fig. 4.54).

Malin (1984, 1985, 1986, 1992) made a series of abrasion rate observations in Victoria Valley, Antarctica, by exposing three types of natural material (basalt, non-welded tuff, sandstone) for sand drift. The abrasion targets 2.5 cm in diameter and 0.5 cm thick were mounted on an aluminium framework that held them at nominal heights of 7, 14, 21, 35 and 70 cm above mean ground level. The targets were faced N, E, S and W in eleven different places. Maximum abrasion occurred at a height of 20–25 cm (Malin 1985: 15). The average rates of abrasion range from about 30 (basalt and sandstone) to 500 micrometres per year (tuff) during the season 1982–83 (Malin 1985: 15). There is a wide range in year-to-year variation in abrasion rates. Abrasion rates determined from single-year observations were probably not representative of long-term averages, although they indicated in general the order of magnitude of abrasion. More abrasion (67–75% of the 5 years' total) occurred during 1984 than in each of the subsequent 4 years (Malin 1992: 27–28). At some sites and in some directions, all abrasion occurred during that single year (1984) when sediment transport velocities were much higher (Malin 1992).

Abrasion measurements of artificial walls in the Sør Rondane Mountains, Antarctica, was carried on by Matsuoka *et al.* (1996). The results were that the abrasion of an asbestos board facing the snowladen katabatic wind suggests that snow particles can polish the windward face of rockwalls. Maximum abrasion took place at 30–40 cm above the ground.

Malin (1992: 28–29) also observed abrasion by wind drifted ice at the Allan Hills ('blue ice'), Victoria Land. It was lower than on sand drift sites with an average amount of 0.001 grams per square centimetre per year (dolerite) and ten times more for tuff. That means about 3 and 80 micrometres, respectively.

Two sedimentological factors appear to play important roles in determining where abrasion occurs: the availability of loose, wind-transportable debris and the small-scale configuration of the surface (Malin 1992: 28). The most abrasion is recorded when surfaces are mostly or partly covered with loose material (Malin 1984, 1985, 1988). 'Surface configuration probably contributes indirectly to enhanced abrasion in two ways. First, sand saltating across rock-strewn, armoured surfaces attains greater kinetic energy (primarily because of higher rebound heights), and even a small amount is more effective in abrading. Second, coarse sand and gravel are often the only wind-transportable particles exposed on such surfaces (owing to previous exportation of finer fractions), which require higher winds to move and, once in motion, have greater kinetic energy owing to both larger mass and higher velocities of transport' (Malin 1992: 28).

Miotke (1979, 1982) determined by a laboratory wind-tunnel experiment that the ventifacts can be formed within a few decades or, at most, a few centuries. Malin's (1986: 19) observations from Victoria Land, Antarctica, do not support these approximations, but he proposed 100 times or greater interval to be more likely.

In Alaska the observed ventifacts (Fig. 5.6) are close to river valleys which are the sources of the blast sand and silt abrading the stones.

5.1.7 *Experimental abrasion in laboratory*

Abrasion of sandblasting has been investigated in the laboratory to find out the wearing rate of pebbles (Kuenen 1960). In the experiment strong wind (25 m s^{-1}) was used, which is hardly ever experienced over long times in most places in the Arctic, and a second run used a wind speed of 11 m s^{-1}. The pebbles were quartzite and the wind drifted material was crusted quartz with grain size fractions from 0.10 to 1.2 mm, sand and a mixture of both. The data showed that the increase of wind speed reduced the required time for a certain amount of abrasion 67 times. Increase of sand size from 0.25 to 1.0 mm reduced the required time to one third. To produce ventifacts thousands of hours of strong wind is needed. Kuenen (1960: 447) estimated that it should take 70 years to produce a ventifact in interior Holland in present conditions. In a cold environment much less time is needed because of stronger winds and less vegetation and because drift snow is also available. Kuenen concluded that the finest mineral material is so light that it has very little abrading effect.

With wind tunnel studies it has been shown that silt drifted by wind also causes abrasion on rocks producing microscopic and macroscopic features on ventifacts (Whitney & Dietrich 1973; Dietrich 1977a, 1977b; Whitney 1978; Schlyter 1995a,b). Dietrich (1977a,b) could show with laboratory experiments that the low temperature requirement, which is based on the assumption that a missile must have a hardness equal to or greater than that of the target if the target is to be abraded, should be revised. He could with silt-sized ice crystals at an air temperature from −10 °C to −25 °C (hardness between 2 and 3.5) abrade, in a 10 minutes run in the wind tunnel, the edges and corners of calcite, synthetic fluorite and periglase. Dietrich tested also chalk dust and silt-sized calcite and barite particles as abrasive material in much longer runs – 455 days and 50 hours, and found them also causing abrasion. He concluded that 'erosion caused by collisions may be dependent upon the kinetic energy of the projectile and the bond strength of the target but also that such erosion may involve projectiles softer than the targets'. We should keep in mind that Dietrich used wind velocities (6.2–10.2 m s^{-1}) well below blizzard or gale force and the silt-sized (0.002–0.06 mm) particles are much finer than normal snow crystals in nature. One more thing which differs in the wind tunnel experiments from the natural conditions is that in the tunnel more dust can be pointed against the target, while in nature the dust will rise higher up from the Earth's surface because the wind velocities are greater higher up and that encourages the suspension load at higher elevations in the atmosphere.

The relative momentum of an airborne grain of sand with mass M is 29.3 M times greater in air than in water. The corresponding kinetic energy is $(29.3)^2 M/2$ or approximately 430 times greater in air than in water (Pettijohn *et al*. 1987: 311). This greater kinetic energy of windblown sand explains the stronger abrasion by wind than by water transport that Kuenen (1960) and others have reported. On the basis of experiments, Kuenen (1960: 442) estimated sand to lose weight 100 to 1000 times faster by aeolian rather than water transport. He also suggested that the much smaller absolute viscosity of air means that cushioning in air prior to impact is also minimal. The greater rounding of some aeolian sands has been attributed to the higher kinetic energy of saltation transport by wind. The greater kinetic energy of wind-driven sand is also a reason for the ventifacts.

'Effective mass' takes the surface area of impact into consideration and thus it is to a large extent dependent upon shape as well as size and specific gravity of the projectile. Angular relationships at impact relate to the angle between the lattice and exposed surface of the target as well as to the angle at which the projectile hits the target (Dietrich 1977a: 245).

5.2   Rock varnish

So-called case hardening is coating of a rock surface with a cement or desert varnish by evaporation of mineral-bearing solutions. Desert varnish is commonly associated with warm and arid regions (e.g. Merrill 1898; Hunt 1954; Watson 1989) but case hardening is found all over the arid regions including Polar deserts. Several reports of desert varnish close to glaciers can be found in the literature (e.g. Glasby *et al*. 1981; Lindsay 1973b; Selby *et al*. 1973). An early observation of rocks with desert varnish from western Ötztaler Alps, Austria, is reported from a rather humid environment which is seasonally dry (von Zahn 1930). In Svalbard the feature is found on blocks and rock outcrops, preferably on deflation surfaces on raised beach ridges and on exposed mountain ridges (Åkerman 1980: 247). The blocks with well developed ventifact shapes and polished surfaces may have a desert varnish which differs from the ordinary polished surfaces in colour and granular structure. The surface is very smooth.

Desert varnish forms a thin coating on rock surfaces. Crusts vary in colour from brown to dark red, purplish and near black (Tedrow 1977: 81). The coat of varnish is usually composed of about 70 per cent clay minerals and some oxides and hydroxides of iron and manganese admixed with detrital silica and calcium carbonate (Dorn & Oberlander 1982). Crusts are formed of iron and manganese, which apparently move outward along large temperature and moisture gradients. Fe^{2+} and Mn^{2+} oxidize to Fe^{3+} and Mn^{4+} upon reaching the zone of the rock surface, with subsequent stabilization (Tedrow 1977: 82).

Why do we consider this matter together with abrasion? The formation of desert varnish seems to be a chemical and physical process (Cooke *et al.* 1993: 47) but the polishing of the varnish might be a result of aeolian abrasion (Klute & Krasser 1940). Desert varnish is also called rock varnish (Cooke *et al.* 1993: 46) which illustrates the idea that it can be produced also in other environments than deserts.

Rock varnish has been observed in Antarctica (e.g. Täubert 1956) not only in the valleys but on the Polar plateau on elevated regions of southern Victoria Land, too (Glasby *et al.* 1981). From there they presented a detailed study of rock varnish. According to Glasby *et al.* (1981) the desert varnish 0.01–1.0 mm in thickness in southern Victoria Land contains mainly of iron oxide and is formed on the higher altitudes because of increased relative humidity with clouds. The formation of desert varnish is found especially on the leeward side of the boulders which reflects the lack of wind erosion (Glasby *et al.* 1981: 395). This author has recognized rock varnish on stones and boulders (Fig. 5.2 and Plate 4) on Basen, Queen Maud Land, Antarctica, close to the Finnish and Swedish base. On Basen the crust is brown and purplish or dark brown and often the rocks are ventifacts with typical wind worn facets. Solar radiation heats the rock surface 25–30 °C above freezing and the crust development may be related to this.

Rock varnish probably forms very slowly, as indicated by the thickness of 0.1 mm formed over periods of up to 4 million years (Glasby *et al.* 1981: 395). This slow process on Basen is not possible because of very strong year-round abrasion by drift snow.

5.3   Frosted quartz grains

Cailleux (1942) pointed out that surface morphology and texture will give information about the sedimentary processes which deformed the grains. In each sedimentary environment several processes are working in parallel or in sequence and all gradually modify the surface and form of quartz grains.

Abrasion caused by the process of transport results in the mineral grains becoming rounded. From the degree of wear found on grain edges it is possible to determine whether the grains have been transported great distances before being deposited. Quartz is very hard, subject to irregular splitting and resistant to weathering so that it is the best of commonly found minerals on which to conduct roundness tests (cf. Bagnold 1941: 7).

When sand grains are transported by wind and they creep or saltate on the sand surface, they dash against each other continuously. This quickly wears soft minerals which decrease in size compared to the harder minerals, and even feldspars vanish relatively fast in this process. The hard, most resistant sand grains are left over, mainly quartz grains which will be well rounded after a long transportation, and many minerals which do not survive rigorous transportation find their place in silts or muds rather than sands (Pettijohn *et al.* 1987: 25).

At the same time, when their edges are rounded their surfaces become etched. The surface is no longer clear but frosted or matt and the grains are opaque. This is a characteristic feature for wind drifted quartz grains and that is how they differ from the water transported and rounded quartz grains. This is one kind of micro abrasion on the sand grain level. The surface texture of a sediment particle can demonstrate the process history of the particle (Krinsley & Doornkamp 1973).

Grains originated by frost shattering (e.g. Lautridou & Seppälä 1986) and glacial breakage are angular. During wind transport the quartz grains are worn matt and rounded. Fluvial and coastal processes leave them shiny. The average proportion of unpolished sand grains in Quaternary deposits is 80% in Fennoscandia and in southern France and Brittany, about 40% in Poland, northern Germany, Denmark, the Netherlands and in the North Sea. In Fennoscandia the proportion of frosted aeolian quartz grains is about 10%, in the Netherlands and Jutland, northern Germany and Poland 40% and reaches its maximum of 80% on the middle Vistula River. This does not mean, however, that the quartz grains were totally rounded during the Pleistocene periglacial aeolian transportation. The sources of much of the sand deposits drifted by wind during the Ice Age and late glacial times are in the Devonian Old Red sandstones and Triassic Bunter sandstones which are abundant in aeolian sands and also in sandstones of Jurassic, Cretaceous and Tertiary origin (Cailleux 1942; Troll 1948).

Old sedimentary rocks such as sandstones and well rounded, old fluvial or coastal deposits as a source of surficial aeolian material will be misleading when interpreting the recent distance of transportation by means of roundness. The grains have already, at the beginning of aeolian transport, been shaped by multiple other wearing processes. This is a risk in central Europe and in the Keewatin area, Canada, where a lot of Mesozoic and Tertiary sandstones exist everywhere with old alluvial deposits.

However, it has been shown that freezing and violent changes of temperature can cause the same kind of frosted surface in quartz grains as aeolian drifting (Kuenen & Perdok 1962). Frosting was noticed on the surfaces of some very fresh, unworn quartz grains in sand dunes in Finnish Lapland and it was regarded as frost action (Seppälä 1971a: 60). In Lapland the original weathered material was transported by the ice sheet, glaciofluvial processes, and then by wind but only for short distances. Quartz grains are still very angular and unworn but some surfaces are frosted.

Aeolian quartz grains are easy to identify with a scanning electron microscope (SEM) (Krinsley & Margolis 1971; Margolis & Krinsley 1971). They have been submitted to an aeolian episode of high energy collisions and that means that their surfaces are covered by a high density of crescent-shaped impact features. Even with low energy transport the impact features are very characteristic, showing very sharp edges without the slightest smoothing (Le Ribault 1978: 323). These characteristic surface features on aeolian quartz grains are also called 'upturned plates' – parallel ridges about 0.5 to 10 µm or more in diameter. Plates appear to be spaced from 0.1 µm to about 1 µm apart (Krinsley & McCoy 1978: 250).

Wind-worn quartz grains with characteristic surface textures have been produced in experimental wind tunnel and abrasion chamber studies in laboratories (Kuenen 1960; Kaldi *et al.* 1978; Lindé & Mycielska-Dowgiallo 1980). These observations can be used, for example, in palaeogeographical studies. In these studies it came out that the original grains lose their weight quite rapidly during the aeolian transportation, and abrasion produces silt size particles which in nature can form part of loess deposits (Smith *et al.* 1991).

5.4   Abrasion of peat layers

On the mires we do not normally meet traces of deflation unless there is for some reason a drying-out of the vegetation (Samuelsson 1926: 169). Even though peat is a fibrous material it can be removed by wind in certain circumstances. Erosion of peat has been called 'turf exfoliation' (a German term; Troll 1944). When the peat layer itself is exposed, is rather dry or frozen and it has formed a steep wall, then, especially, the wind-drifted ice crystals are able to erode it in significant amounts. It is abrasion which is directed to peat. We find examples of this process on palsa mires in Fennoscandia where deflation creates large areas of bare dark brown peat on palsas (e.g. Åhman 1977; Seppälä 1988, 2001) (Plate 5). In the Schefferville area, Québec, during two successive winters approximately 18 cm of surface peat was eroded from a palsa surface (Cummings & Pollard 1990: 101). In Finnish Lapland strong winter winds have removed in some cases more than 40 cm thick layers of peat from the palsa surface (Seppälä 2001). Sometimes the peat cover of palsas has been totally removed and only the frozen silt core remains (Åhman 1976, 1977; Allard *et al.* 1986). These features have lost their thermal insulation and cannot be very stable in the present climate. They will thaw rather rapidly.

On the fell summits in Finnish Lapland small peat calottes occur with signs of strong erosion caused by wind drifted snow (Seppälä 1972b; Luoto & Seppälä 2000). If the blanket bogs in mild climates, for example, in Wales, dry they soon will be eroded by wind and form peat bluffs. G. Samuelsson (1910: 247–248) reported from Scotland that 'denudation of moors is principally due to the erosion of the wind, only to a rather small extent to that of the water'. In Iceland deflation of peat is a common phenomenon (Samuelsson 1926: 170).

When the deflation has worn the sand layers underneath the peat cover, blocks of peat slide down to the deflation basin because of undercutting. This causes steep sand walls covered by peat. In Iceland these cliff-like deflation walls (bluffs) are very common features (Samuelsson 1926, Fig. 5). Cailleux (1939: 61 and Plate V: Fig. 1) called these steep walls yardangs as in loess regions. Ashwell (1966: 536) described the present stage in the following way:

> The present situation is that islands of turf still remain on parts of the plateau, some of them only a few square yards in extent, others more widespread, but heavily indented by trenches of erosion. The soil under the turf is fine grained, predominantly of silt size, with very few included stones. Where the turf has been removed the landscape is often a gently rolling plain, covered by a surface litter of boulders, cobbles and pebbles, many of which are rounded, and are pitted by wind erosion. It is often stated that the soil has been removed and the underlying moraine exposed, but in very many parts of the south central part of Iceland removal of the top litter of stones shows that the loessial soil lies below.

In the Canadian Western Arctic the wind deflates earth hummocks and produces fluting in thick peats (Bird 1974: 713). Although peat cover is thin in the Arctic it somewhat protects the underlying mineral soil against deflation. This is why sharp erosion edges and steps 10–20 cm high can be often found. Their formation may have started by meltwater drainage or by frost and then developed further by deflation. Schunke (1986: 68–69) calls them *Rasenkanten*.

5.5   Practical meaning of abrasion

In the Arctic and Antarctica abrasion by wind-drifted snow is a much bigger problem than sand blasting in temperate and hot environments, where it is easier to protect buildings and other structures against abrasion. In dry and cold conditions snow moves freely in air and huge amounts of material are continuously inpacting everything which rises above the snow surface. Mawson (1915: 123–124) described abrasion in Antarctica in the following way: 'The abrasion effects produced by the impact of the snow particles were astonishing. Pillars of ice were cut through in a few days, rope was frayed, wood etched and metal polished. Some rusty dog-chains were exposed to it, and, in a few days, they had a definite sheen. A deal box, facing the wind, lost all its painted bands and in a fortnight was handsomely marked; the hard, knotty fibres being only slightly attacked, whilst the softer, pithy laminae were corroded to a depth of one-eighth of an inch.'

Mawson's (1988: 74) observations from Antarctica were also somewhat quantitative: 'Wind abrasion . . . has in 5 days cleared the label off a Sunlight soap box and abraded the wood, unequally, now in ridges about 1/32 inch [equal to 0.8 mm] off hardest parts and 1/8 inch [equal to about 3 mm] from soft.'

Samuelsson (1926: 113) had found in Green Harbour, Spitsbergen an old Dutch wooden memorial cross on which the letters were oil-painted and they had been protected. All the other parts were eroded to a depth of 1 cm. Similar observation were made on Kapp Thordsen, Spitsbergen where a claim post was eroded by wind. On it the text was engraved and then it was painted with tar. The layer of tar was thicker on the letters and after 70 years of niveo-aeolian erosion the text was embossed and all the other parts had eroded about 1 cm deep (Seppälä 1987a).

The same abrasion by wind can be seen on the wooden walls of Svenskehuset, the oldest standing house in Spitsbergen (Seppälä 1987a) and on many huts on the Norwegian mountains. Knots and harder parts of the tree rings are embossed.

Mawson (1915: 123–124) pointed out also the troubles which abrasion of snow surface causes on travelling in Antarctica: 'The effect of constant abrasion upon the snow's surface is to harden it, and, finally, to carve ridges know as sastrugi. Of these much will be said when recounting our sledging adventures, because they increase so much the difficulties of travelling.

Even hard, blue ice may become channelled and pitted by the action of drift. Again, both névé and ice may receive a wind-polish which makes them very slippery.' (see Plate 14)

All these examples indicate that in cold environments the abrasion is a significant factor, not only from the geomorphic point of view, but it has a notable practical meaning for constructions, too.

Plate 1 Hoar snow on spruces (*Picea abies*) on the Riisitunturi fell, Kuusamo, Finland (ca. 66° 15′ N). Trees have caused turbulence and wind channels on the snow. Photo by P. Havas.

Plate 2 Wind abraded rocks from Iceland. Photo by J. Käyhkö.

Plate 3 Wind abraded sandstone from Iceland. 1973.

Plate 4 Rock varnish on a basalt stone (some 5 cm in length) from Basen, Queen Maud Land, Antarctica. Photo by D. Krinsley.

Plate 5 Wind abraded palsas north of Erttehvarri, western Utsjoki, Finnish Lapland. July, 1972.

Plate 6 Tongue-like sand accumulation on the leeside of a deflated sand dune, Hietatievat, Finnish Lapland. July, 1973.

Plate 7 Late winter view of the deflated summit of an esker on the slope of fell Peldoaivi. Other parts of the region are still covered by snow. April 26, 1969.

Plate 8 Deflation scours in Iceland. July, 1973. Photo J. Käyhkö.

Plate 9 Aerial photograph of a vast deflation and sand dune area at Kobuk River, NW Alaska. (15–131 NASA JSC 386 JUL 78 ALASKA CIR 60.)

Plate 10 Arctic riverbeauty (*Epilobium latifolium*) on a small aeolian sand accumulation in Iceland. Photo by J. Käyhkö.

Plate 12 Dust storm on the flood plain NW of Askja, Iceland. Photo by J. Käyhkö.

Plate 11 Satellite image with oriented lakes on the coast of the Beaufort Sea, Prodhoe Bay area, Alaska. August 27, 1985.

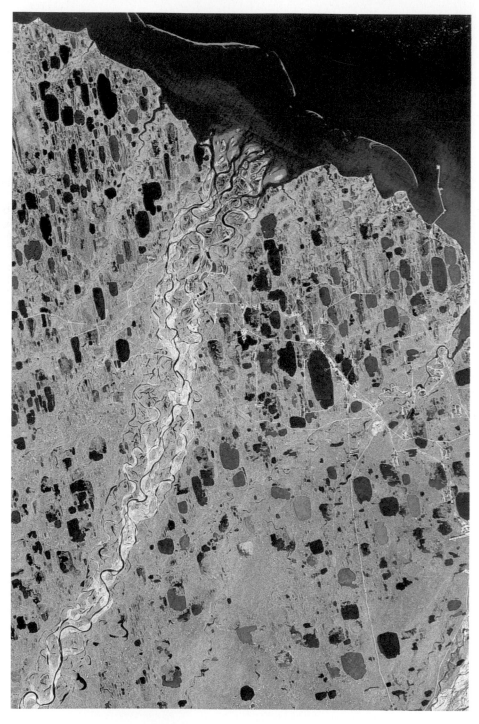

Plate 13 Cornice in Tarfala valley, northern Sweden. Photograph by W. Karlén, 1982.

Plate 14 A snow scoop at the edge of a nunatak in Heimefront Fjella, Queen Maud Land, Antarctica. Photo by P. Lintinen, 1994.

6

Deflation

According to Bird (1967: 240) landforms produced entirely by deflation are few in the Canadian Arctic. 'The most widespread are shallow depressions, rarely more than three meters across, from which about 1 cm, and never more than a few cm, of fines have been blown away to leave a stony, and generally vegetation-free, residue. These patches are found throughout the Canadian Arctic. Although rarely studied they may be important, as without a plant cover the surface is left unprotected from other geomorphic processes.' The task is now to improve somewhat this rather narrow view of deflation and deflation forms.

6.1   Definitions

Wind erosion removing loose fine-grained particles from the ground surface is called deflation, which produces degradation forms of different size. David (1977: 21–23) classified aeolian erosion forms in the following categories: *wind/pit*, or *blowout hollow* is a circular or more frequently oval-shaped depression formed by deflation of sand through a breach in the vegetation cover (Fig. 6.1) or other form of surface protection, such as a soil pan. In some areas the surface outline of a blowout is rather irregular. The inside of the hollow is spoon-, bowl-, or funnel-shaped depending on the size of the feature and on a number of other factors. The smaller forms are called wind/pits which are most commonly circular (Fig. 6.1) and the larger ones are called blowout hollows. In the following text we use the term *blowout*. Large blowouts we shall call deflation basins (Fig. 6.2). They are normally surrounded by parabolic or blowout dunes. A special form of blowout hollow is a *dune crevasse* which is a depression produced by deflation on a formerly stabilized dune (Fig. 6.3). Because of the nature of the sand in the dune, dune crevasses are generally elongate in plan view with pointed ends, and funnel-shaped in profile with steep inside slopes. A *deflation depression* is the blowout hollow associated with a dune. It is a morphological element of a dune: the depression from which the sand in the dune ridge was blown out and which is surrounded by the dune, specifically the head, wind, and

Fig. 6.1 Secondary deflation relief in a vegetated deflation basin of a parabolic sand dune. Kiellajoki, Finnish Lapland (Seppälä 1971a: Fig. 13).

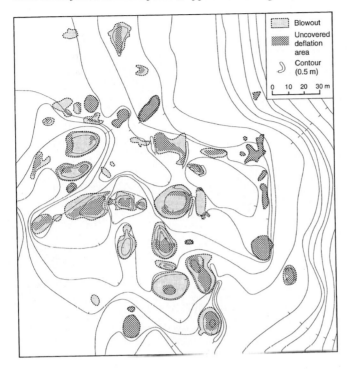

Fig. 6.2 Deflation basin in a parabolic dune on Hietatievat sand dune area, Finnish Lapland. July 1965.

Fig. 6.3 Dune crevasses (white areas) in stabilized periglacial sand dunes in Lake Petsimjärvi area, Finnish Lapland. Aerial photograph published by permission number 41/2004 of Topographic Service of FDF (Seppälä 1971a: Fig. 15).

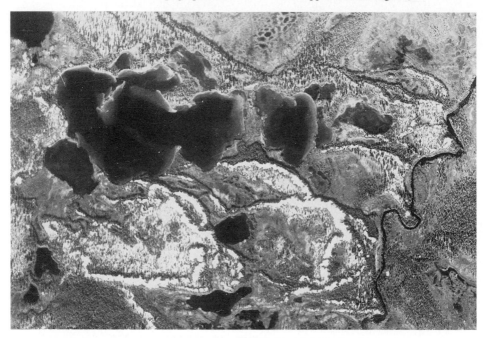

Fig. 6.4 A butte-type deflation remnant on the sand dune field at Hietatievat, Finnish Lapland. Buried soil representing the old stabilized dune surface is covered by redeposited aeolian sand during the deflation. August 26, 1975.

back-ridge. The *deflation area* is according to David (1977: 23) a large surface from which sand was deflated into dunes. It may in fact form from the merging of a large number of individual blowout hollows and deflation depressions when the dunes migrate a certain distance downwind from their place of origin. The floor of the deflation area may be formed by non-sandy deposits, and in that case it forms a featureless plane though it will always have a very thin layer of aeolian sand on it.

In Lapland on relict periglacial sand dunes strong deflation has left butte-like erosion remnants (Fig. 6.4) and steep edges in which can be seen also the initial dune surface as a podzol horizon buried by redeposited aeolian sand (Fig. 6.5). On these erosion remnants mountain birch (*Betula pubescens*) and juniper (*Juniperus communis*) often grow. Pissart *et al.* (1977: 2468–2469) have found butte-like erosion remnants at Sachs Harbour on Banks Island, Arctic Canada, which grow willows (*Salix arctica, S. alaxensis, S. richardsonii*).

Deflation requires, besides strong winds, certain conditions: (1) suitable grain size composition of mineral material for wind transport, (2) rather well sorted material without too big grains, (3) unvegetated soil surfaces, scattered or patchy vegetation, and (4) low-lying ground water table.

Sand deposits of various origins with similar grain size composition to aeolian sands are especially sensitive for deflation. Ideal objects for deflation are the old

Fig. 6.5 Remains of deflated sand dune with the original palaeosoils with charcoal buried by redeposited aeolian sand at Lake Pöyrisjärvi, Finnish Lapland. June 19, 1973.

vegetated sand dunes which consist of aeolian sand. If the vegetation cover has been destroyed or limited by factors such as a forest fire, or trees felled by storms, clear cutting, sheep or reindeer grazing or some natural hazard such as insects damaging the forest (Seppälä & Rastas 1980) then deflation can start and gain the upper hand over the vegetation which otherwise keeps the surface stable.

Deflation seems to be extremely active close to and just outside of the northern boreal forest limit in the zones of sporadic and discontinuous permafrost (Fig. 2.18). When we move polewards to the continuous permafrost zone where the soil moisture is higher because of frozen ground effects on infiltration (permafrost table restricts the melt waters penetrating deeper and keeps the active layer wet) we do not find so many active deflation features in spite of the absence of trees. For example, in Ungava Peninsula in Arctic Canada some 500 km north of the forest limit the ground surface is mostly covered by a thick layer of moss. Exceptions are of course the very arid Arctic regions, the Polar deserts in northernmost Greenland and Ellesmere Island or Dry Valleys and nunataks in Antarctica where the ground surface has no moss cover.

At present in the Subarctic and Arctic environments an investigator easily gets an impression that many of the geomorphological features formed during the dramatic events of glaciation and deglaciation are in a state of stagnation and decay. The main eroding factors are meltwaters, heavy rains, thawing of permafrost, mass wasting and deflation.

6.2   Activation of deflation

Vegetation tries to fix the growing substrate. Most plants suffer from moving sand, not liking to be covered by soil or to have their roots exposed. If the ecological conditions change unfavourably for the vegetation then it cannot resist the force of wind. Aridity is the main limiting factor for growth in hot climates. In cold environments relatively low precipitation can be enough for plant growth because of the low evapo-transpiration. However, cold and dry air is a fatal combination for most plants. They will be frost dried during the long winter without liquid water, if they do not get enough protection from snow covering them. Kihlman (1890) considered dry and strong winter winds as the most critical for the trees at the forest limit on the Kola Peninsula. Any impairment in climate may cause deflation. For example, less precipitation, decreasing relative humidity, drought, thinner snow cover, colder winters, cooler and shorter summers, increasing wind speed etc. may cause the death of vegetation. Natural hazards: mass-expansion of insects (Seppälä & Rastas 1980) and fires in Subarctic environments (Seppälä 1971a, 1981, 1995a; David 1977; Filion 1984; Kotilainen 1991) suddenly and unexpectedly destroy vegetation and support deflation locally and decisively.

When studying the role of wind for denudation in Mackenzie Plains and Central Otago, South Island, New Zealand, Zotov (1938) pointed out the close relationship between frost action, deflation, plant cover and animal life. After plant cover protecting the soil has been depleted by man-made fires, wild animals, such as rabbits and deer, and overgrazing, 'winter frosts are sufficiently severe to pulverize exposed soil and to lift up small plants' (Zotov 1938: 227A). In the high country frequent frosts throughout the greater part of the year, assisted by extensive heating and drying of exposed soil together with drying wind commence the erosion (Zotov 1938: 240A). From West Spitsbergen Åkerman (1980: 258) lists factors disturbing vegetation cover in Arctic conditions: desiccation cracking, frost cracking, thermokarst processes, frost heaving, needle ice formation, ground ice formation, fluvial processes, with human activity, animals etc. also activating deflation. Åkerman (1980: 258–266) also gives small examples of these processes supporting the deflation in West Spitsbergen.

Animal tracks on the vegetated sand surfaces can be the starting point for deflation. Reindeer in Lapland (Seppälä 1984; Käyhkö & Pellikka 1994; Käyhkö 1997) (Fig. 6.6), sheep in Iceland (Thorarinsson 1962; Ashwell 1966), caribou in Alaska and muskox in Arctic Canada tramp their tracks and cause deflation. Bird (1967: 241) reported even Parry's ground squirrel (*Citellus parryi*) and the arctic fox (*Alopex lagopus*) contributing to dune development in Arctic Canada. The former may be directly responsible for blowouts by destroying the vegetation and

Fig. 6.6 Reindeer on a deflation basin at Hietatievat sand dune area, Finnish Lapland. July 1977.

burrowing in the sand. They also attract the Barren Ground grizzly bear (*Ursus richardsoni*) which tears up the vegetation over a wide area looking for squirrels. At the north tip of Seward Peninsula on coastal dunes of Cape Espenberg, Mason (1990: 122) observed burrowing activities of arctic fox and ground squirrels. He found fox dens at least every 0.5 km laterally across some dune ridges.

In Spitsbergen flocks of ptarmigan (*Lagopus* sp.) searching for food under snow and tearing off the vegetation cover contribute to deflation (Czeppe 1966: 122). In West Spitsbergen sometimes eider ducks (*Somateria molissima* L.) dig their nests in the dunes and thereby initiate wind erosion (Åkerman 1980: 241).

The most recent factor which destroys vegetation is the human being moving with all kinds of off-road vehicles on the surfaces of fine sand deposits. The vehicle tracks stay unvegetated very many years in these severe growing conditions and they play an increasingly important role in deflation activation. This may also cause some economic impact because large areas of the best growths of *Cladonia* lichen, on which reindeer feed, are damaged. Åkerman (1980: 264–265) made some measurements of the deflation impact of a tractor track in West Spitsbergen and noticed up to 40 cm annual increase in the length of some erosion scars (Åkerman 1980: Fig. 4.33).

One factor which increases deflation and may be a combination of several climatic elements is the drop of the ground water table. This has been identified on the Subarctic dune fields in Lapland (Seppälä 1984, 1995a; van Vliet-Lanoë *et al.* 1993).

If we think in an opposite way and summarize the main factors which prevent deflation in cold environments we can list the following: favourable growing conditions for vegetation, thick protecting snow cover, high ground water table and low wind activity. No change in climate is needed if deflation reaches the ground water table or a coarse grain layer under the deflated sand deposits. Then the deflation will stop and succession of vegetation will take place on the former deflation area.

6.3   Blowouts in the Subarctic

The form of blowouts in sand dunes (Figs. 6.3) depends on the original shape of the sand deposits which have been eroded. The blowouts can be elongated basins which follow the original ridge. Only some remains of original dune edges could be left (Figs. 6.5 and 6.7). On a flat surface blowouts can be rather round in shape (Fig. 6.1). They resemble bomb craters with a somewhat higher rim wall around them. Blowouts are surrounded by low birch forests (*Betula pubescens*), often like park tundra. Some large deflation basins in Subarctic Lapland are over 10 m in depth and they cover several hectares in area (Fig. 6.8). Normally the slopes of the blowouts are steep (about 35°, which is

Fig. 6.7 Large deflation basin with remnants of original sand dune, Hietatievat, Finnish Lapland. June 1988.

Fig. 6.8 Some 10 metres deep deflation basin at Kiellajoki sand dune area, Finnish Lapland. July 18, 1967.

the slip angle of aeolian sand). Sand has moved over the rim shoulders from the basin and is covered by vegetation and this often produces oversteepened slopes dipping over 55°. From the blowout small wind channels (dune crevasses) lead out and at their outermost ends tongue- or delta-like sand accumulations form (Plate 6). By measuring the orientation of these channels and redepositions the direction of the effective winds can be determined (Fig. 6.9) (Seppälä 1971: Fig. 21).

Suitable material for deflation has been deposited by glaciofluvial, glacio-lacustrine, and fluvial transport. Glaciofluvial deltas, eskers and valley trains can be reformed by deflation. On these features in Subarctic and Arctic conditions it is rather common to find barren sand surfaces caused by deflation (Fig. 6.10 and Plate 7). On eskers and glaciofluvial deltas the blowouts often follow the upper edge which is exposed to the strongest winds. The depth of blowouts on glaciofluvial deposits depends on the thickness of the fine sand layer. This mate-rial often has coarser sand grains, stones and even bigger particles which then are enriched on the surface during the deflation, and because wind cannot transport them further, they form a pavement which protects the underlying sand deposits

Fig. 6.9 Transport directions and orientation of redeposition forms (arrows) and axes of relict parabolic dunes of the Kaamasjoki-Kiellajoki dune field (wind rose), Finnish Lapland. Redrawn after Seppälä (1971a, Figs. 4 and 21).

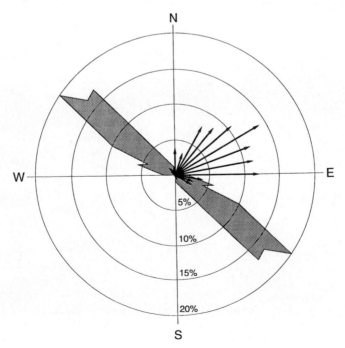

from erosion and prevents deepening. These surfaces can still stay a long time uncovered by vegetation even when they have reached their mature stage and maximum size, because there is some sand and snow movement on the surface back and forth and wind will carry the seeds away from the blowout surface.

David (1977) in his comprehensive dune inventory of Canada mentioned only very few large deflation areas. The most extensive active sand patches in Canada are in the Archibald Lake dune area in the Lake Athabasca region (59° N), Alberta and Saskatchewan, which is 457 square miles in total area (Rowe & Hermesh 1974). Some of the active patches appear as large blanket sand surfaces forming parabolic dunes, too. Very extensive present-day wind action has been reported from the eastern coast of Hudson Bay (Filion & Morisset 1983). There, old sand dune relief in boreal forest and forest tundra zone is largely deflated.

6.4 Deflation in Iceland

The two factors, that Iceland is surrounded by the ocean and the lack of forests, cause the frequent and violent gales to be a morphological and ecological factor of very great significance (Nielsen 1933: 245). Thus there is an abundance of suitable material for deflation together with favourable climatic and terrain conditions and Nielsen (1933: 247) regards the region west of Vatnajökull as perhaps the most pronounced Arctic wind erosion region in the northern

Fig. 6.10 Deflated esker complex at Lake Pöyrisjärvi. Reindeer like to follow the crests of eskers. June 19, 1973.

hemisphere (Fig. 6.11). Moraine material and glaciofluvial deposits provide transportable material for wind. Rapid weathering of volcanic rocks produces large quantities of fine material, the so-called palagonite dust, and as well occasionally after the eruptions enormous quantities of volcanic ash (tephra) are available (Nielsen 1933: 247).

Wind action in Iceland has been very strong during the Holocene because of severe climate, scattered vegetation and great abundance of volcanic material with fine particles (Cailleux 1939). Wind erosion phenomena characterize large parts of the desert regions of the interior of Iceland (e.g. Sapper 1909; Samuelsson 1925, 1926; Nielsen 1933).

Deflation has for centuries been a serious problem in Iceland (Thorarinsson 1962) (Plate 8). Large regions in Iceland have been eroded totally during the last centuries. A metre-deep or more covering of loess-like earth has been removed and there has remained only stone and block covered pavements. Winds have undermined the vegetation layer along almost vertical escarpments (Plate 8).

Fig. 6.11 Regional deflation intensity in Iceland after Ólafur Arnalds. More or less marked erosion can also be observed in low-lying terrain, where it is predominantly due to overgrazing. Source: Bernes (1996: 105).

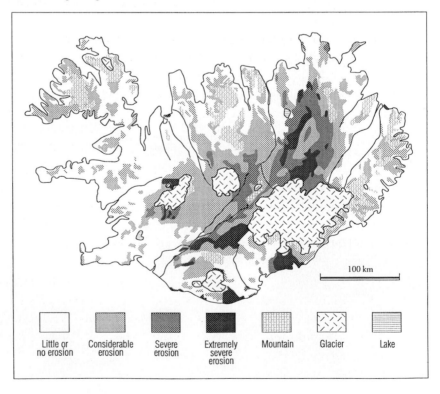

Thus the loess-covered lava fields form the subsistence of a herd of sheep and cattle, and the deflation in these regions is nothing short of a national catastrophe as a number of farms have been laid waste and others are threatened (Nielsen 1933: 250).

The principal agent of the erosion is the sand that sweeps along the surface and undercuts the finer more coherent loose earth. People have tried to stop this process of destruction by building stone fences at right angles to the prevailing winds in order to catch and hold the sand (Nielsen 1933: 250).

In past times people cut trees for structural timber and firewood and progressively destroyed the balance of vegetation both in central Iceland and in the lowland areas (Ashwell 1966). After the Second World War this problem was actualized by the rapid increase in the number of sheep. Deflation in Iceland takes place often in finer loess-like volcanic ash layers (Cailleux 1939) but also fine sand and even coarse sand. The vegetation cover in Iceland has several times been destroyed by volcanic eruptions and this is considered as a main activating factor of deflation. Climatic change after the colonization of Iceland plays its role in the deflation history. Clear cuts close to the Arctic forest limit did not become reforested because of the more severe climate.

In front of the Hofsjökull, central Iceland, the recent moraines carry a continuous cover of vegetation, while the older moraines farther away from the glacier practically lack vegetation and undergo wind erosion. The moisture of the ground is very great near the glacier but decreases with distance. It is evidently the strength of the wind and the cohesion of the soil, not the temperature, that determine whether vegetation can get a foothold or not. With decreasing moisture deflation reaches its greatest intensity on the plateaus of the interior, and here the most extensive deserts are to be found. Strong cold winds sweep down from the glaciers, carrying away the silt and sand from moraines and glaciofluvial surfaces and exposing and polishing the block pavement (Samuelsson 1925; Nielsen 1928). Often the surfaces are paved with almost uniform stones not more than 5 cm in diameter (Nielsen 1933: 248) (Fig. 5.4). Strongly deflated regions in Iceland exist on the highlands south of Longjökull Glacier (Samuelsson 1926, Fig. 30) and to the west and north of Vatnajökull Glacier (Fig. 6.11).

In front of Hofsjökull the recent moraines are covered by a continuous vegetation, while the older moraines farther away from the glacier are mostly uncovered and undergo wind erosion (Nielsen 1928). The reason seems to be that near the glacier the ground is very wet but the moisture content decreases with distance. Wind velocity and the cohesion of the soil may determine whether vegetation can develop or not. With decreasing moisture deflation reaches its greatest intensity on the plateau of the interior of Iceland and there the most extensive deserts are found. In Iceland wind erosion has produced a number of surfaces that in all essentials correspond to the hamada phenomenon in the temperate, subtropical and tropical deserts (Nielsen 1933: 250).

6.5   Wind channels in New Zealand

A special type of deflation quickly develops wind channels some 50 cm in depth and of varying intervals, on high exposed ridges of Kaimanawa Mountains, North Island, New Zealand (Zotov 1940: 262B). Where the prevailing NW wind blows directly up over the slope the wind channels merely widen with time and destroy all vegetation. The finer soil, rhyolitic tuff, is blown away, while the coarser material creeps downhill. Usually, however, the wind sweeps over the ridges and summits along a more or less spiral path (Fig. 3.13). The direction of wind channels tends to approach right angles to the maximum slope. The action of the frost on the exposed ground of the channels repeatedly disturbs stony matter (Zotov 1940: 262B).

6.6   Arctic blowouts

Blowouts in the Arctic are much smaller as Bird (1974) points out. The reason why Subarctic regions are more sensitive for present-day deflation than the

Fig. 6.12 Edge of Great Kobuk Sand Dunes, Alaska. Photo by E. Koster.

Arctic regions is unknown. Black (1951) quoted David M. Hopkins' observations on Seward Peninsula, northwestern Alaska, about oval blowout areas undergoing active deflation on the Bering coast. They are only about 60 m wide and 150 to 300 m long. One area is indented about 2.5 m below the general level of the surrounding terrain, but others only 'a few inches'. This is typical for permafrost regions.

In Alaska vast areas are covered by aeolian sand and silt. Mainly they are vegetated but at least three large deflation areas can be found: at the lower course of the Koyukuk River (a tributary of the Yukon River), central Alaska, and two areas at the southern side of the lower course of the Kobuk River, northwestern Alaska (Plate 9 and Fig. 6.12) (Karlstrom *et al.* 1964). All have the same characteristics of vast sand seas with moving sand dunes, strictly delimited from the surrounding boreal forests where blowouts are too small to be identified on the aerial photographs, although extensive parts of the river valley adjacent to the active and stabilized dune fields are covered by vegetated aeolian sand sheets and loess deposits (Koster & Dijkmans 1988). The Great Kobuk Sand Dunes (Fig. 6.12) are the largest, occupying approximately 62 km^2 (Kuhry-Helmens *et al.* 1985). The active layer in the forests surrounding the dune field is only a few decimetres thick and the unfrozen sand layer of the open sand area was observed at several localities to be at least 1.5–2 metres thick (Koster & Dijkmans 1988). This fact and the very sharp edge of the dune field give us reason to speculate that the deflation and aeolian processes in general are activated probably by forest fire, even though no charcoal layers have been reported so far.

6.7   Deflated sand and silt surfaces

Fluvial accumulations such as flood plains, alluvial fans and deltas are often uncovered sand and silt surfaces in the Polar regions and therefore subjects for deflation. Nickling (1978) has intensively studied the aeolian processes on a relatively large delta (7 km^2) of the Slims River, Yukon. There he found shallow scour pits resulting from deflation. The maximum local relief of the delta sediments is less than 1 m but wind transports large amounts of sediments deflated from the delta surface.

Fluvial erosion can be the activating factor destroying the vegetation cover on glaciofluvial, glaciolacustrine or fluvial sediments. Along many rivers crossing these type of deposits can be found barren sand and silt surfaces exposed to deflation. One typical example is Rivière Deception in northern Ungava Peninsula, Québec, Canada. The river has cut its course through a glaciofluvial delta which originated during deglaciation. Currently the about 30 m high river bank and the terrace surface above are deflated (Seppälä *et al.* 1991). Deposits are mainly well sorted sands, and strong winds cause small sand storms on the delta scarp drifting sand clouds over the river on the tundra.

Another modern example of terrace scarp deflation is provided by Swett and Mann (1986) from eastern Greenland where deflation on the numerous terraces of Vibekes Elv is drifting sand and loess and also depositing aeolian forms. Vibekes Elv is entrenched 150 to 200 m through glacial, glaciofluvial and glaciolacustrine deposits which consist of sorted and poorly sorted clay- to boulder-size materials. Approximately 35 km² of mostly unconsolidated sediments at the angle of repose are exposed along the terrace scarps. A deflated and abraided armoured pavement veneers virtually all the terrace surfaces that are not covered by low vegetation or water.

In Scoresby Sound, east Greenland the south shore especially is wind eroded, while in Isortokfjord, west Greenland, the southern shores are more vegetated and the reason is that local winds mainly disrupt the north shore (Samuelsson 1926: 88).

Normally the short-term but very effective spring floods in the Subarctic and Arctic transport the loose material that slides down slopes, even blocks, away from the river banks and cut new fresh exposures of fine sediments for deflation. This is also the case at Vibekes Elv. Fluvial erosion of the thick gravel deposits has produced vast exposures adjacent to the river. As wind deflates the finer particles (sand, silt, and clay) from these slopes at their angle of repose, the pebbles, cobbles and boulders roll downslope and accumulate at the base of the slope rather than remaining as a lag of the steeply inclined surfaces (Swett & Mann 1986). River meandering and lateral cutting remove the coarse clasts and the process continues. Fluvial erosion, mass wasting and deflation together slowly demolish the deposits pilled by deglaciation.

The unvegetated sand surfaces of outwash plains (sandurs) are certainly touched by deflation. Meltwaters from the glaciers regularly bring new material onto the surface of the outwash plains and this dries and will be drifted by wind. Since the glaciofluvial processes are still active the small blowouts and aeolian accumulations are short-lived and immediately during the next flood are destroyed by fluvial erosion and accumulation, if sand is not deposited outside the sandur on its edges (e.g. McKenna-Neuman & Gilbert 1986). Deflation taking place on sandurs and valley trains is an important process on loess formation further away from the material sources.

Gilbert (1983: 164) has found in the fjords of Baffin Island that deflation of sediment from sandurs during summer may be reduced because the surface is wetted by the higher discharges, but in September and through the winter the surface is dry and deflation is unrestricted. 'Even where the surface froze before drying, sublimation may occur to a limited extent to leave a surface layer of dry sediment that may be easily eroded.' On the ancient outwash plains from the last deglaciation nice examples of deflation can be found, for example, in Swedish Lapland (Seppälä 1972a).

One of the factors producing favourable sand and silt surfaces for deflation is the advancing glacier, pushing fine sediment deposits in front of the ice-margin and forming glacier-pushed ridges and push moraines. Often the removed sediments are glaciofluvial outwash sediments, glaciolacustrine deposits, former fluvial sands and silts, tidal flat sediments or proglacial fan deposits. An early description of wind erosion on push moraines in Spitsbergen was given by Gripp (1929: 230–231). He observed the falling winds coming down from the glacier and causing dust clouds, but also some warmer contrary winds in the opposite direction bringing sediments up on the glacier and forming so-called cryoconites (ablation pits) on the glacier surface.

This author has seen this type of push moraine ridge complex at the Taku Glacier, close to Juneau, southeastern Alaska (Fig. 6.13). The material was glaciofluvial silt which formed a large tidal flat at the end of the fjord. This bottom material was recently pushed by the advancing Taku Glacier. Formed

Fig. 6.13 Push moraine of Taku Glacier, SE Alaska. Katabatic wind to the left. July 1971.

silt ridges were vegetated, dry and easily drifted by wind. Aeolian deflation and redeposition features could be seen among fluvial erosion phenomena. Strong winds falling down from the glacier drifted material also to the water.

Riezebos *et al.* (1986) made observations of aeolian activity on a similar but still bigger glacier-pushed ridge complex at the glacier Holmströmbreen in Ekmanfjorden, West Spitsbergen, where the glacier about a century ago obviously surged and pushed up a proglacial fan and a regressive sequence of fjord muds and freshwater tidal flat sediments (Gripp 1929). The ridge complex is about 5 km long and about 800 m broad and reaches up to about 85 m above sea level. It consists of several parallel ridges separated by troughs.

Many features on the surface of the ridges reflect strong aeolian erosion (Riezebos *et al.* 1986). The authors came to a general conclusion that at present on the push-moraine ridges of Holmströmbreen, degradation processes occur similar to the planation of the Dutch glacier-pushed ridges during the Pleistocene. This was accomplished relatively rapidly by aeolian processes which formed the cover sand fields.

6.8   Frost action and deflation

Troll (1948: 16; 1973:4) used the term 'gelideflation' for the frost effected deflation in periglacial conditions in Subarctic and mountainous environments. Often it is connected with needle ice formation which makes the surface soil susceptible for deflation (Zotov 1940; also Czeppe 1966: 122). McGowan (1994) reports from New Zealand that needle ice activity on the exposed glaciofluvial deltas during late spring and autumn was observed to fragment surface crusts in source areas and resulted in the formation of surfaces highly susceptible to deflation. Needle ice was also responsible for the reactivation of relic pavement surfaces at Lake Tekapo. Needle ice is not a very common feature in Polar regions (Schunke 1986: 69).

Old solifluction terraces with signs of very strong deflation on the slope of Reuterskiöld mountain, Spitsbergen, were reported by Samuelsson (1921). Their upper flat surface is unvegetated and covered by small stones without fine material and the steep lower edges are totally covered by *Andromeda* and *Dryas* vegetation. According to Samuelsson similar deflated solifluction terraces exist in high fells in Lapland.

Frost action can produce several features which can be deflated. The surfaces of well-formed unsorted and sorted soil polygons are uncovered and therefore suitable subjects for deflation. Often the frost-active soil such as till contains so much fine fraction that wind can only keep its surface uncovered but cannot remove large amounts of material. This has been observed on several sites. The origin of these patterned ground fields is often connected with locally thin snow cover affected by wind and the erosion takes place in the winter time (Fig. 6.14). Schunke (1986, Abb. 18) has provided a very good picture of this type of deflation surface on fine sediment polygons at Flakkerhuk, east Greenland. Schunke does not explain the barren surface as a result of deflation. From interior Jameson Land, east Greenland, the thick layers of glacial, glaciofluvial and glaciolacustrine deposits and signs of aeolian erosion and abrasion have been reported but no aeolian deposits (Möller *et al.* 1994).

Nichols (1969: 82–84) described from Inglefield Land, north Greenland, several heavily wind-eroded earth hummock fields. The hummocks, 30–60 cm in diameter and similar in original height, have been truncated so that the original upper curved surfaces have been removed and flat erosional surfaces of loess have been formed. Parallel grooves are formed on the erosional surfaces. Small undercut clifflets less than 2 cm in height are also present, and occasionally a granule has protected the loess in its lee. A rim of vegetation which was resistant to erosion commonly surrounds the truncated surface. Apparently the hummocks were covered with vegetation at the time the flat surfaces were cut. Nichols then discussed different possible origins for these hummocks and came to the conclusion that they were formed by wind erosion. One argument is for example that: 'The spaces between the hummocks are not eroded, nor are they filled with

Fig. 6.14 Frost heaved patterned ground on the slope of fell Peldoaivi, Finnish Lapland. April 26, 1969.

material from the hummocks'. Nichols does not mention the winter winds as the most probable deflating factor. The spaces in between the hummocks are filled by snow and the surfaces are exposed to wind and moving ice crystals.

6.9   Seasonal variation of deflation

Most of the year the deflation basins in Arctic and Subarctic environments are mainly covered by snow and only the steep slopes are somewhat exposed to winds. The spreading and intensity of the landscape effects of gales are limited and this means that the main part of deflation takes place during the summer (e.g. Nielsen 1933: 245). The other very important limiting factor is seasonal frost which makes the sand layers unmobile and hard as rock (Seppälä 1971a: 28). The uncovered sand surfaces, such as outwash plains, deltas and alluvial terraces and deflation areas, thaw and dry somewhat in strong wind and this gives loose sand for transportation and it has great significance for deflation. In different regions it seems to be different seasons of year which are favourable for aeolian processes. The timing of aeolian activity during the year depends on the site characteristics. Kobendza & Kobendza (1958) observed that deflation on sand dune areas takes place in winter in middle latitudes with some snow, as in Poland.

In snow-rich climates deflation can be active only during the summer. Most of the year the deflation basins and blow-outs are covered by thick snow except on the upper parts of their steep edges, which are somewhat uncovered but also they are cemented by frost. Very little loose, dry sand is produced by thawing and drying by sun and wind. This is the case, for example, in Finnish Lapland some 69° N latitude (Seppälä 1971a: 28).

According to Fristrup (1952–53) in central and southern Peary Land, Greenland, the strongest wind erosion occurs during the winter. That region is very arid. The total precipitation is insignificant, hardly above 100–125 mm. During the winter the total quantity of snow is so small that the land is not covered by a continuous layer of snow and large areas exposed to the wind are free from snow.

In West Spitsbergen the period December to April is entirely dominant concerning wind transportation of mineral particles (Åkerman 1980: 241, 277). Wind drift of mineral particles was observed also in June and July upon dry braided river plains and on outwash plains but not measured (Åkerman 1980: 277). According to Czeppe (1966: 122) strong winds (>10 m s^{-1}) occur in Hornsund, SW Spitsbergen from October till March when the flat surfaces are either wet, frozen or covered by snow, and that limits deflation to protruding snow-free edges, ridges and block fields. Riezebos *et al.* (1986) described the ice-pushed ridge complex blown free of snow and significant quantities of windblown sand on an old snow surface up to 500 m from the source ridges. They concluded that 'it seems likely that only during prolonged dry winter periods, when any interstitial ice binding the surface sediment layers has been removed by sublimation, do strong winds achieve these effects'.

On Ellef Ringnes Island and Amund Ringnes Island in Arctic Canada (78° N lat.) the aeolian action varies greatly between summers; an early snow-melt and low rainfall promote wind action which is restricted to poorly vegetated or bare areas of unconsolidated sand or silt without clay (Hodgson 1982: 28). No aeolian processes are active in wet summers.

In southern Spitsbergen Migala and Sobik (1984) made deflation measurements in September and October and their opinion is that in these oceanic conditions an intensive sublimation in a cold period, temperature slightly below freezing and dew-point lower than 0 °C are needed to activate deflation.

On eastern Baffin Island, Arctic Canada, winters are much drier than the summer period but still snow cover on proglacial sandurs may also block or retard deflation. There it has been found that deflation occurs preferentially during dry cold autumn and winter months (Gilbert 1983; McKenna-Neuman & Gilbert 1986; McKenna-Neuman 1989, 1990a). Bare sand surfaces are typical of winter conditions since snow cover is never more than several centimetres deep and seldom covers the entire surface of the outwash plains (McKenna-Neuman & Gilbert 1986). Also the existing niveo-aeolian deposits give evidence of sand drift during the winters. Koster & Dijkmans (1988) came to the same conclusion for northwestern Alaska by studying dune forms and niveo-aeolian sediments.

On north central Banks Island, Canadian Arctic Archipelago, the timing of aeolian activity during the year depends on the site characteristics. Where the snow cover is incomplete in winter, either because a rugged microtopography resulted in terrain elements protruding above the surface of the snow or because sublimation of snow exposed the ground surface, aeolian transport can be year-round. In areas with a flatter microtopography, summer transport took place but deflation was negligible over the winter (Lewkowicz & Young 1991).

On Western Sverdrup and adjacent islands, Arctic Canada, the most favourable areas for aeolian processes are below the marine limit, particularly wide fluvial deposits and strandline flats without clay (Hodgson 1982: 28). There, deflation of deltaic sediments containing plant debris strata which resist erosion gives residual pillars or clumps up to 2 m high; there can also be deflation of the finer fraction from gravelly sediments to leave a granule or larger size lag on the surface (Hodgson 1982: 30).

The spring floods of Colville River, northern Alaska, carry in three weeks up to 10 cm thick layers of fine sand onto the flood plains (Walker 1967). For most of the year these deposits are too moist or too covered by snow for effective sand removal by wind, which takes place in summer until early winter and again in late April and early May when an early melt season provides an uncovered source of sand for the wind (Walker 1967: 11).

In Victoria Valley, Antarctica, Calkin & Rutford (1974: 200) assumed that most of the sand movement takes place in the warmer months, from November to February, and it is probably rare from March to October, when up to a metre of snow or a surface ice matrix remain on the dune field, or when the dunes are moistened from below.

If the winter transport is important, then much snow can be found in sand deposits. It causes moisture in sand layers which in the form of liquid water increases the value of the threshold shear velocity of sand movement (Belly 1964; Calkin & Rutford 1974: 204) and decreases aeolian sand drift during the summer time.

6.10   Sand transport and deflation measurements

Deflation studies in cold environments have been mainly descriptive. Direct measurements of this process found in literature are rather scattered. To measure the amount of wind drifted material and the deflation rate in remote places is a difficult task. The measurements can be divided into short term measurements and long term observations. In short term measurements we study the occasional sand movement during one or several recorded storms in one or a few seasons. Short term transport measurements in cold environments have not been made in the winter time, however, this is in many cases the most important drift season. The winter measurements are extremely difficult because all kinds of traps will be filled by snow instead of mineral material. It is even difficult to record the precipitation in winter time because the rain gauges do not catch the total amount of snow fall because of turbulence around the gauge. Another annoying factor is the drift snow packed in the gauge. With long term deflation observations we mean repeated measurements within longer periods such as several years' time including both summers and winters. Then we do not know the weather conditions exactly or when the sand in fact was moving. In long term measurements the investigator does not sit beside the study site to await storms.

With traps we are collecting the material moving along or above the surface but we do not record the net deflation, and that is a problem when interpreting the results of measuring transport rates. Drifted material cannot be easily related to deflation landforms. The transport can be quite significant without much net erosion.

Butterfield (1971:126) in New Zealand states the matter in the following way: 'Wind erosion per se cannot be measured directly. Conventional methods of

erosion measurement such as peg displacement, painted stripes or surface lowering fail to differentiate between soil eroded by run-off, rain splash or other
agencies, and that removed by wind. Eroded material must be caught either during the transportation or subsequent deposition. In either case the inference for
erosion itself is indirect.' This is true if we are not measuring the deflation in
a blowout or deflation basin without outlet and which has a bowl shape, from
which only wind can transport the material out. Especially, meltwaters of snow
have to be taken into consideration. In short term measurements we can also
be sure about the eroding factor if we are staying in the field during the whole
observation period.

6.10.1 *Measuring techniques*

Applied techniques of measurements are the same as those used in temperate and subtropical climates when studying wind drift. They are mainly made
by trapping the moving material with different equipment especially constructed
for this purpose. Morgan (1986) divided the traps into horizontal and vertical
ones. The horizontal traps are at the ground surface, buried boxes or jugs or
containers or flat plates. With them we are able to collect creeping and saltating
material. Vertical traps are tubes or columns or boxes with separate compartments
rising above the sand surface. The vertical traps also collect the suspended load
from the air flow.

Horizontal sand traps. Owens (1927) designed a simple horizontal sand trap
(Fig. 6.15). It consists of a box 34 cm high, and 67 cm long (see Pyc & Tsoar
1990: 324). The box has vents on both upwind and downwind sides to allow free
entry and exit of air. The internal baffles cause sand deposition in the trap.

Horikawa & Shen (1960) used a rectangular box 2.4 m long with 18 compartments. The difficulty was how to orientate the long axis parallel to the wind

Fig. 6.15 Horizontal sand trap constructed by Owens (1927).

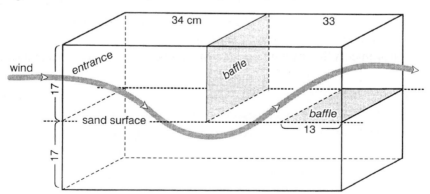

direction. In Spitsbergen, Piotrowski (1983) used rectangular containers 0.12 m² in area and filled them with water. The upper edge and the water surface were placed at the ground surface. Water will avoid the turbulences caused by an open space and the trapped material stays in the container, even the suspended one. Piotrowski's traps work well only on fully horizontal sand surfaces. They should be placed when it is calm, and the sampled material is difficult to get out according to Käyhkö's (1991) experience.

Käyhkö (1991) used open rectangular plastic jars 10 by 10 cm and 13.5 cm high with water in the bottom, or empty ones, which were buried into sand and could get reasonable results. This technique is not sensitive to the changes in wind direction but the jars cause scouring effects at their edges, as do all the other types of horizontal traps, too.

Tsoar and Yaalon (1983) used 7.5 by 7.5 cm plastic plates coated with vaseline and placed on the sand surface. They collected especially fluorescent dye-traced moving sand grains which became stuck on the plate. According to our knowledge no one has used this technique in cold environments.

Vertical sand traps. Bagnold (1938) designed a vertical collector which was 76 cm high and 1.3 cm wide to minimize interference with the airflow. The collector contained baffle plates inserted at an angle of 40° to prevent grains bouncing out of the collector. Horikawa and Shen (1960) tried to improve this basic design by providing an exit port for the airflow. The collection efficiency even now was relatively low, generally ranging from 20 to 40% (Knott & Warren 1981). A Bagnold-type trap can be equipped with a fin to enable it to rotate around an axial pole so that it always faces into the wind (Pye & Tsoar 1990: 326–327).

Belly (1962: 17) constructed a vertical trap (Fig. 6.16) collecting particles from different levels in a wind tunnel. Almost immediately after the beginning of a run, a scour takes place around and below the trap. According to him this scour seems to remain steady and therefore does not influence the measurements. For runs of long duration, however, the platform becomes undermined, and this is probably a cause of error, for the surface creep does not enter the mouth of the trap. According to my knowledge Belly's type of sand trap has not been tested in the cold environments, but the same problem is noticed with other traps at the sand surface.

Butterfield (1971) introduced a trap which was used on the mountains in New Zealand. That fixed trap consists of five circular collecting elements bolted vertically above one another between dexion strips. The whole structure is mounted between steel posts ('waratahs') driven into the ground and guyed against strong winds by wire rope. The use of slotted dexion allows each collecting element to be easily raised or lowered, but in general these are logarithmically spaced from 5 mm above the mean ground surface to 150 cm. Each collector is constructed from a PVC plastic pipe 7 cm in diameter, one end of which is sealed with a 120 mesh stainless steel gauze. The length of each collector is 15.5 cm. Particles entering the orifice of the collector strike the gauze and drop vertically into a detachable plastic jar. Only at the highest wind velocities is there significant loss of fine particles through the gauze.

Åkerman (1980: 274–276) constructed sediment collectors with wind-vanes which face automatically against the wind (Fig. 6.17). One type was placed

Fig. 6.16 Vertical sand trap constructed by Belly (1962).

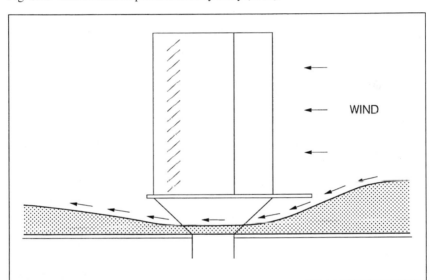

Fig. 6.17 Sediment collectors constructed by Åkerman (1980).

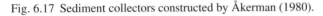

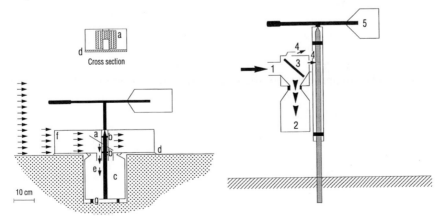

at ground surface and the other type 0.1, 0.5, 1.0, 2.0, 5.0, 10.0 and 20.0 m above the ground (Åkerman 1980: 279). They were emptied once a week during the summer and daily during the winter. However, the trap was placed upon a surface which is swept bare of snow all the winter. I am a little sceptical with the winter measurements because traps could be filled in a few moments by snow blown by strong winds. Åkerman measured in the winter time the relation of mineral particles to the amount of snow collected. It is really hard to estimate the turbulence effects these traps caused.

Rutin (1983: 90–92) constructed sand traps which collected sand from one direction only. The sand trap consists of a wooden housing 60 cm in height with 9 or 21 plastic trays standing on each other. The opening of each tray is 97 mm in width and 68 mm in height. The upper half of the tray is a flat plane sloping slightly towards the inner collector. The plane forces the sand particles into the lower collecting tray with volume equivalent to about 650 grams of dry sand.

McGowan (1994) used sediment traps similar to those designed by Butterfield (1971). They consist of a PVC drainpipe T-joint extended to 265 mm in length with a 60 mm diameter orifice. A perspex fin attached to the rear of the trap enables the trap to be self-orientating into the airstream, while a detachable clear

Fig. 6.18 Simple sand trap in operation (Seppälä 1974).

sample jar suspended below collects dust particles filtered from the airstream by an internal gauze filter with a mesh size of 212 micron. Traps of this design were tested by Butterfield and found to be 92% efficient for wind speeds less than 15 m s^{-1}. McGowan (1994) deployed some traps with 77-micron filters to minimize the loss of fines during windstorms. No significant difference was observed in the samples collected using either 212- or 77-micron filters. Loss of the fines during windstorms seems to be minimal.

Seppälä (1974) used similar jars to Käyhkö (1991; see above) but two vertical plates at right angles to each other were fitted into the instrument so that the jar was divided into four sectors (Fig. 6.18). The plates rose 11.4 cm above the jar to stop saltating sand and guide it into the jar, which was buried into the sand so that the upper edges were at the sand surface. The jar and its plates caused some turbulence which at first probably filled the jar too much, but later on the small wind channel guided the creeping sand grains around the trap. In west winds also the southern and northern sectors of the jar were filled, and a strong west wind also partly filled the eastern sector, which is evidence of turbulence.

Leatherman (1978) developed a very simple vertical trap and it was modified and enlarged by Rosen (1978). It consists of a 10 cm diameter, 100 cm long PVC tube which is half-buried in the sand. The upper end has two slits which extend 46 cm down from the top. Air enters the front slit (6.5 cm wide) and leaves through the back slit (10 cm wide), which is covered by 60 μm mesh to trap the sand grains. Sand is collected in a removable tube liner which has a fine mesh base to allow free drainage of water. Käyhkö (1991) used among others the Leatherman-type traps in Lapland. A good review with pictures of sand traps is presented by Pye & Tsoar (1990: 324–330).

Lee (1987) constructed a sand trap with registering balance which will give the opportunity to correlate the sand movement with wind velocity, gustiness and other weather characteristics. This technique is especially good in very short term measurements combined with very detailed current and turbulence measurements.

6.10.2 *Short term measurements*

Some attempts at short term deflation measurements in cold environments can be found in the literature. All these quantitative measurements of deflation are very local and limited in time. I think that in this phase of research they, however, can give us some idea of the meaning of deflation for the landform evolution, but the total view is still to be created.

Sand trapping techniques can be used for short term deflation measurements. We can also place marking pins in the deflation basin and indicate the present surface on them, and after a certain period or a known storm observe the new position of the sand surface (Seppälä 1974; Käyhkö 1991).

These measurements have been carried on at Hietatievat, Finnish Lapland (Figs. 6.7 and 6.19) by Seppälä (1974) and Käyhkö (1991) but they counted the sand transport passing a gate (g cm^{-1} h^{-1} or kg m^{-1} s^{-1}). Seppälä used in his calculations the diagonal of the trap jar as the width of the gate and Käyhkö used the mean value of the diagonal and the side of the jug.

At Hietatievat sand transport was measured in June, 1973, on a large blowout 7600 m^2 in area in an old sand dune (Seppälä 1974). The maximum rate of sand transport was about 25 g cm^{-1} h^{-1} when the wind velocity averaged 6.9 m s^{-1} with gusts exceeding 10 m s^{-1} at 2 metres height. For longer periods (11–12 h) with lower wind speeds (mean 6.0 m s^{-1}) the sand drift rate ranged from 3 to 10 g cm^{-1} h^{-1} with the highest rates in the centre area of the blowout and decreasing towards the sides. The results gave an impression that one or just a few periods with strong winds can keep the deflation basins and blowouts uncovered and deflation continuously active.

Käyhkö (1991: 102) tested the vertical and horizontal traps at Hietatievat and found out that the vertical trap used by Seppälä (1974) could catch only 24% of that amount of drifted sand as the same trap jar without plastic plates rising 10 cm above the sand surface when the average wind speed was 4.1 m s^{-1} measured 55 cm above the sand surface. Käyhkö (1991: 113) measured with lower wind speeds the maximal sand movement rate 8.4 × 10^{-4} kg m^{-1}s^{-1}.

One method to measure the magnitude of the aeolian transport especially used by Polish investigators is to measure the weight (g) of the trapped material per area (m^2). It is a useful technique when sand accumulation takes place but it is rather difficult to transfer to the geomorphic meaning as to how much a surface has been eroded or how much a blowout has expanded. This type of measurement gives an idea of how much material is passing a certain area. The site could be in balance with deflation and accumulation and still on it will be drifted quite big amounts of material. One dm^3 of sand weighs about 2000 g. In two areas at Oscar II Land, Spitsbergen, Piotrowski (1983) carried on aeolian transport measurements for about 1.5 months and the maximum value was 1397.7 g m^{-2} at Kaffiöyra for a sandy-gravel beach. On the outwash plain and marine terrace the aeolian transport was much less, only several percent of the crest zone values (max. 4.5 g m^{-2}).

From Lewis River Sandur, Baffin Island, Church (1972) gave an estimate that 1.6 kg sediment was transported down the valley per metre width of sandur in July 1965. 70% was transported in three days and the effective wind speed was 70 km h^{-1} (>19 m s^{-1}).

Migala & Sobik (1984) noticed blowouts and niveo-aeolian phenomena predominantly on alluvial deposits close to glacier Werenskioldbreen, Spitsbergen, which become enriched every year with fine-grain mineral matter carried by ablation waters. Observations were made in late summer and autumn. According to them drying of sediment by evaporation and sublimation played an important role. Congelation initiated the change in the internal structure of the deposits, which were rather silty. This was caused by ice segregation or needle ice and micro ice-vein formation. Before freezing the sediment was not suitable for wind drift. From the central part of the temporary proglacial lake 14.5 kg m^{-2} of sediment was drifted away in about two weeks' time. The bottom sediments were eroded about 11 mm vertically.

When thinking of sublimation in the High Arctic we have to keep in mind that there are several months without direct sun radiation, and therefore the possible sublimation is only caused by latent heat transported by wind. On the basis of experimental research on sublimation of frozen sand samples McKenna-Neuman (1990b) made some approximations of the possible production of loose sediment on an outwash plain. The rate of particle generation (kg m^{-2} s^{-1}) was calculated by dividing the weight of sediment in the collection pans by the test surface area adjusted for the diminishing block dimensions.

In the early autumn when surface temperatures are just below zero centigrade the greatest rate of loose material production through sublimation appears (McKenna-Neuman 1990b: 334). She assumed that the outwash sediments are near saturation at that time and then the regression model predicts that a 1 km^2 area could produce approximately 7.1×10^5 kg loose sediment in one day in Pangnirtung Pass, N.W.T. Coarse aeolian sediments adjacent to the sandur could produce 3.5 times more loose material and this amount is equivalent to a volume of 1640 m^3. However, when the temperature drops the amount could be just 0.5% of that in the autumn. Sublimation does not continue in a sufficient way if the loose sand staying on the surface is not drifted further and new frozen material uncovered. An illustrative example of the protecting effect of the dry sand is the appearance of dirty cones on glaciers. 2–3 cm thick layer of sand and gravel prevents or slows down the further ablation, the mineral material on the surface stays rather dry because it has small capillarity and high porosity, the underneath ice stays unmelted and sublimation does not take place.

This is the weak point in McKenna-Neuman's (1990b) application of laboratory results to the natural conditions. Each time during the experiment when a sand grain was released from the cementing ice by sublimation it dropped off from the test block, and from the system. Loose sand was not producing an insulating layer on the test block, which means in the natural conditions a constant sand drift and continuous transport of material outside the sublimating surface. However, the experiment supports clearly the field observations that sublimation is fundamentally important for autumn and winter aeolian transport. However, coarse grains enrich during the winter deflation and produce an immobile surface pan, and that also decreases the sublimation rate in the natural conditions.

6.10.3 *Long term measurements*

Seppälä (1984) in Lapland used 100 cm long wooden pins 3 to 4 cm in diameter which were driven into sand at a depth of 50 to 70 cm. 20 pins stood in intervals of 20 m in four lines, and the longest line was 140 m in length. The position of the sand surface was marked on each stake. The vertical position of the sand surface was observed four times over three years. The accuracy of the measurements was ±0.5 cm. The blowout was the same as in the short term measurements (Seppälä 1974) and it covered an area of 7600 m^2. The average amount of deflation ranged annually from 1.6 to 2.5 cm of vertical sand surface lowering. This means from 121 to 190 m^3 or 242–388 tons of sand moved from this single blowout annually (Seppälä 1984: 44–45). In these three years the maximum deepening at one observation point was -19 cm. The total vertical deflation in the three years, on average, was 6.2 cm counted from all the measurement points. The average values for one year are 2 cm, 157 m^3, and 314 tons. This was the first time these types of value of deflation in Subarctic conditions could be presented. Later Käyhkö (1991) measured at the same dune field the deflation rates with repeated levellings of many more points than Seppälä (1984) and got a value of -3.0 cm for average net deflation in about two years' period.

Lewkowicz & Young (1991) in Arctic Canada used 300 mm, long wooden pins (diameter 7 mm) equipped with metal washers (diameter 50 mm) and they were inserted to depths of 150 mm, but they faced some difficulties because such short pins could not be relocated when large amounts of erosion or deposition had occurred. In sand deposits and with thin wooden pins we do not have troubles with the impact of frost heave (Seppälä 1984: 43; Lewkowicz & Young 1991: 200). They considered the measurements of deflation and deposition with the accuracy to ±1 mm. On Banks Island in Arctic Canada there were recorded in the same area detectable changes ranging from maximum vertical erosion of 158 mm over two years to maximum deposition of 48 mm in a single year (Lewkowicz & Young 1991: 208). They do not make any calculations of the total transportation or deflation for their study areas.

In the literature we could find very few approximations of the annual total deflation based on field observations. From the short term measurements at
Hietatievat, Finnish Lapland (Seppälä 1974) compared with the long term measurements at the same site (Seppälä 1984) we can conclude that a few storms or
just one single strong wind period could cause most of the annual sand transport
and keep the deflation basin uncovered and the process active.

With normal levelling with long enough intervals on the blowouts we can
get a total view of the magnitude of the deflation and accumulation. The best
method is that the measuring points are marked and we have the reference points
outside of the deflated area. When working carefully the vertical accuracy of
levelling is normally ±0.5 cm. Käyhkö (1991) used this technique at Hietatievat,
Finnish Lapland, in the years 1988–1990 (Fig. 6.19) and found the deflation and
accumulation rates very similar to Seppälä (1984) in the same region in the
period 1974–1977.

The sand accumulation on the edges of the blowouts can be measured with
a bio-indicator introduced by van Dieren (1934) and Kobendza & Kobendza
(1958). Some plants do not suffer if they are covered by drift sand. Every year
after some sand accumulation they grow a new leaf node at the sand surface
and by measuring the distances between the nodes we get an idea of the vertical
sand accumulation at the site. The plants used by Seppälä (1974) at Hietatievat
for accumulation measurements were: *Carex* sp., *Descampsia flexuosa, Juncus
trifidus* and *Solidugo virga-aurea*. Even 12 years backwards could be measured,
and the maximum accumulation at those sites was 3.6 cm and the mean annual
accumulation rate was 2.1 cm.

Fig. 6.19 Deflation rate at Hietatievat, Finnish Lapland, measured by levelling. Drawn
after observations by Käyhkö (1991: Fig. 68).

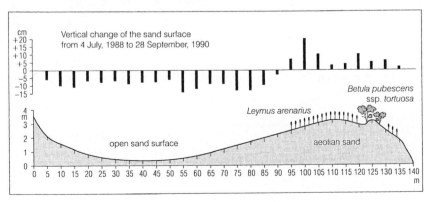

Modern techniques such as the electronic tachymeter (e.g. Nikon) based on an infrared beam and prism reflector have been used recently in the deflation surveys. The advantage is that in a short time hundreds and even thousands of points can be measured from the small area with great accuracy (in distance ± 10 mm/1 km and ± 1 mm in height) and all the values can be stored in digital form and evaluated further with a computer. When we have two sets of data with a time interval we can calculate the volume of the moved material and the mass change. The results of deflation measurements from Hietatievat, Finnish Lapland, do not differ much from the previous measurements (Käyhkö 1991) even though the measurements were made in different years. The mean vertical lowering of the sand surface on the studied deflation basins in different places in Finnish Lapland was 6 mm yr^{-1} (ranging from $+145$ to -184 mm) in 1993–94 (Käyhkö 1997: 155).

McKenna-Neuman & Gilbert (1986) used a time lapse camera on eastern Baffin Island, taking one photograph at noon each day on 16 mm film in the winter time for about 7 months to study the role of snow cover in shielding surface sediments from winter deflation. This technique can also be used to record the short term variation in aeolian relief, but its accuracy is rather limited. A time lapse camera cannot be used in the High Arctic north of the Arctic Circle in the winter time because of the lack of daylight.

Fig. 6.20 Low level vertical aerial photograph of deflation basins at Hietatievat, Finnish Lapland. June 30, 1973. Photo by R. Larsson.

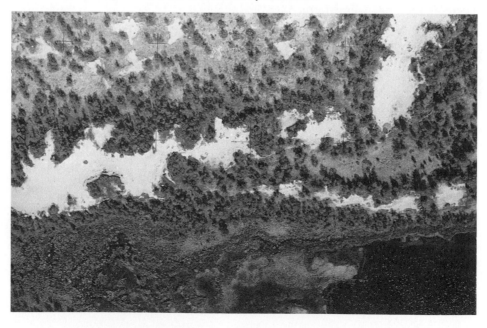

Aerial photographs are rather practical tools when doing long term observations of deflation. On them the deflated and revegetated areas can be easily
detected and mapped with reasonable accuracy. The quantitative stereomapping
of amounts of drifted material is difficult to recognize because the uncovered sand
surface has a very high albedo, almost like snow, and therefore surface details
are difficult to identify (Plate 9). Special 'home-made' low level photographing
has been done in Finnish Lapland using a small plane and a Hasselblad MK70
camera (Fig. 6.20). Those pictures have been used for stereomapping of deflation
areas, on a scale of 1:2000 and with 0.5 m contour interval.

Resolution of satellite images is often too small to identify the small deflation areas, for example in Lapland. Especially in the numerical interpretation of
satellite data we shall meet great difficulties when trying to locate only blowouts,
because several other bare sand surfaces and some other things have the same
reflection characteristics, too, e.g. patterned ground, road cuttings, gravel pits,
beach sands, house yards etc. Also lichen-covered (*Cladonia*; *Stereocaulon*) surfaces and some clear cut areas give very similar radiation to blowouts. (Personal
communication from J. Käyhkö 1995.)

An attempt to locate and quantify deflation areas with Landsat TM mapping in
Lapland was made and it was found that the mixed pixels formed a problem and
the results probably underestimate the total area (about 1000 ha) with aeolian
activity (Käyhkö *et al.* 1999).

6.11   Cessation of deflation

The end of deflation has not been much debated in the literature, however, most of the cold-climate dune fields and sand sheets are vegetated or only
partly in an active stage (Koster 1988). The active deflation phenomena are so
striking in the Polar regions that the investigators have not paid much attention
to the cessation of deflation. The only clear exception is the vast pavements in
cold environments which are mentioned by several authors (e.g. Högbom 1912;
Samuelsson 1921, 1925, 1926; Fristrup 1952–53; Tedrow 1966; Schunke 1986).
In many cases deflation is encouraged by other geomorphological processes as
described above, and new erodable material is always available and that means
that the cessation of deflation cannot yet be noticed.

When we have more vegetation covering the ground surface and the succession can take place in a relatively short time, as in Subarctic Fennoscandia, then
it is more obvious that the old deflation features get some interest, too. Once
deflation has begun, it may cease for any of five reasons (Seppälä 1971a, 1984):
(1) the fine sand is carried away and deflation reaches till or glaciofluvial deposits underneath, leaving coarser or finer material on the surface, e.g. gravel
or silt (see e.g. Cailleux 1972: Photo 1); (2) deflation reaches the level of the

ground water (e.g. Ohlson 1957); (3) the ground-water level rises; (4) a blowout becomes so deep that the wind can no longer reach deflating velocities at the bottom of the blowout (deflation continues, however, at the edges of the basin so long as there is suitable material available); (5) the climate becomes wetter and/or warmer, so that conditions favouring plant growth are improved. All these factors encourage the growth of vegetation in the basins and limit the power of wind.

6.12   Pavements

In periglacial conditions wind has produced vast earth surfaces covered by blocks, cobbles and pebbles (Figs. 5.2 and 5.3). These so-called pavements have been described by several scientists from Spitsbergen (e.g. Högbom 1912; Samuelsson 1921, 1926; Riezebos *et al.* 1986), Iceland (Sapper 1909; Samuelsson 1925, 1926; Nielsen 1933; Thorarinsson 1962), west Greenland (Jensen 1889), Peary Land, north Greenland (Fristrup 1952–53, Nichols 1969), Canadian Arctic (Bird 1967) and Antarctica (Péwé 1960; Calkin & Cailleux 1962; Nichols 1966, 1969). Also many of the stone and block concentrations covering the summits of mountains and fjelds, for example in Lapland and Labrador, result from deflation combined with frost heaving of blocks from coarse till to the ground surface. Of course heavy rains and melt waters are also washing the fine fractions from the surface, but on the flat tops of fells their role is not sufficient while strong winds are very frequent.

Fine sandy till surfaces in a glaciated environment can also be successfully attacked by deflation if they do not contain too much fine fractions or coarse grains. Several examples of deflated till surfaces can be found in Finnish Lapland on the fell regions (Fig. 6.14). It seems to be obvious that deflation on till surfaces has been activated by frost activity which is at first destroying the roots of plants. These blowouts are normally only some tens of square metres in area. Bird (1967: 241) postulates that in the Canadian Arctic 'most widely distributed is the deep sandy till which frequently has blow-outs but rarely develops well-formed dunes'.

Pavement forms of old till deposits have been described from east Greenland by Schunke (1986: 67–68). He calls them by a German term 'Steinpanzer' (stone armour), and the origin is surface washing, frost heave, deflation and probably snow drift, i.e. the same geomorphic processes working on the fell summits.

6.13   Succession of vegetation on deflation surfaces

Numerous papers have been published on the vegetation development on coastal dunes because it has a practical meaning for stabilization of aeolian material (e.g. van Dieren 1934; Chapman 1964; Pluis & Winder 1990). Very little information is available of the succession of vegetation on deflation surfaces in Polar regions (e.g. Tobolski 1975). From West Spitsbergen Åkerman (1980: 237) mentioned as colonizing pioneer plants on aeolian sand surfaces *Puccinellia* and *Carex* sp., and after the dunes are stabilized the most common succession plants are *Dryas octopetala* and *Saxifraga* species. *Dryas* is a very effective colonizing plant on new sand surfaces in cold environments because it has on its roots some bacteria which are able to fix the nitrogen from the atmosphere. In Iceland small aeolian sand accumulations are often colonized by Arctic river-beauty (*Epilobium latifolium*) (Plate 10 and personal communication by Jukka Käyhkö 1995).

Our experience from Finnish Lapland (Seppälä 1971a: 30, 1974, 1982b) is that the stabilization of deflated sand surfaces starts with scattered vascular plants, e.g. *Empetrum hermaphroditum* and *Arctostaphylos uva-ursi* (Tobolski 1975), which are originally from the edges of blowouts, and they are very soon followed by some lichens, e.g. *Solorina crocea*, which are able to fix nitrogen from the air, and some blue-green algae. In the next phase several other lichens (e.g. *Stereocaulon pascale*) and mosses (*Polytrichum piliferum* and *Dicranum fuscescens*) start to colonize the deflation area. One of the first species starting to grow on blowouts is *Festuca ovina*. On the sides of deflation areas the succession is somewhat different because accumulation of sand takes place there also. There we are able to find pure growths of perennials such as *Descampsia flexuosa, Juncus trifidus, Solidago virga aurea, Leymus arenarius* and *Carex* sp. which survive even when they are partly covered by drifted sand. Partly *Archtostaphylos alpina, Vaccinium myrtillus, V. uliginosum* and *V. vitis-idea* grow on the sides and they slide down to the basin following undercutting. If the deflation has reached the ground water table then willows (*Salix* ssp.) colonize the basin very fast (Seppälä 1995a).

On the same deflation area in Hietatievat, Finnish Lapland, we can find blowouts in several different succession stages. Some have almost no vegetation, others are partly anchored by vegetation and a few are totally covered by a vegetation mat. Often in the vegetated blowouts concentric vegetation zones can be seen, following the directions of the edge of the blowout.

Shrubs such as *Juniperus communis* (Fig. 2.20) and some willows in Lapland (Ohlson 1957) and willows (*Salix alaxensis*) in the Arctic Coastal Plain of Alaska along the Meade River (Rickert & Tedrow 1967; Tedrow 1977: 302) form sand accumulation centres which can form small sand dunes. Sand is covering their stems but still they are able to continue growing by extending their branches above the sand surface. The pillow-like growths of plants such as *Silene acaulis, Diapencia* sp. and *Empetrum* sp. are deflated on the foreside and growing on the lee side (Figs. 2.28 and 2.29). From these tussocks the

Fig. 6.21 Roots of some 50 years old birch (*Betula pubescens*) laid bare by deflation on the Kaamasjoki-Kiellajoki sand dune area, Finnish Lapland. July 17, 1968. (Seppälä 1971: Fig. 18).

Fig. 6.22 Pines (*Pinus silvestris*) on a sand dune close to Lake Kotkamajärvi, Inari, Finnish Lapland, felled by W winds. July 25, 1972.

prevailing direction of deflating winds can be measured. Deflation may dig roots of trees barren (Fig. 6.21) and the next strong storm may fell the weakened trees (Fig. 6.22).

In Northern Alaska near Atkasook, *Leymus mollis* ssp. *villosissimus*, often accompanied by *Tanacetum bipinnatum* ssp. *bipinnatum* and *Mertensia drum-mondii* (Lehm.) G. Don, is the dominant plant of herbaceous active sand dune vegetation (Komárková & Webber 1980: 461, Fig. 10). On willow-vegetated dunes along the Meade River the soil is colonized by *Eriophorum vaginatum* (Rickert & Tedrow 1967; Tedrow 1977: 302).

On the Arctic coastal plain of Alaska is a 50–70 km wide zone of aeolian deposits with ancient and active sand dunes. There quasi-stabilized dunes are covered by grasses and willows and by lichens, especially *Alectoria nigricans* (Everett 1979: 213).

6.14   Ground water table and deflation lakes

If the ground water table is close to the surface it keeps the ground surface wet and prevents deflation. In present-day deflation areas in Lapland the ground water table is normally at least 60 cm below the surface in the middle

Fig. 6.23 Oblique aerial view of heavily deflated parabolic sand dune at Lake Pöyrisjärvi, Finnish Lapland. Notice the dark wet area where the deflation has reached the ground water table in the middle of the deflation basin somewhat below the centre of the picture. June 30, 1973.

Fig. 6.24 Vegetated deflation basin at Hietatievat dune area, Finnish Lapland. June, 1974.

Fig. 6.25 Deflation lake dammed by relict parabolic sand dunes on Kaamasjoki-Kiellajoki dune field, Finnish Lapland. July 19, 1967. (Seppälä 1971a: Fig. 11).

of the summer. Deflation may reach the ground water table (Fig. 6.23) and that supports the colonization of pioneer plants (Fig. 6.24).

If the deflation basins are filled by water they form small ponds or lakes (Figs. 6.25 and 6.26). In northern Fennoscandia on several sand dune areas

Fig. 6.26 Bathymetric maps of two deflation lakes on Kaamasjoki-Kiellajoki dune field, Finnish Lapland. Contour interval 50 cm. The deepest measured points are 4.25 m and 3.9 m (Seppälä 1971a: Figs. 7 and 9).

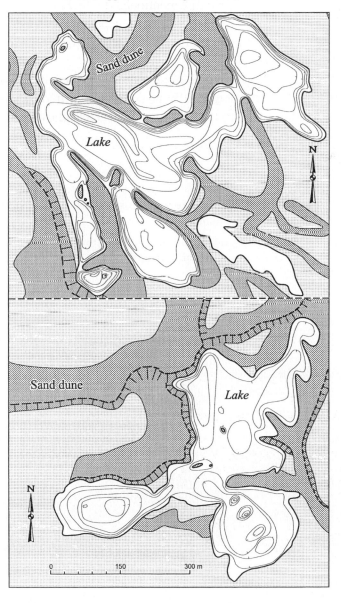

the deflation basins are lakes at present (Seppälä 1971a, 1972a). The ground water table rose some 5 to 6 m after the dune ridges were formed and it filled the deflation areas of parabolic dunes which are now partly subaquatic. This took place some 4000 years BP in Finnish Lapland (Seppälä 1971a). The same phenomenon might be found around the circumpolar area (Payette & Filion 1993) but it has not yet been carefully studied.

At Fagnant Lake, south of Great Whale River, western Québec, is found a dune field where old deflation areas are filled by peat moss (Filion & Morisset 1983: 89). Such peat-filled dune fields are mentioned for the James Bay Territory (Dionne 1978). They show that after the main aeolian period finished the ground water table rose and peat formation started in the deflation depressions (Filion & Morisset 1983: 89).

During a very warm and dry summer some shallow (<2 m in depth) deflation lakes dry up, and sand at their bottoms can be somewhat drifted by wind. This took place, for example, in July 1972 on the dune field at Kaamasjoki river basin, Finnish Lapland (Fig. 6.27).

Fig. 6.27 Seasonally dried deflation lake on Kaamasjoki-Kiellajoki dune field, Finnish Lapland during a very warm and dry summer. July 27, 1972.

Locally dunes can dam and control the rise and fall of the ground water table without special changes in regional climate, as has been investigated in Nebraska Sand Hills where sand dunes blocked surface drainages and the water table was raised as much as 25 m (Loope *et al.* 1995). We do not think this is common.

7

The question of oriented lakes

7.1   General characteristics

Oriented lakes form groups of lake basins with a common long-axis orientation. Some lake basins are entirely the result of thawing, others are merely modified by it (Washburn 1973, 1979: 271). They are conspicuous features of Arctic and Subarctic lowlands in both tundra and taiga regions (Hopkins & Kidd 1988: 790). Oriented lakes are found in vast areas of the Arctic Coastal Plain in northern Alaska (Fig. 7.1 and Plate 11), in the North West Territories of Canada east of the Mackenzie Delta and on Banks Island (Bird 1967: 211–215) (Fig. 7.2), and according to French (1976: 123) also in the Siberian Arctic coastal plains, as well as in temperate climates with various origins (Price 1968a, 1968b). Locally, from 20 to 50% of the surface is occupied by oriented lakes (Carter 1987: 615). At an elevation of less than about 100 m, these lakes may occupy from 15 to more than 50 per cent of the coastal plain between Sagavanirktok (148° W) and Utukok (162.5° W) rivers in Alaska (Brewer *et al.* 1993). Porsild (1938) in his classic paper on pingos of the Mackenzie Delta was perhaps the first to examine the 'lake-filled country' of lakes and former dried lakes. Cabot (1947) was the first to describe oriented lakes of the Arctic in detail. Oriented lakes, similar to the Arctic ones, have been described from SE United States already much earlier according to Cooke (1934) and Price (1968a).

Most of the Coastal Plain of Alaska is a cold desert. According to Ryan (1990) mean annual precipitation there is typically around 150 mm, of which about half falls as snow (around 100 cm per year) (Fig. 2.8). Winds blow almost continually, with the shape of the lakes usually indicating the predominant wind direction during the ice-free season. Strong winds at Barrow, Alaska, a major oriented lake district (Fig. 7.1 and Plate 11), blow frequently from the west and east sectors (Black & Barksdale 1949: Fig. 3; Carson & Hussey 1962: Table 1, Figs. 4 & 5).

Oriented lakes in permafrost regions are originally thermokarst features resulting from melt-out of ice within unconsolidated permafrost sediments (Harry & French 1983). In general they are also called thaw lakes (Hopkins 1949: 119;

Fig. 7.1 Oriented lakes on the Arctic Coastal Plain, northernmost Alaska. Redrawn after Carson & Hussey (1962: Fig. 2). Reproduced with the permission of © The University of Chicago Press.

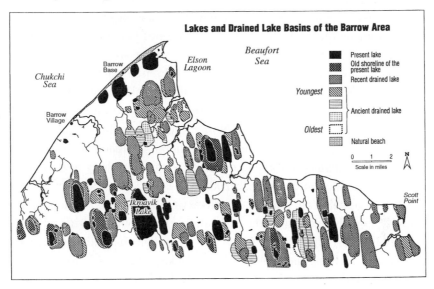

Fig. 7.2 Major areal extent of the oriented lakes in Alaska and Arctic Canada. Sources: *Base Map of Northern Alaska*, US Geological Survey, October 1944 (Black & Barksdale 1949: Fig. 1), Bird (1967: Fig. 57) and Carter *et al.* (1987: Fig. 24).

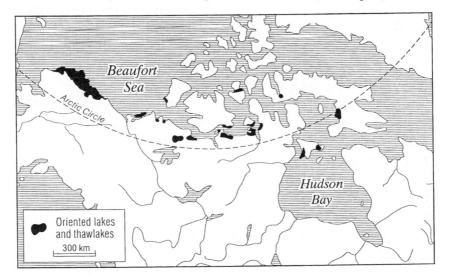

Washburn 1979: 271) but their formation seems to be in close connection with aeolian processes and the general shape of larger lakes is the result of the influence of wind (Fürbringer & Haydn 1974). Bird (1967: 211–216) and Washburn (1973; 1979) make a distinction between thaw lakes and oriented lakes. The origin of both might be the same but thaw lakes are not yet shaped so much by wind. In Alaska they are especially well developed in areas with aeolian sand and silts (Hopkins & Kidd 1988). According to French (1976: 123) fluvial terraces, outwash plains and Arctic coastal areas with silty alluviums and high ice contents are the most favoured localities for widespread development of thaw lakes.

The shape of oriented lakes in North America is often elliptical, having axial ratios around 1.5 with their long axis generally aligned N-NW, perpendicular to the predominant summer winds (Ryan 1990). The long axis is normally less than 3 km with an area less than 200 ha (2 km^2). The shapes of the lakes of the Alaskan Coastal Plain may be described as elliptical, cigar-shaped, rectangular, ovoid or egg-shaped, triangular, irregular, or compound, having any combination of these shapes. The ratio of length to width ranges from almost 1:1 to more than 5:1 (Black & Barksdale 1949: 111). Mackay (1963: 47–50) divided oriented lakes according to their shapes and central deeper basins into four categories: (1) lemniscate, (2) oval, (3) triangular and (4) elliptical. Equations were set up for each type.

Many lakes show narrow oblongs three to four miles long, apparently elongations of the original polygons (Cabot 1947). The most rapid development of an oblong lake would take place by the joining of two adjacent lakes (Cabot 1947). The lakes and ponds in the Colville River include remnants of the oriented lakes among lakes of other origins such as intra- and inter-dune lakes (Walker 1993).

The thaw lakes are relatively shallow with maximum depths generally ranging from 3 to 4 m (Ryan 1990) and one 5 m deep is relatively deep (Brewer *et al.* 1993). The littoral margins are often less than 0.3 m deep (French 1976: 124). Sublittoral shelves form the lake bed adjacent to the long sides of the lakes and are separated by a central deep that commonly extends to both ends of the lake (Carter 1987: 615). The ice thickness of thaw lakes in the winter time range from 2 to 2.5 m (Ryan 1990). Most of them freeze to the bottom during the winter (Walker 1993). In connection with syngenetic ice wedges, thaw lake basins as deep as 20 m are known (Carter 1988: 710; Hopkins & Kidd 1988: 791). If the basin deepens sufficiently to allow liquid water to persist under winter ice, the heat anomaly intensifies, and a basin of thawed ground develops beneath the lake bottom (Hopkins & Kidd 1988: 791).

These small lakes are slowly migrating and draining into one another (Hopkins 1949: 127), or into the Arctic Ocean, because of thermal erosion of permafrost and especially ice wedge polygons (Mackay 1988; Brewer *et al.* 1993). Wallace

(1948: 179) found in eastern Alaska from tree-ring counts that annual bank recession was 6 to 19 cm. This means that a lake of 90 metres radius would form in approximately 1600 years or, at a rate of 19 cm per year, only 500 years would be required (Wallace 1948: 179). Migration rates of about 30 cm per year have also been recorded (Tedrow 1969: 339).

Wallace (1948: 179) called the thaw lakes cave-in lakes. Lakes are located in the forest covered sand dune areas of the floodplains of the Nabesna, Chisna and Tanana rivers, eastern Alaska. In their origin he distinguished three events: the deposition of fine, frost sensitive sediments, the volume of which changes with freezing and thawing, followed by development of permafrost, and finally, a change in the thermal balance to cause thawing of permafrost. The lakes are not formed until after the flood plains are covered by vegetation. It is apparently the change from a good insulator, the vegetation cover, to a good accumulator and distributor of heat, such as the lake water with relatively high specific heat, that enlarges the lake basins. The effect of removal of the vegetation cover has been demonstrated in many places where caving has taken place in ground that had been cleared for construction purposes. In nature the initiation of a lake could result from such an event as the overturning of a tree by wind, with resulting uprooting of the vegetal mat (Wallace 1948: 180). Wallace did not point out any ice wedges or massive ice permafrost in connection with cave-in lakes. They might not be real oriented lakes even though they have a thaw origin.

7.2   Formation processes of oriented lakes

A thaw lake cycle and the melting deposits in Alaska have been described by Britton (1957: 117–123). In the literature no observations of a thaw lake cycle from Canada have been found. An idealized cycle of the development of a typical thaw lake at Imuruk Lake area, western Alaska, was presented by Hopkins (1949) (Fig. 7.3). At the first stage a locally deep thaw is initiated (1) by disruption of the vegetal cover by frost heaving and thawing collecting in a small water pool, (2) by accelerated thaw beneath pools occupying intersections of ice wedge polygons or (3) by accelerated thaw beneath pools in small streams. Pools of water, once formed, enlarge themselves by thawing and caving at their margins. Escarpments can be 1–3 m high. Thaw proceeds most rapidly at water level. Large blocks are undercut and occasionally collapse. This process resembles the alas formation by thermokarst in Siberia according to Czudek and Demek (1970: Fig. 9). The difference is that in Alaska the thaw lakes are not in a forested region and in Siberia wind probably does not play a great role in the alas formation.

Cabot (1947) mentioned that strong wave action by summer winds could have undermined the cliff walls of soft sand and gravels. Floating turf mats with living plants were seen on some lakes. Wave erosion becomes an important process after the lake has attained a diameter of about 30 m. The bodies of clear ice of ice wedges quickly melt out and their sites are marked by gullies up to 3 m deep and 60 m long. Melting ice wedges also form the trenches for drainage channels (Hopkins 1949). Underground drainage of lakes is known also from permafrost regions (Mackay 1958: 56). Ice wedges may also cause piping (Seppälä 1997c).

After drainage the lake bottom is soon colonized by *Carex*, *Sphagnum*, and *Salix*. Perennially frozen ground forms again in the lake bottom. Evidence of newly formed ice wedges is lacking in the bottom of drained thaw lakes (Hopkins 1949: 127) or they may also persist beneath the lake bottom, covered by less than half a metre of lake sediment (Billings & Peterson 1980: 417; Hopkins & Kidd 1988: 791). Cabot (1947: Figs. 5 & 6) illustrated rounded polygon areas which probably represent drained thaw lakes. The succession of vegetation of drained oriented lakes in the wet coastal tundra of Alaska has been studied (Britton 1957). Billings and Peterson (1980) examined and modified Britton's (1957) thaw-lake cycle hypothesis giving an interesting contribution. Pioneer species are *Dupontia fisheri* and *Arctophila fulva*, while *Eriophorum angustifolium* is characteristic throughout all of the terrestrial part of the cycle. *Carex aquatilis* is the most successful competitor and dominates the vegetation for about 2000 to

Fig. 7.3 Cyclic development of a thaw lake at Imuruk Lake area, western Alaska. Redrawn after Hopkins (1949: Fig. 2). Reproduced with the permission of © The University of Chicago Press.

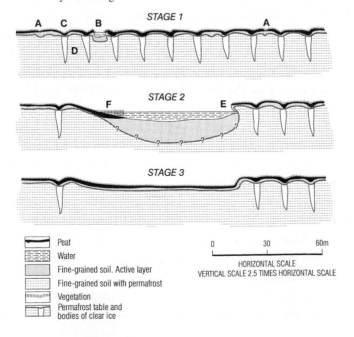

3000 years. It disappears locally when thaw ponds developing from low-centre polygons become too deep.

Deviating from the above observations, there are lots of ice wedges everywhere on drained oriented lake bottoms in Canada (Mackay 1958: 77, Fig. 18) At Illisarvik, an experimentally drained lake, ice wedge formation was the first event. Although Illisarvik is not an oriented lake, wedges occur everywhere on drained oriented lake bottoms in Canada (Ross Mackay, personal communication 1995).

After new perennially frozen ground has formed, the floors of some drained lakes become the sites of smaller new thaw lakes. Several present lakes are surrounded by escarpments marking former shores of ancient, larger lakes (Hopkins 1949: 127).

7.3 Origin of oriented lakes

The orientation of the thaw lakes of the Arctic Coastal Plain became the subject of considerable discussion (Black 1969b: 132–133). The concept of their origin seems to be still a somewhat unanswered question. Cabot (1947: Fig. 6) pointed out lakes on oval polygon areas in various stages of development. One had begun to drain itself before being completely thawed. Wahrhaftig's (1965: Plate 2: Fig. 2) idea is that 'melting of the top of the ice wedges produces the shallow trenches. Crack systems tend to both parallel and intersect perpendicularly the gradually receding shorelines because of stress differences generated by horizontal temperature gradients. Random polygons form in areas of more nearly uniform stress.'

Cabot (1947) discussed that the long axis of oriented lakes is at right angles to the prevailing winds of today but is the same as that of summer winds from the Cordilleran icecap that prevailed during interglacial periods which then generated the wave erosion forming the lakes. Black and Barksdale (1949: 117) also concluded that the oriented lakes were formed by former predominant winds parallel to the long axis and that they gradually extended downwind. These winds were presumably connected with the major climatic changes during the Pleistocene. This seems to be impossible. Glaciation ended more than 10 000 years ago, yet the lakes are in balance with present conditions. 'When I last saw Black before he died, I believe he had changed his mind.' (Ross Mackay, personal communication 1995).

Then arises the question of the age of these lakes, which is difficult to determine because of much redeposited material (Hopkins & Kidd 1988). Carson (1968) dated several lacustrine benches and found the majority to be less than 3500 years old since drainage. He believes that the number and size of thaw lakes reached a maximum during the hypsithermal about 3500 to 4000 yr BP. The life span of a thaw lake is relatively short – typically persisting 2500–3000 years, but several large lakes appear to have existed for 4000 to 5000 years (Hopkins & Kidd 1988: 791). The lake formation in Alaska and Arctic Canada seems to me to be a continuous process, and when the small, recent second or third generation lakes (cf. Hopkins & Kidd 1988: 792) show a similar orientation, then the present day conditions have their impact on their formation. And the dominating factor seems to be wind.

Livingstone (1954) pointed out that the observed rapid rate of modification of the lake shores required that a change in wind direction such as that postulated by Cabot (1947) and Black and Barksdale (1949) must have occurred during the last few centuries. Otherwise, evidence of this previous wind regime would have been obliterated by erosion. He suggested instead that the lake forms resulted from modern processes.

Livingstone (1954) first called attention to a possible mechanism whereby currents account for the elongation of the lakes at right angles to the present winds. Water moves towards the downwind shore, setting up a return flow around the lake ends and causing erosion. Rosenfeld and Hussey (1958) pointed out that the problem is more complicated than the simplified approach of Livingstone, and that his hypothesis cannot apply equally to small ponds, a few metres long, and to large, kilometres long lakes. Black (1969b) also criticized that Livingstone's hypothesis cannot be applied equally to the very shallow lakes and those with a deep central elongate basin.

Rosenfeld and Hussey (1958: 285–287) proposed the possibility of fault and joint patterns controlling the orientation. NNW-SSE and ENE-WSW bedrock lineaments could be responsible for the lake orientation in Alaska and Canada. Rosenfeld and Hussey (1958: 286) based their idea on the tendency of many small streams to be oriented at right angles to each other with one set parallel to the direction of lake elongation. Reasoning that permafrost would react competently to stresses, they thought it likely that a master fracture pattern had developed in the surficial deposits and controlled the maximum development of ice wedges. Then thawing of these fracture controlled ice-rich zones would give rise to oriented thaw depressions. This is somewhat doubtful because bedrock is under thick sediments in many areas with oriented lakes. Fürbringer & Haydn (1974) based the size and orientation measurements of thaw lakes in northern Alaska on satellite imagery. The statistical evaluation of these data supported the idea that there is an evident correlation not only between the orientation and the size of larger water bodies but also between the orientation and the prevailing lineations which are restricted to the smaller lakes. They consider that the orientation of the smaller water bodies is evidently controlled by morphological and structural features. Zones of maximum ice wedge development parallel to lake elongation have not been observed in the northern Alaskan lake district, and no fracture systems are known to occur in the unconsolidated deposits (Carter 1987: 616). The Arctic Coastal Plain of Alaska is underlain by Upper Cretaceous and early Tertiary sedimentary rocks essentially flat-lying with dips less than 5°. These rocks have gentle folding with large low domes and anticlines and broad, shallow basins and synclines. About 50 m of unconsolidated sand, silts and clays and peat cover the rock surface (Black & Barksdale 1949: 108). Bedrock is far too deep in many areas to control the orientation.

A hypothesis has also been presented that the effects of greater thaw in a N-S direction, as a result of greater insolation at noon during the Arctic summer together with wave action, could impact on the lake orientation (Carson & Hussey 1959), although later measurements (Carson & Hussey 1962) showed that directional variations in insolation were of insufficient magnitude to cause the lake orientation.

Already in aerial photographs of sand dunes and oriented lakes from Alaska (Black & Barksdale 1949: Plate 3B) we can see that small newly formed ponds are elongated initially to the prevailing winds, similar to sand dunes, and only as the lake grows does elongation perpendicular to the prevailing winds develop (Carson 1968; Carter 1987: 616). For example, on the satellite image of the Prudhoe Bay area (Plate 11) not many sand dunes can be seen, probably because the Arctic Coastal Plain, even though it consists of vast amounts of permanently frozen fine sediments, is mainly covered by mosses and scrubs.

Black (1969b: 131) proposed that 'the concept of their origin is relatively simple . . . In permafrost regions thaw of ground ice which comprises more volume than pore space in unconsolidated sediments results in thaw depressions that may become lakes.' The ice content of permafrost in the upper 5 to 10 metres of permafrost, where fine-grained materials are present, commonly contains 30 to 90 per cent of ice by volume. The permafrost is supersaturated, i.e., the mineral and organic matter are a suspension in ice. The essential requirement for thaw lake development is the presence of substantial quantities of ground ice (Hopkins & Kidd 1988: 791). Thaw of permafrost may start after the disruption of the vegetal mat or active layer, by gravity movements, wind or water work, or even animals, becomes sufficient to induce thaw. The thaw continues until an insulating blanket of material is reestablished or until the supersaturated permafrost is destroyed (Black 1969b: 134)

Several questions arise for this kind of explanation: what is destroying the thick vegetation mat (e.g. Black & Barksdale 1949: 107), gravity movements and water erosion are minor factors on flat surfaces without slopes ('Oriented lakes occur chiefly in lowland areas . . . on featureless plains' according to Black & Barksdale 1949: 108), winds are strong but they do not wear moss without sand or snow drift, animals dig hollows in the active layer with random orientation and their interest is that these diggings stay uncollapsed. Lemmings run along the ice wedge furrows and their paths may cause local thawing but why oriented?

The major concern has been with the possible cause of the orientation of the basins. The initial thought that the orientation was by wind oriented at right angles to that of today is disputed by all who have dealt with the problem (cf. Black 1969b: 139).

The initial cause of the thaw lake or depression may be the quite random melting of ground ice, subsidence of the ground, and the accumulation of water in the depression (Hopkins 1949; Black 1969b: 134; French 1976: 123). Hopkins and Kidd (1988: 790) put it in the following way: 'Ponding of surface water to some critical depth is required to initiate the thawing of ice-rich permafrost that produces a thaw lake'. They have seen incipient thaw lakes developing over ice wedge intersections. A growth of palsa may disrupt drainage and cause ponding (Hopkins & Kidd 1988: 791). It has been suggested that thaw lakes originate by basin coalescence caused by breaching of low-centre ice wedge polygons (Rex 1961: 1022; Harry and French 1983: 456). Once formed, a tundra pool tends to warm and thaw the ground beneath its bed and beyond its banks, because water has a low albedo and a high heat capacity. The water is often brown, darkened by humus. As ground ice melts, the pond deepens (Hopkins & Kidd 1988: 791).

Black (1969b) summarized how Carson and Hussey (1959) reviewed five possible hypotheses for the lake orientation and concluded that each alone was not enough, but that a composite would suffice. The five are: (1) wave action from winds parallel to elongation during earlier time, (2) present winds which produce wave-current systems which scour at right angles, (3) present winds which distribute sediment on east and west shores, insulating them from thaw, (4) orientation produced by thaw during maximum insolation at noon, and (5) lakes controlled by north-south trending ice wedges which formed in the north-south components of a right-angle fracture system. They conclude that oriented ice-wedges might develop in the rock fracture system which has been shown above as irrelevant (Carter 1987); and that maximum insolation would be more effective in melting the north-south trending wedges than the complementary set, and this idea they later dismissed themselves; that the depressions so oriented would be perpetuated and enlarged by thaw and wind (wave) oriented sediments deposited on the east-west shores. Others (Carlson *et al.* 1959) also suggest that preferentially oriented ice wedges play a role in the orientation of the lakes.

Carson & Hussey (1960) provided additional data on the hydrodynamics in three lakes near Point Barrow, Alaska, including conditions when the lakes are ice free and when an ice cover is present. Their measurements show that the highest rate of erosion is going on at the ends of the elongated lakes by long-shore currents as predicted by the hypothesis of Rex (1961). Rex applied hydrodynamic theory and the principles of shore erosion and sediment transport. He utilized the conclusions of Bruun (1953) that maximum littoral drift on a curved beach is at an angle of 50° between the deep water angle of the waves and a line normal to the shoreline, with minimum littoral drift occurring where the angle is 0° (the nodal point), or directly on the downwind side. The resulting equilibrium configuration of erosion is a cycloid. Lakes formed under the influence of mutually opposing winds of equal strength should have opposite shores of cycloid form.

Carson and Hussey (1962) summarized their field data which support the circulation hypothesis of Rex (1961) and reject their earlier postulations of structural control by ice wedges. They point out that circulation in the smallest oriented ponds resembles that suggested by Livingstone (1954), but is too slow to erode. However, wave action on the lee sides of those ponds is sufficient to erode. As lakes increase in size, a gradual change in circulation systems was observed by Carson and Hussey (1962). They determined that the predictions of Rex (1961) for the zones of maximum current velocities and littoral drift are valid for lakes with a fetch of greater than 550 m. The ideal form and orientation of the Alaskan coastal plain lakes are produced by endogenic aqueous forces depending primarily on the bimodal wind distribution, represented by two approximately opposed vector resultants from the east and west (Carson & Hussey 1962). They calculated vector wind roses. Because they did not mention anything about using exponential values of wind speed when comparing the effectiveness of winds of different speed, this lack gave to Price (1963) a reason to comment: 'aerodynamic theory and experiment seem to show that winds should be compared by the cubes of their speeds', as has been done when analysing sand dune behaviour (e.g. Price 1960).

Mackay (1963) developed a mathematical model to relate lake shape to resultant wind vectors and the square of the velocity, and attempted to analyze the equilibrium forms of lakes that might be produced by winds of today. 'If it is assumed that winds from each of the 16 compass directions recorded in climatic records tend to develop curved bays, approximately cycloidal, then an oriented lake may be viewed as the summation or integration form of 16 cycloids, each of different size.' He assumed that the diameter of the generating circle of the cycloid is equal to the resultant for a given wind direction. The computed shapes and actual shapes of lakes from the Mackenzie Delta and the Barrow, Alaska, regions agreed nicely. However, Mackay emphasized that the precise mechanism of lake orientation remains unexplained. Harry and French (1983) showed that by using data for storm winds, lake shapes calculated with Mackay's (1963) model closely simulate the D-shape of oriented lakes on Banks Island.

An important factor in lake elongation is the development of sublittoral shelves; growing basins quickly establish wave-generated equilibrium bottom profiles off downwind shores that rapidly adjust to changes in lake depth (Carson & Hussey 1960; 1962). Sublittoral shelves insulate permafrost and dampen incoming storm waves, and limit the rate of expansion parallel to prevailing winds, without restricting expansion perpendicular to prevailing winds. In the brief period prior to the development of sublittoral shelves, lake expansion occurs parallel to the prevailing winds (Carson 1968).

7.4   The possible role of snow in the formation of oriented lakes

When reading all these different theories of the development of oriented lake we already know much of their form and the summer conditions. The winter conditions on ice-wedge polygon fields are not described in detail in the literature. The initial question is why ice wedges and depressed centres of polygons in certain directions do start to melt easier than others and form the lake basins in certain directions more frequently than in others. Melting may start as a random event (cf. French 1976) but then it will develop in certain prime directions during the formation of the lakes. One possible explanation of thermokarst hollows, which should be checked in the field, is the drift snow accumulation parallel to strong winds and perpendicular to winds with lower speed. Ice wedge furrows and raised ridges on both sides which are perpendicular to the dominating strong winds cause snow dunes on the surface and collect more snow on the centres of polygons than those which are parallel to or in other directions to the wind, which are cleaned of snow. Snow insulates the ice wedges and they are then warmer when the thawing season starts and the active layer will develop deeper in these places and create thermokarst. The other ice wedges stay cold and unthawed. This means that the thermodynamics of the active layer and the upper part of the permafrost affected by insulating snow could have a decisive role. When the initial melting has once started then it continues because the thermokarst hollows collect water and continuously more snow and keep the basin warmer than the surroundings. Water and saturated soil have higher thermal capacity and this keeps the basin unfrozen for a longer time and deepens the active layer. In the depressed polygon centres water accumulates. It starts to drain along thawing ice wedge furrows in a certain direction determined by the accumulated snow and the active layer underneath gets thicker. Snow surface may also undulate because of turbulences and this causes certain intervals in between the lake distribution. Drift snow accumulation could be the possible explanation for the orientation of the thaw lakes and their relationship to the wind directions. When destruction of ice wedges once starts it proceeds rapidly to depressions and then further to ponds and lakes (cf. Black 1969b: 137). Then a secondary matter is the wave erosion and other factors impacting on these ponds during the summer time. They may explain the oval form and further growth of the lakes when their development continues. This means that both winter and summer winds have a decisive role in the formation of oriented lakes. If this hypothesis is valid then it could be proved with a simple field experiment. On a ice wedge polygon field snow fences are built in different directions and the reaction of the ice wedges to the increased amount of snow is observed. Both the vertical and horizontal temperatures of the permafrost should be recorded on the experimental site and in control localities in wide areas.

Mackay (1978) built snow fences at Garry Island, Canadian Arctic, in 1974 and at Inuvik in 1980 (Mackay, personal communication 1995). According to his observations, when the snow is trapped on ice wedges they stop cracking.

Fig. 7.4 Small oriented lakes on the marine terrace with ice wedge polygons, Harrowby Bay in Liverpool Bay area, N.W.T., Canada. Redrawn after Mackay (1956: Fig. 4).

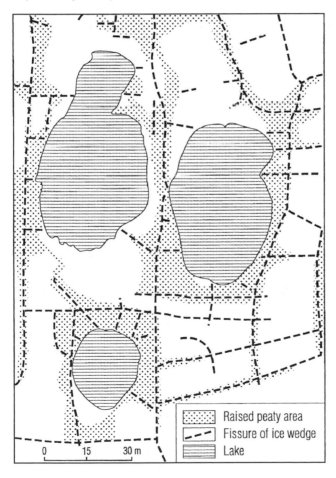

He has measured the ground temperatures in these sites and concluded: 'Insofar as I can determine, the small thaw ponds develop, at each site, because of local random events.' Numerous factors seem to be involved and the initiation of the lakes can start in many different ways.

7.5   Wind is the key factor

In general the oriented lakes are fascinating complex features. Their formation is an example of multiple factors proceeding together simultaneously in geomorphology. It has been revealed that the oriented lakes are in response to modern winds. (a) The axes of large lakes several km long and small lakes 50 m long are parallel (Mackay 1956: Fig. 3; 1958: Fig. 17). (b) When an old oriented lake drains in part, a new oriented lake can form with the axis parallel to the old (Mackay 1957: Fig. 18). (c) The shorelines of many oriented lakes are partly erosional and partly depositional, i.e. in balance with present currents. (d) Oriented lakes as small as 30 m in length can be growing in areas where oriented lakes several km long are also growing (Mackay 1956: Figs. 2, 3 and 4) (Fig. 7.4). (e) The oriented lakes of Alaska, east of the Mackenzie Delta, and Baffin Island all occur in tundra areas where there are no trees to decrease the strength of the wind and the sediments are easily eroded. (f) The strongest winds in summer blow in essentially opposite directions (Mackay, 1963; Mackay, personal communication 1995).

8

Accumulation

Wind drifted material deposits on the ground when the wind speed drops to the threshold of the cessation velocity (Figs. 4.1 and 4.7) and the drift stops. For different modes of transport: suspension, saltation and creeping the thresholds are different. The coarsest grains creeping on the ground surface move slower and they stop first, and the very small grains transported with high speed high up in the atmosphere settle last. Deposited grains form different kinds of accumulations with and without stratification. The smallest aeolian bedforms are ripples which are waves on the sand surface. A single ripple has differences in grain size, and sorting of grains produces the initial ripple formation (Seppälä & Lindé 1978).

8.1   Ripple marks

Ripples occur on all sand surfaces on which aeolian drift exists and the wind speed is not too high. Ripples can also disappear fast if the wind conditions change. Ripples are related to the sand movement more than accumulation, but when the wind slows down then the ripples persist. We seldom find ripples in aeolian sediments though they are common in fluvial and lacustrine deposits.

In general the wavelengths of ripples range from a few centimetres to tens of metres, and their height from 1 to 30 cm. The big and more stable ripples, which are formed in long lasting transportation, are called megaripples. Ripples have an asymmetric profile with gentle windward slopes (from 8 to 13°) and lee slopes up to 30° (Werner *et al.* 1986).

Ripples are usually formed of sand and occasionally of gravel size particles, but in Antarctica very strong winds also form much bigger ripples of very coarse grains (Fig. 8.1). In ice-free areas some 200 km around the McMurdo Station, Victoria Land, Antarctica, the aeolian pebble ripples are asymmetric in profile, and subparallel in plan, with some convergence and divergence, resembling miniature transverse dunes. Their height reaches 30 cm, and the intervals between crests is up to more than 1.5 m (Smith 1966).

Large ripples composed of pebbles and cobbles have been described from the Allan Hills, Victoria Land, Antarctica. Denton *et al.* (1984) interpreted fields of these ripples found throughout the Transantarctic Mountains to have been formed by subglacial sheet flow, and used them to infer major ice-flow directions for ice-sheet episodes. Maximum ripple wavelength in the Allan Hills is in excess of 30 m, with an amplitude of more than 2 m; most are much smaller. Mostly the ripples are very thin mantles of debris. Some ripples have well defined internal bedding, but for the most part such features are absent. Malin (1986: 20), however, considers it likely that wind has been important in their formation and modification. He has calculated free-stream wind speeds approaching 70 m s^{-1} (250 km/h) (Malin 1986: 18) and these rare but strong winds may have the potential for transport of pebble-sized materials (Malin 1986: 20).

8.2   Aeolian accumulation among vegetation

As van Dieren (1934) wrote, vegetation is a very efficient stopper of sand drift and it can cause sand accumulation and even start dune formation. As stated above (sections 2.8 and 6.16) some plants survive even if they are covered by drift sand (Fig. 6.20, Plate 10 and Fig. 8.2). Plants which support sand accumulations in Lapland are, for example, *Leymus arenarius, Festuca rubra,*

Fig. 8.1 Very coarse grain aeolian deposit close to Davis Research Station, West Fold Hills, Antarctica. 1990. Photo by K. Nenonen.

Empetrum hermaphroditum, Arctostaphylos uva-ursi, Vaccinium uliginosum, V. vitis-idaea, Juniperus communis and *Betula nana* (Seppälä 1971a, 1974, 1984). Pissart *et al.* (1977: Table 2) give a long list of plants growing on aeolian deposits on Banks Island, Arctic Canada. There are the same species (e.g. *Carex bigelowii*) and similar species (e.g. *Arctostaphylos rubra*) as in Lapland. Aeolian accumulation among vegetation causes deposits with alternating thin layers of sand and organic strata (Pissart *et al.* 1977: Fig. 8) which are typical on the edges of deflation basins. They can be some metres in thickness and give a good data base to study the succession of vegetation on sand surfaces.

8.3   Sand dunes

Aeolian dunes are large accumulations of sand shaped by wind drift. Sand dune formation is dependent upon some basic elements: an adequate source of suitable sediment, enough strong winds, long enough fetch area and at least locally scattered vegetation as well as some reason to subside the transport of sand grains and cause the deposition. Deposition may originate with small obstacles such as scattered vegetation, stones causing surface roughness, moisture differences on the sand surface, uneveness of the sand surface caused by deflation, etc. All kind of irregularities on the sand surface cause turbulences in the

Fig. 8.2 Sand accumulation among *Empetrum hermaphroditum* growth on Hietatievat, Finnish Lapland. July 1965.

wind pattern. Large dune field formation requires a great source of sand which is not very common in cold environments. Suitable sand for wind drift is found on outwash plains (sandurs), braided river valleys, flood plains, deltas, along sandy shores, eskers, kames and some push moraines.

On very few occasions in cold environments we find barchans or crescent-shaped dunes (Plate 9) which are typical desert dunes. This is probably because large sand seas are missing, relative humidity is rather high and it makes the conditions favourable enough for certain pioneer plants binding the sand surfaces relatively fast. The most common type of dune in cold regions is a parabolic dune and dune fields formed of agglomerations of parabolic dunes (Figs. 8.3 and 8.4). Parabolic dunes indicate that during the formation of the dunes they are at least partly vegetated.

Along the Arctic river banks can be found partly vegetated longitudinal dunes which are bordering the vegetated tundra, as described by Rickert and Tedrow (1967: 302) (Fig. 8.5). At the Colville River delta on the coastal plain of Alaska,

Fig. 8.3 Schematic diagram showing varieties and development of parabolic dunes. The arrow indicates the direction of dune-building winds. Redrawn after Smith (1949b: Fig. 6).

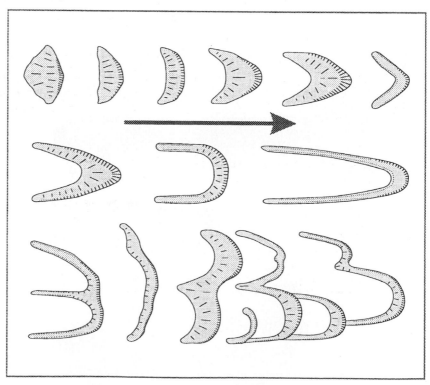

flooding deposits huge amounts of fine and very fine sand which is drifted by
the northeast wind and accumulated as riverbank dunes, especially on the west
bank of the river (Walker 1967). The flood deposits of the ephemeral streams
and their channels are subject to aeolian modification during the arid summer
months on Banks Island, Arctic Canada (Good & Bryant 1985).

Fig. 8.4 Oblique aerial view of parabolic dunes in Lake Pöyrisjärvi area, Finnish
Lapland. June 30, 1973.

Fig. 8.5 Schematic representation of a dune area along the Meade River in the Arctic
Coastal Plain of Alaska. Redrawn after Rickert and Tedrow (1967: Fig. 5).

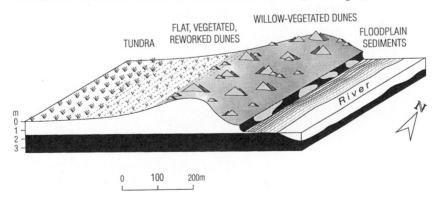

Coastal dunes occur widely along the shores of north and west Alaska (Black 1951). In cold environments they can be rather large but they are not the most common type of dunes (Fig. 8.6). In this text we do not especially touch the coastal dunes because their characteristics are mainly similar to those of temperate regions, their formation is a special process which is in close connection with other coastal processes such as wave action and there is much information concerning them (e.g. Norrman 1981; Pye 1983; Nordstrom *et al.* 1990; Carter *et al.* 1992). We may expect that frost and ice cause some special features on coastal dunes in a cold environment. For example, small kettle holes were formed by melted ice blocks buried in small beach dunes at Kongsfjord, Svalbard. One cold environment speciality is small thermokarst lake-shore dunes which have been described from Tuktoyaktuk Peninsula, Northwest Territories by Ruz (1993). Their formation seems to relate to lake drainage and coastal retreat of 1 m/a as a result of thermokarst caused by continuous sea level rise.

The windward slopes of parabolic dunes incline in general between 4° and 16°, while the inclinations of lee slopes range from 20° to 35° (Fig. 8.7) which is the static friction angle of aeolian sand. If the deposition has taken place among vegetation the maximum dip can be even more – up to 55° (Seppälä 1971a: 31, 1995: 801). The heads of the parabolic dunes are normally higher than the arms.

Fig. 8.6 Large coastal dune on Kapp Weissenfels, Svenskeøya, Svalbard. 1980. Photo by O. Salvigsen.

The internal structure of sand dunes has been studied mainly in hot deserts (e.g. Bagnold 1941; McKee 1966; Pye & Tsoar 1990). Cold environment dunes also show similar characteristics such as foresets and top sets, cross-bed sets, and some ripple strata. When the main deposition takes place on the lee side of the dune ridges then the dips of strata incline steeply downwind. That means in parabolic dunes that dips are to different directions on different arms (Fig. 8.8). The maximum angle of rest for aeolian sand is 35° which determines the maximum dip of dune strata. Open sand at a steep angle is very unstable and therefore slumps, and slip faces on the lee sides of sand dunes are common. Slips can be caused by sudden rains or changes in wind direction. Sand with horizontal lamination occurs in interdune areas. Many kinds of secondary sedimentary structures can be found in sand dunes caused by scour and fill by wind or water, vegetation, seasonal freezing and thawing, buried snow clumps, burrowing animals and seismic shocks.

8.3.1   Sand dunes in the Arctic

It is surprising how few the dune areas are in the Canadian Arctic compared with the dune areas in northern Fennoscandia. Isolated patches of sand dunes are widely distributed in northern Canada (Craig 1965), but their occurrences are so small that, for example, David (1977) in his large inventory of sand dunes in Canada does not mention more than two areas: Fort Simpson – Manners

Fig. 8.7 Height [in m] and inclination of slopes of parabolic dunes on the dune field in Kaamasjoki-Kiellajoki interfluve area, Finnish Lapland. (Seppälä 1971a: Fig. 6).

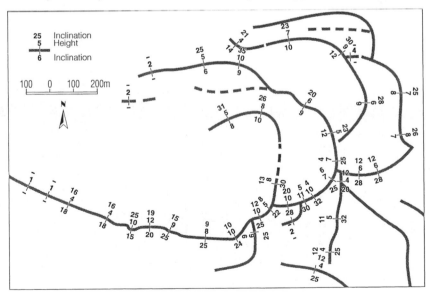

Creek sand hills along the Mackenzie River (some 61° 45′ N) and Trail River sand hills on the east side of the Mackenzie River (66° 06′ N). The western Arctic seems to be more suitable for dune formation because of fine sandy soils and drier, warm summers in a sparse tundra vegetation. 'The largest area of aeolian deposits in northern Canada is between the middle Thelon and middle Back rivers. The dunes in this region are irregular in form and rarely show any preferred orientation.' (Bird 1967: 240). 'The ultimate source of the sand is the Dubawnt sandstone that underlies most of the area; the dunes develop from the surface deposits lying on the sandstone. Larger but more restricted areas are lake shore dunes, rarely more than 8 m high, formed from the exposed sandy beaches as the lake level drops in the summer; such dunes are common on the shores of Aberdeen, Garry, and Pelly lakes.' (Bird 1967: 241).

Additional sand sources for dune formation are esker systems, outwash sands and the beds and lower terraces of many rivers (Bird 1967: 241). In the Tibielik river area some isolated dunes reach 25 m in height (Bird 1967: 241). Many rivers in the western Canadian Arctic have sand dunes, both on an existing delta such as the Mackenzie delta area (Mackay 1963) and on remnants of higher

Fig. 8.8 Internal sedimentary structures of a parabolic dune (Seppälä 1971a: Fig. 27).

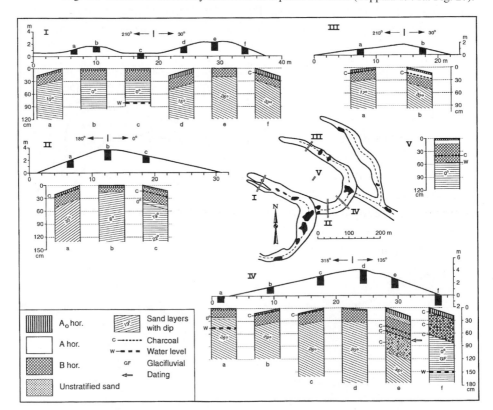

deltas, constructed during the postglacial marine transgression – at the mouth of Hayes River and the coastal zone of the Pleistocene delta of the Mackenzie, for example. The majority of dunes have parabolic or rounded forms; they are invariably low, heights greater than 4 m being uncommon, and generally stable with only occasional blowouts.

Gilbert (1983: 164, Fig. 10) has found sand dunes in connection with sandurs along several fjords on Baffin Island, and described an extensive sand dune area at the head of Maktak Fiord. The sand bank is 3.5 × 0.6 km and reaches about 100 m above the sandur surface.

Very impressive dune fields are scattered along a 100-kilometre stretch of the south shore of Lake Athabasca, northern Saskatchewan, starting west of the William River and extending to the MacFarlane River in the east (Karpan & Karpan 1991: 43). There are 16 600 hectares of sand and this is the largest active sand surface in Canada. Originally sand was deposited by meltwaters from the continental ice sheet in glacial Lake Athabasca. Today's active dunes are sustained by new sources of sand exposed in the river deltas and beaches. Prevailing winds have drifted these large sand masses eastward and formed white sand dunes some 20 m in height (Karpan & Karpan 1991: 43–46).

In Subarctic conditions thoughout the eastern coastal area of Hudson Bay, northern Québec, parabolic dunes are found within the boreal zone and the forest subzone of forest tundra (Filion & Morisset 1983). The dunes show multilobate forms and particular spatial arrangements such as normal-incident dunes, dunes in opposition and hemicyclic dunes due to the effect of multidirectional winds (Filion & Morisset 1983: Fig. 6) (Fig. 8.9). In the shrub subzone of forest tundra and the shrub tundra the dunes show either an imbricate or *en échelon* pattern built by easterly unidirectional winds (Fig. 8.9). The sand dunes formed of original reworked glaciofluvial deposits and secondary marine, lake or river sediments seem to be very much deflated and stabilized by vegetation. The tree line forms the limit between the southern forest-type and the northern tundra-type aeolian systems. Wind activity in open areas is stronger but less varying. In forested sites dunes have a much greater variety in form than in the northern part of the region (Filion & Morisset 1983: Table 4).

In the coastal area of Hudson Bay it has been found that parabolic dunes develop through elongate dunes to lateral residual dune ridges which are parallel to the wind direction when the accumulating amount of sand decreases (Filion & Morisset 1983: 86, Fig. 16). The idea of the development from a transverse dune through a parabolic dune to two longitudinal dunes was first presented by Kádár (1938: 169–171) and then supported, for example, by Landsberg (1956: 177), Galon (1958: 16, 1959: 98, Fig. 6), Kobendza and Kobendza (1958: 169) and Stankowski (1963: 142) (Fig. 8.9).

In Alaska aeolian sand is located mainly outside the last glaciated areas on lowlands and alluvial plains, Fig. 8.10 (Lea & Waythomas 1990: Fig. 1). Sand dunes occur in central Alaska on river banks and in connection with flood plains (Wahrhaftig 1965: 17; Komárková & Webber 1980: maps). According to the geological map of Alaska (Karlstrom *et al.* 1964) large sand dune areas occur: in eastern Alaska along the Tanana River in the vicinity of Tetlin Lake, on the southern side of the Tanana River in central Alaska, three areas west of Fairbanks, at the lower course of the Koyukuk River, a tributary of the Yukon River, in two areas along the Kuskokwim River and two on the south side of the lower course of the Kobuk River and along the Yukon River in NE Alaska. For example, an active sand dune field extends inland for several hundred metres from the Meade River on terrace remnants (Everett 1979: 209; Komárková & Webber 1980: 455). Some dunes form long low ridges parallel to the effective wind direction while others are great wavelike ridges at right angles to the wind. They are asymmetric in cross-section having gentle windward and steep lee slopes. Most of the sand dunes of Alaska are now stabilized by overgrown vegetation, but also active

Fig. 8.9 Parabolic dunes indicating different wind directions (after Aufrère 1931; David 1977; Filion & Morisset 1983) and schematic development of sand dunes from transverse via parabolic to longitudinal dunes (after Galon 1959).

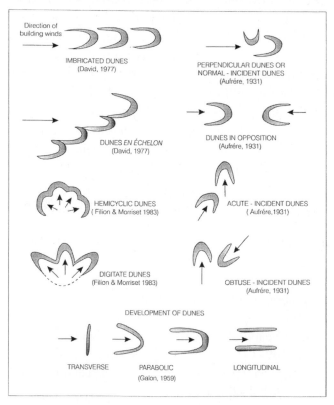

barren sand dunes are found, for example in parts of the Koyukuk Flats and Kobuk Valley (Wahrhaftig 1965: 17) (Fig. 8.10).

In northern Alaska on the river floodplains and terraces between Meade River and Colville River many stabilized longitudinal and parabolic dunes occur (Everett 1979: Fig. 1; Galloway & Carter 1993b: 3). They measured from aerial photographs a maximum dune density of over 275 dunes for a 100 km^2 area 12 km south-southwest of Teshekpuk Lake. West of Ikpikpuk River, there is an average of 135 dunes per 100 km^2 (Galloway & Carter 1993b: 3). The dunes vary in length from 320 to 2500 m. Their width ranges from 15 to 40 m and their height only from 1 to 3 m. The Ikpikpuk dunes in Alaska are composed of large, linear dunes up to 20 km long, 1 km wide and 30 m high (Carter 1993: 79).

Nichols (1969: 80) found in northwest Greenland small shadow dunes associated with the boulders on the surface of a sandy terrace and some 1 m high sand dunes adjacent to an ice-dammed lake. Along the rivers and coast in Peary Land

Fig. 8.10 Glaciated areas, alluvial plains and distribution of upper-Quaternary aeolian sands in Alaska. Arrows indicate the effective wind directions. Compiled after Lea & Waythomas (1990: Fig. 1) and Wahrhaftig (1965: Fig. 6).

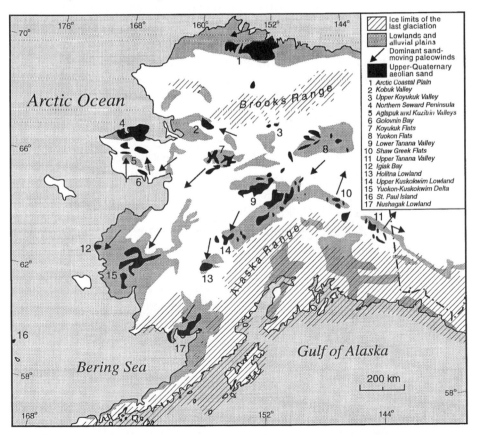

he also reported small dunes some 2–3 m high. Still larger dunes are present around Søndre Strømfjord (Fristrup 1952–53: 60). Although most of Inglefield Land is devoid of vegetation, the lack of major aeolian deposits gave Nichols (1969: 80) occasion to conclude that 'the work of wind is of very minor importance' but even he mentioned a number of different indicators of wind action. Only great sand resources can cause large sand dunes though the winds are strong.

Åkerman (1980: 235–241) describes small sand dunes (less than 1 m high) on top of and on the slopes of a raised beach ridge on the west coast of Spitsbergen.

8.3.2   Sand dunes in Antarctica

On the northern side of lower Victoria Valley, Antarctica, is a field of whaleback mantles, each about 0.5 to 1 km long (Cailleux 1963: 101–105; Calkin & Rutford 1974: 195) which are less than 10 m thick and have gentle slopes of less than 3°. These surface sands are less sorted, coarser, and contain some small pebbles showing very strong wind. They are followed by a dune belt with finer sands showing dissipation of the carrying power of wind.

In South Victoria Land, Antarctica, to the east of Lake Vida, barchan type sand dunes 9 to 15 m high and 60 to 90 m long associated with longitudinal dunes have been formed under the influence of the local up-valley wind (Webb & McKelvey 1959: 127). This entire sand area is underlain by permanently frozen silt. Where sand dunes encroach upon the delta drainage system during the summer flow period their flanks are quickly cut away. Dune and sand areas are located between glaciofluvial gravel and boulder moraine areas (Webb & McKelvey 1959: 123). Sand dunes contain inclusions of snow layers and ice-cemented layers up to 10 cm thick, which suggest that some sand is moved when the dunes are snow covered (Calkin & Rutford 1974: 196). Snow and ice layers stabilize the sand dunes at least during the cold period. Conditions during the cold periods are favourable for the upward movement of moisture from the subsurface ice and snow (Everett 1971: 429; Calkin & Rutford 1974: 196–198). Strong summer sublimation probably allowed some sand movement after the meltwater period in the dune area in the middle of November (Calkin & Rutford 1974: 200) (cf. the separate chapter on sublimation).

In Victoria Valley the dune field is 3.5 km long and less than 1 km wide and it consists of many partial barchan forms (compound and single), which together form transverse dune ridges between 70 and 600 m long (Calkin & Rutford 1974: 196). The slip faces of the dunes are oriented NW and dip from 31.5 to 34.5° at the top and 31° at the base. At the wind velocities of more than 20 knots ($10 \ \mathrm{m \ s^{-1}}$) avalanching of sand is rapid, and the slip faces then consist of a series of sandflow tongues whose surfaces dip 30°. On the windward side slopes vary from 10 to 20° (Calkin & Rutford 1974: 196).

8.3.3   Sand dunes in other cold regions

The dune distribution in Iceland is also rather limited. Nielsen (1933: 250) gave the impression that there are no sand dunes, and according to Nielsen this is undoubtedly connected with the fact that the aeolian activity is a temporary phenomenon, being restricted to a short period of the year, and that throughout the whole of the remainder of the year the surface of the ground is exposed to other forces which level the rudimentary dunes formed during the sand storms.

At Hraun, on the SW coast of Iceland, some active sand dunes are moving inland and made of fine volcanic particles. At Gunnarsholt north of Myrdalsjökull sand dunes 5 to 8 m high are found, and in some places ashes are heaped by wind into dunes 1 to 4 m high (Cailleux 1939). Wind action upon volcanic ashes and lapillis is more frequent in Iceland than in other countries, probably because the existence of vegetation cover is checked by the severe climate (Cailleux 1939: 60).

In Chapter 6 about deflation we considered several sand dune areas in Fennoscandia (e.g. Klemsdal 1969; Seppälä 1971a, 1972a, 1973a, 1995; Sollid *et al.* 1973; Bergqvist 1981) (Fig. 8.11) where parabolic dunes were mainly formed some thousands of years after the deglaciation and after that their development has been mainly destructive. Dunes were evidently derived from fine sand of eskers, kames, deltas, valley trains, ice-dammed lake deposits and outwash plains (Figs. 8.12–8.14). Therefore the dunes are mainly located at the lower levels in the topography. For example, in northern Sweden the altitudes of dune areas vary from 155 to 620 m a.s.l. (Seppälä 1972a: 88).

Parabolic dunes exist in clusters which also have the form of a parabola (e.g. Solger 1910; Högbom 1923; Galon 1959; Seppälä 1971a, 1972a). The longest arms of individual parabolic dunes in Fennoscandia are seldom over 1300 m (Seppälä 1971a, 1972a). The broadest dune ridges are 150 m wide and the widest parts are found at the nose of the parabola (Seppälä 1971a: 15). The highest dunes found in Finnish Lapland are about 12 m in height but normally they range between 4 and 6 m (Seppälä 1971a: 16) (Fig. 8.7).

The distance between successive dune ridges following the direction of the wind parallel with the axis of parabolic dunes is usually 100–200 m (Seppälä 1971a: 15, 1972) (Fig. 8.12). In some cases the ridges have an interval of only 40–60 m.

Nice examples of the relationship between the eskers and dunes can be found in Fennoscandia. If the esker runs parallel with the effective wind direction then the dunes locate on both sides of the esker (Seppälä 1972a: Fig. 3) or in the downwind end of the esker (Seppälä 1972a: Fig. 4) (Fig. 8.13). If the esker is in a transverse position against the wind direction the dunes are found only on the leeward side of the esker (Seppälä 1972a: Fig. 6) (Fig. 8.14).

One rather unusual example of dunes derived from a valley with an esker should be mentioned from Finnish Lapland. These are long dunes climbing up the slope of a fell. One of the dunes crosses the ridge and falls downwards on the other side (Seppälä 1993) (Fig. 8.15). This type of climbing dune has been reported from desert environments (e.g. Evans 1962; Pye & Tsoar 1990). Tsoar (1983) made a wind tunnel model of echo and climbing dunes. How well the model corresponds with the Lappish dunes is questionable because they have parabolic characters (Fig. 8.15) and that should mean vegetation cover during the formation.

Many of the sand dunes in the Arctic and Subarctic environments are parabolic in form, which means that vegetation has partly anchored their arms during the deposition and thus aeolian activity occurs especially during the summer time. Inner structures of parabolic sand dunes show very clear stratification and few disturbances which means that the deposition has happened without snow (no

Fig. 8.11 Distribution of inland dunes in northern Fennoscandia (Klemsdal 1969; Seppälä 1972a, 1973a, 1996; Aartolahti 1976; Sollid *et al.* 1973; Bergqvist 1981; Tikkanen & Heikkinen 1995). Shaded area indicates the region with palsas (Seppälä 1988: Fig. 11.1).

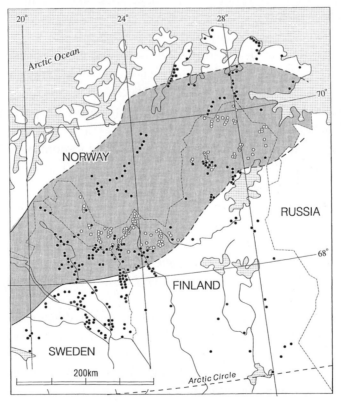

Fig. 8.12 Geomorphological map of dune fields in northern Sweden. Key to symbols:
1. sand dune, 2. glaciofluvial deposits, 3. till field, 4. alluvial deposits, 5. mires,
6. lake and river, 7. glacial drainage channel, 8. steep river-cut slope. Redrawn after
Seppälä (1972a, Fig. 11).

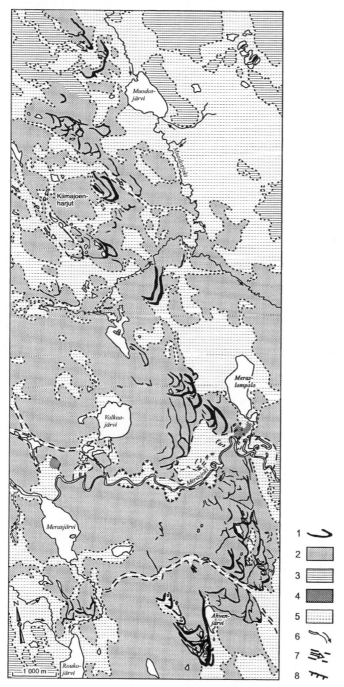

niveo-aeolian features). David (1979) explained the parabolic shape of dunes to be the response to the moisture in the sand. His idea is that the sand surface dries fast, and sand is present in great quantities on deflation basins where the wind speed is highest and funnelled onto the converging arms of the dune. From these depressions the wind removes the dried sand and drops it over the crest on the slipface just in the front of the depression, providing the rest of the slipface with little or no sand. The slipface in front of the depression receives more and more sand and projects further ahead in a curved shape and the parabolic form is the result. The weak point in this theory is that the water table in the deflation basins is closer to the surface than in the crests (Fig. 6.23). Moisture might play a role at the beginning of dune formation and it is important for the vegetation, and when the parabolic shape has started to form then the sand on the arms parallel to the wind direction moves slower and gives more favourable conditions for plants to grow.

8.4   Dune migration

Dune movement is measured in several survey techniques. Calkin & Rutford (1974) used a marker survey in Victoria Valley, Antarctica. They inserted bamboo poles near the toe of the dune slip faces and remeasured these poles later. The slip faces advanced on average 4.8 cm a day during the summer time. The effect of winter westerly winds is unknown.

A second technique used in Victoria Valley, Antarctica, is photogrammetry. Three sets of vertical aerial photographs with intervals of more than 2 and almost

Fig. 8.13 Geomorphological map of a dune field near Vuottasjaure, northern Sweden, where an esker was parallel with the dune forming winds. Key to symbols: 1. sand dune, 2. esker, 3. glaciofluvial deposits, 4. till field, 5. mires, 6. lake and river, 7. glacial dainage channel, 8. steep river-cut slope. Redrawn after Seppälä (1972a, Fig. 4).

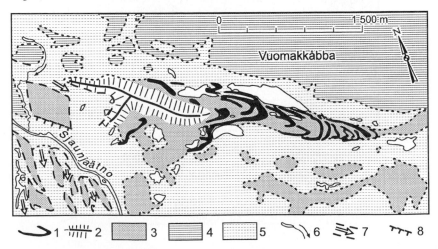

7 years were used to measure dune movement. The difference in crest positions, as measured for the dune at right angles to the initial position, was taken as a measure of dune movement. In the two years period the mean movement was 13 m, with a range of −6 to 62 m and a standard deviation of 8.5 m. Errors in measurements may approach 4 m (Calkin & Rutford 1974: 199). The mean of the following seven year period was only 8 m, with a range of −14 to 28 m and standard deviation of 10 m. The irregularity of mean movement rates over the 10 years period is clearly evident.

According to these measurements the counted daily rates of sand dune movements, assuming that the sand was moving four months a year, are 4.7 cm for the short period, and for the seven years period one obtained a mean of 1.1 cm a day. At such rates the dunes could have migrated 3 to 6 km from Victoria Lower Glacier margin to their present positions in 500 to 5000 years. However, the age and place of the original sand source is uncertain (Calkin & Rutford 1974: 200).

Fig. 8.14 Geomorphological map of a dune field near Lake Temmingijärvi, northern Sweden, where an esker was perpendicular to the dune forming winds. Key to symbols: 1. dune, 2. esker, 3. glaciofluvial deposits, 4. till field, 5. mires, 6. lake and river. Redrawn after Seppälä (1972a, Fig. 6).

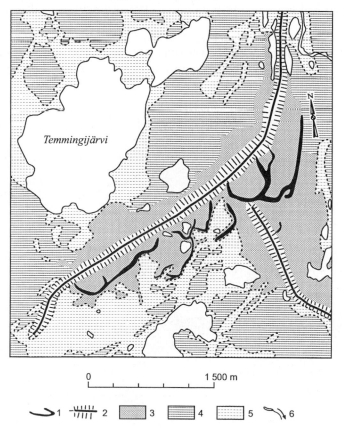

The migration rates of sand dunes, as also the long term deflation rate measurements, are always the net rates because different winds from different directions partly compensate the former transportation and contribute to recycling the sand. This might be the case in Victoria Valley, too, because the summer winds are easterlies and the winter winds westerlies.

8.5   Loess formation and dust storms in Polar regions

Cold environment loess deposits are known in Greenland, Alaska and Antarctica (Westgate *et al.* 1990). We shall take some examples of them.

Loess deposits have been reported from southwest Greenland. In the vicinity of the ice margin (*'som framme vid israndens närheten täcker landskapet och särskilt de flesta höjderna'*) – 'they cover the landscape and especially most of the hills' (Nordenskjöld 1910: 25).

Loess-like deposits about 60 cm thick were found above the highest stream terraces and in places where post-glacial streams have not existed in Inglefield Land, northwest Greenland (Nichols 1969: 80). The material was derived by wind from the valley train during the winter months, when there was little or no meltwater, and when the valley train sediments deposited the previous melt season were dry and not covered with snow. The orientation of the ventifacts

Fig. 8.15 Climbing and falling sand dunes in northernmost Finland. From Seppälä (1993: Fig. 1). Reproduced with the permission of © The Geological Society Publishing House.

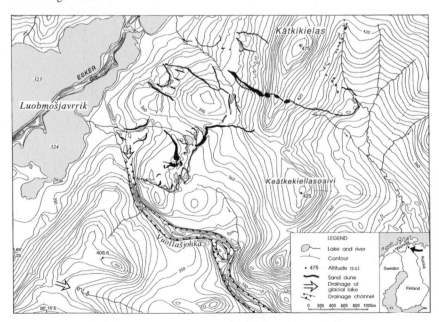

suggests that part of the loess driven from an uplifted dissected delta surface was probably blown out into the bay. Böcher (1949: 25–28) has described more loess in the Søndre Strømfjord, west Greenland.

Bryant (1982) identified loess deposits in the Adventdalen valley in Spitsbergen on the basis of granulometric and micromorphological analyses. Some loess has also been found from the Ny-Ålesund area (Gallet *et al.* 1998). They are small accumulations of multiple sedimentary processes which have started by glacial grinding, then undergone aquatic sorting and finally been drifted by wind. Older loesses have also gone through some chemical weathering.

Loess-like deposits have been reported from south-central Yakutia (Péwé & Journaux 1983). According to Konishchev (1987) in northern Yakutia thick layers of loess-like deposits occur, with high ice content on river banks and valley slopes. These loess-like icy silts contain mainly quartz, feldspars, amphiboles, pyroxenes, biotite and muscovite and no calcite. The origin is somewhat problematic but it seems to be redeposited eluvium with subaquatic origin more than aeolian. This is a strange matter if the very large streams of Siberia with high spring floods and wide flood plains do not produce any suitable deposits for aeolian transport and loess formation. Signs of at least Quaternary aeolian deposits are Cailleux's (1969: 198) observations of numerous fresh wind-worn quartz grains from the middle course of the River Ob (85 to 90 per cent) and from the flat Lena River basin (75 to 90 per cent) about which Cailleux concludes that 'the wind action here has been as strong as in the periglacial belt of Europe'.

Smalley (1966) proposed that glacial grinding provided loess material and the idea was supported by Boulton (1978: 796), and Boulton went further by suggesting that most of it is produced in the basal zone of traction, which he believed to be a uniquely glacial environment in which large forces at 'non-inertial' shear contacts produce fine-grained wear products. Glaciers produce rock flour direct from solid rocks with the drifted mineral material by crushing it. Rock flour on the rock surface under glaciers produces a hard compacted concrete-like layer on the rock surface and this protects the rock surface and drifted particles from further grinding. We should keep in mind that only small portions of the total glaciated areas have been rock based, even in the erosional regions of ice sheets with roches moutonnées as in Finland (Fogelberg & Seppälä 1979). Mainly, ice sheets have moved on a sediment base formed by earlier weathering materials, fluvial deposits and tills. All of them contained loess-size silts suitable for aeolian transport. These deposits formed from preglacially weathered materials drifted by ice sheets to central Europe, sorted by glaciofluvial drainage and dried before the succession of vegetation, and were also transported from the outwash plains and flood plains of major rivers such as the Rhine by wind and deposited among vegetation as loess (Fig. 8.16). The North American loess is essentially a wind-drifted glacial material – its origins are directly linked to the northern ice-sheets (Smalley & Smalley 1983: 61).

'Ohne Frost, kein Löss' stated Carl Troll and this seems to be essentially true. Cold conditions produce loess and the loess deposits formed especially during the cold phases of the Quaternary (e.g. Smalley & Smalley 1983: 64). According to Troll (1948) loess is purely a product of gelideflation which is associated with the higher latitudes of the northern and southern hemispheres and produced by periglacial deflation and accumulation. This somewhat categorical opinion is no longer accepted since much of the loess deposits originate from Pleistocene glacial sediments.

Frost weathering is able to produce silt size material and can be considered as a major mechanism for supplying loess material (e.g. Brockie 1973; Lautridou & Ozouf 1982).

Loess-like deposits about 60 cm thick were found above the highest stream terraces and in places where post-glacial streams have not existed in Inglefield Land, north Greenland (Nichols 1969: 80). The material was derived by wind from the valley train during the winter months, when there was little or no meltwater and when the valley train sediments deposited the previous melt season were dry and not covered with snow. The orientation of the ventifacts suggests that part of the loess driven from an uplifted dissected delta surface was probably blown out into the bay.

The North American loess is essentially a wind-drifted glacial material – its origins are directly linked to the northern ice-sheets (Smalley & Smalley 1983: 61) (Fig. 16.1). Rutter *et al.* (1978: 314–315) reported that thin loess deposits occur widely on flat or gently sloping surfaces in the glaciated parts of central

Fig. 8.16 Distribution of loess in central Europe. Redrawn after Haesaerts (1985).

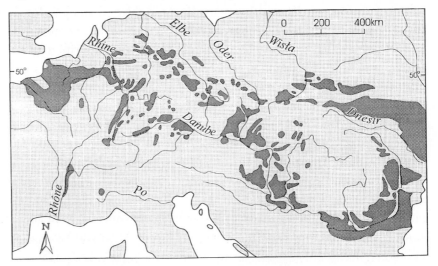

Yukon up to about 300 m or more above the floors of major valleys. Loess is similarly disposed in the unglaciated areas adjacent to the Yukon River and its tributaries, such as the Pelly, Stewart, Klondike and White Rivers that carried glacial meltwaters beyond ice frontal positions during successive glaciations.

8.5.1   Loess formation in Alaska

Loess formation in Alaska is a present-day process which has a long history and has produced thick layers of fine silts covering large areas (Westgate *et al.* 1990). Loess is the most widely distributed superficial deposit in the nonglaciated parts of Alaska and Yukon Territory (Taber 1943: 1471). Silt with some organic material is called muck in Alaska (Taber 1943: 1471; Wahrhaftig 1965: 17). The silts in Alaska might have been formed and deposited in several different ways. Taber (1943: 1473) mentioned glacial, volcanic, organic and weathering origin and residual, aeolian, fluviatile, lacustrine and marine deposition. Recently the aeolian origin has been considered the main factor. As a typical combination of processes forming silt and loess deposits is the following. Large parts of the lowlands of Alaska have been unglaciated. They have been formed by the deposition of sediments brought in from adjacent highlands and glaciated areas by great rivers and wind. Most of these rivers which flow through the lowlands had their sources in ice sheets during the Pleistocene, and many of them still head in glaciers (Wahrhaftig 1965: 16). They brought and still bring great quantities of sand, gravel and silt originally produced by frost shattering and glacier grinding. When the rivers with sediment load enter the lowlands the coarse material is deposited, first forming fan-shaped outwashes which then are followed by bare flood plains of the braided rivers. On the outwash plains and flood plains the rivers carry silt and sand. The glacial rivers are subjected to daily and annual floods (Wahrhaftig 1965: 17). During the periods of low water when the fine deposits have dried on the lowlands, the wind blows great clouds of dry silt and sand from the river beds, partly among the vegetation. The sand accumulates as sand dunes and the fine material settles to mantle the plains and lower slopes of the surrounding hills as loess. Great dust storms forming loess in Alaska are common at the beds of the Delta, Donjek, and White Rivers (Wahrhaftig 1965: 17). Loess deposits are often rather thick in Alaska and it is supposed that dust storms must have been very frequent and extensive during the glacial time, but the loess formation continues today (Plate 12).

8.5.2   Dust storms

Dust storms are common features in many places in cold environments. From Iceland they are well known. Trautz (1919: 122) described them from the outwash plain, Sprengisandur, on the northern edge of Vatnajökull glacier.

South of Langjökull glacier, Iceland, a grey-yellow dust storm is often observed, a so-called '*mistur*' (Icelandic) (Samuelsson 1926: Fig. 30). In the area south of Hofsjökull icecap a dust storm rose to 2000–3000 m over the plateau and fine dust spread out over a tremendous areas (Nielsen 1933: 247–248). Dry north winds give rise to a peculiar atmospheric condition in south Iceland and far out over the Atlantic, characterized by a soft, yellowish-brown gleam as a consequence of the refraction of the dust particles (Nielsen 1933: 248).

The complete lack of vegetation cover on the Skeidarársandur south of Skeidarárjökull, Iceland, is due to the deflation of volcanic ash which took place daily in the summer season according to Bogacki (1970: 279). South winds from the ocean are humid but the dry and stronger north winds from the glacier carry volcanic dust from the glacier surface, outwash plains and deltas.

Deltas in the western Canadian Arctic and northern Alaska contain quantities of silt, and brown clouds of dust are whipped up by the wind from vegetation-free areas in the spring and deposited over considerable distances (Bird 1974: 713). Drainage of meltwaters from the glaciers continuously feeds the outwash plains (sandurs) with fresh uncovered sediments. Gilbert (1983: 164) tells about winter travellers on the sandurs in the fjords of Baffin Island who met violent sand storms carrying sand 1–2 m above the ground in sufficient concentrations to reduce the visibility to a few metres. Probably the term sand should not be taken seriously in this case, but it was blowing silt. At Maktak Sandur the dusting of sediment spread over the sea ice to a distance of 15 km attests to the importance of winter transport. According to Gilbert (1983) the terrestrial aeolian sediment drift from the sandur surfaces is an important part of the marine sedimentation in the fjords of the Canadian Arctic.

Once particles have been set in motion it is necessary to distinguish between dust storms and sand storms. The small particles of dust are kept in suspension by upward currents of the air movement. It is a suspension phenomenon and the dust cloud may rise to a height of 1000 m or more. In an erosion desert, where there is very little dust, the initial dust cloud is blown away and, soon after the wind exceeds the threshold strength, one can observe a thick low-lying sand cloud driving across the country. It has a clearly marked upper surface about 0.9–1.2 m above the ground (Nickling 1978; Raudkivi 1990: 51). The problem is to explain how large grains get up to heights of 0.9–1.2 m and what keeps them there. Measurements show that the upward eddy currents near the surface are not sufficiently strong to keep these grains in suspension or to lift them up, particularly since the vertical component of turbulence is much reduced by the presence of the solid boundary, the ground.

Dust storm genesis over exposed lake shorelines and braided river beds in alpine environments at the Lake Tekapo Basin, New Zealand, with regard to *Föhn* windstorms, has been recently studied by McGowan (1994).

8.6   Niveo-aeolian material

The term niveo-aeolian was proposed by Van Straelen in 1946 in Belgium (Cailleux 1968) and can be defined as a mixture of snow and wind transported mineral particles and the forms and microrelief generated by them (Cailleux 1974). It also contains different kinds of organic particles and vegetative detritus. Normally it is annual, snow melts in summer and leaves unmelted particles on the earth surface, but also perennial niveo-aeolian deposits occur, for example in Antarctica (e.g. Calkin & Rutford 1974). Niveo-aeolian deposits seldom form thick layers or landforms. They can form part of sand dune formation, as observed in the field in the Great Kobuk Sand Dunes, Alaska (Koster & Dijkmans 1988) (Fig. 8.17) and on so-called cliff-top dunes along Mountain River, Northwest Territories, Canada (Bégin *et al.* 1995: 399–400). In Poste-de-la-Baleine, on the east coast of Hudson Bay, niveo-aeolian deposits are annual with the snow melting down every year. They contribute to the nourishment of the first sandy raised beach and of some sand dunes (Rochette & Cailleux 1971; Cailleux 1972, 1976).

Hétu (1992) described the effects of wind on rockwall dynamics of Mount Saint-Pierre, Gaspé, Québec, where the summit plateau received a large amount of aeolian sediments originating from the shale rockwall. The layer of debris

Fig. 8.17 Great Kobuk Sand Dunes, Alaska with wet niveo-aeolian sand. Photo by E. Koster.

covering the snow after one storm had a mean thickness of 11.4 mm and the total volume of accumulation was about 13 m^3, formed of rather coarse grains, 2–4 mm in diameter, with many even as long as 10 mm. According to radiocarbon datings this cliff-top aeolian sedimentation has continued for a thousand years with a mean rate of 1.8 mm yr^{-1}.

Niveo-aeolian deposits can be formed in two ways: (1) falling snow with strong wind is mixed with wind drifted sediments, or deposited snow is removed together with sediments and redeposited, or (2) snow is deposited and on top of that drifted sand or silt without simultaneous snow fall (Jahn 1972: 98). Observations in the Søndre Strømfjord area, western Greenland, indicate that the niveo-aeolian deposition takes place mainly in the early winter when the snow cover is still thin (Dijkmans 1990). Later in the winter the sand is deeply frozen.

Thin layers of aeolian sands called periglacial coversands (e.g. Ruegg 1983) found in western Europe, North America and Russia (Koster 1988) form an unstratified surficial blanket and are at least partly niveo-aeolian in origin.

Cailleux (1968: 60) defined the niveo-aeolian deposits and sand dunes of Victoria Valley, Antarctica (Table 8.1).

When snow is covered by aeolian sand (Fig. 8.18) it thaws slowly and causes special dissipation, compression and tensional structures in the sediment (Ahlbrandt & Andrews 1978) (Fig. 8.19).

8.6.1   *Quantity of niveo-aeolian material*

Czeppe (1966: 122–123) measured in 1957–58 in Hornsund, SW Spitsbergen the amounts of debris from snow. He took snow samples equal to

Table 8.1. *Comparision of niveo-aeolian deposits and sand dunes in Victoria Valley, Antarctica (Cailleux 1968: 60).*

	Niveo-aeolian forms	Dunes
General slope	Weak, 0 to 10°	Strong, 5 to 31°
Median length	Big, 500 to 2000 m	Small, 200 to 600 m
Crests	Well rounded	Angular
Perceptible substratum	From place to place	Only in interdunes
Small pebbles 3 to 5 mm long	Present, rare nests	Nil
Volcanic particles 1 mm, black	Mixed with the rest	Sorted, by trails
Big ripples, wave-length 0.8 to 1 m	Frequent	Absent here
Interstratified layers of snow	Present	Never seen
Contraction cracks	Frequent	Nil or very unusual
Crevasses at the foot and subsidence	Frequent	Nil

one litre of water. In these samples the mineral material content ranged from 0.2 to 0.85 g. He also took 5 cm thick surface snow samples which had mineral material up to 2.0 g per litre of water. The mineral content was mostly (87 weight %) coarse sand (0.3–2 mm in diameter), up to 12% of fine sand (0.3–0.06 mm) and finer fractions less than 1%. The amount of aeolian depositions was in general about 100 g m^{-2} during the four winter months, which corresponds to 300–400 g m^{-2} yr^{-1} as a rough approximation. This means that aeolian transport carries considerable amounts of material in Hornsund, although it does not create any relief forms.

From West Spitsbergen Åkerman (1980: 242–244) made quantitative measurements of the amount and distribution of niveo-aeolian deposits. Snow was sampled with a plexiglass tube. The sample was melted and the amount of niveo-aeolian material was measured. The median value corresponded to 0.1 g l^{-1} of snow and ranged from 0.05 to 0.45 g l^{-1} snow. (Note that Czeppe's measurements above were expressed as g l^{-1} of meltwater.) Åkerman (1980: 244) obtained an average of 60 g niveo-aeolian deposits per cubic metre of accumulated snow during the period from November to May, with an average snow density of 0.6 (range from 0.44 to 0.72). With the assumption that about 10% of the surface of the deposit was swept bare and deflated, he concluded that approximately

Fig. 8.18 Exposure of niveo-aeolian sand with snow layer on the lee side of a sand dune on Great Kobuk Sand Dunes, Alaska (see Plate 9, Figs. 6.12 and 8.17). Photo by E. Koster.

270 g of fine material was removed by wind from each square metre of bare ground during each winter. The figure was based upon a snow depth of 0.5 m.

In Antarctica, McMurdo region, Victoria Valley, niveo-aeolian deposits, made of alternating layers of wind-driven snow and sand, 0–2 m thick, are perennial, and they cover five times larger areas on the north side of Lake Vida than sand dunes (Cailleux 1968: 59, 1972). However, the active layer in Victoria Valley is only 20–30 cm, and it is very difficult to make exposures in permafrost, Cailleux (1968: 60) showed that niveo-aeolian coversand deposits are less thick than sand dunes (Cailleux 1968: 61) because some scattered boulders of moraines from underneath outcrop. Aeolian sand is found there in many places scattered among

Fig. 8.19 Schematic formation of special sedimentary structures in niveo-aeolian sand when a snow layer thaws. Redrawn after Ahlbrandt & Andrews (1978: Fig. 12). © Elsevier. Reproduced with permission.

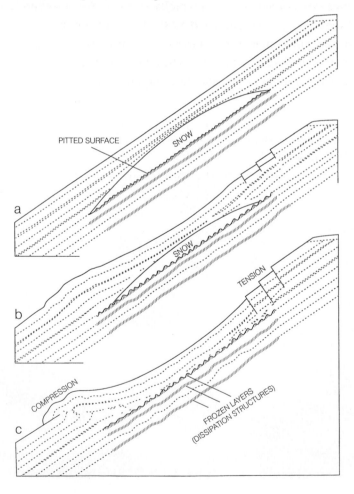

gelifracted stone fragments, showing strong aeolian activity in spite of a lack of transportable material. Niveo-aeolian coversands pass to dunes by transition when the amount of material increases. The aeolian sand sheet, 1 to 3 m thick, is underlain largely by compacted snow.

Bogacki (1970: 283) described some niveo-aeolian material from Skei-darársandur, Iceland, where it was collected on the lee sides of slopes, and in an example the inter-bedding of snow (50 cm) had aeolian sand-silt layers 20–25 cm in thickness. Melting caused a pitted surface topography. Some of the pits were 2–3 m in diameter and 70–80 cm in depth and looked similar to features described by Cailleux (1972) and Koster *et al.* (1984).

Snow lenses up to 25 cm in thickness were preserved throughout the year in sand dunes in Wyoming at an elevation of 2043 m. Sliding of sand on top of the snow layer during the melting has covered the snow (Steidtmann 1973: 797). This type of deposition may explain large-scale deformations in dune sand.

Jahn (1972) reported considerable amounts of niveo-aeolian deposits in the mountain and submountain areas of the Sudetes, southern Poland. He measured on the mountains up to 1.6 kg 'dirt' in a cubic metre of snow and on ploughed fields the amount of aeolian material ranged from 2 to 15 kg m^{-3} of snow (Jahn 1972: 104–105). Responsible for this deflation are the strong *Föhn* winds from the south. The deflation in the winter time in Poland can reach great values and has also a practical meaning. In February 1956 a heavy deflation of loess and sand was recorded in southeastern Poland. The amount of material on the snow surface locally reached the value of 10 000 tons km^{-2} (Strzemski 1957) and removal of mineral material in the Sudetes was 4000 tons km^{-2} (Jahn 1972: 109).

9

Wind directions interpreted from field evidence

9.1   Small indicators

The directions of effective winds can be observed from different small erosional signs in the field. In section 5.2 ventifacts were described. They may show two or three main effective wind directions, but because the rock surfaces are not always perpendicular to the wind direction it will be difficult to measure the effective wind directions from the polished rock surfaces. If the regional wind is very uniform then on one side the rock is abraded (Fig. 5.7) and it is easier to interpret the exact wind direction.

The sharp edges between facets on large ventifacts are often at right angles to the prevailing wind direction. The lengthened spatula-shaped cupules on blocks are mostly directed parallel to the wind direction, and their deeper end is always windwards (Cailleux 1939: 61). Small ventifacts are unsure indicators of wind direction because they can be turned during the erosion process.

In the summer, winds in the eastern part of Victoria Valley, Antarctica, are predominantly from the east, as indicated by sand dunes. The convective winds are related to the removal of snow and to valley heating. They might be largely inactive during the winter. Ventifacts in the valley show predominant cutting by westerly winds even though they are sporadic and the sand dunes may be stabilized in autumn and winter by the time the strong westerly winds penetrate there (Calkin & Rutford 1974: 193).

The wind directions read from the facets or the microforms and other abrasion features on rock surfaces cannot be determined with a higher accuracy than 15–30°. The initial form of the block and its orientation plays an important role in facet formation (Åkerman 1980: 251).

Lichen cover on rocks and trees is often worn off from the windy side. For example, lichens are found on boulder pavements close to the glaciers in north Greenland. 'They do not, however, uniformly cover the exposed parts of the fragments. They are common on the side away from the glacier, but are usually absent on the side toward the glacier. The side toward the glacier is probably blasted by wind-driven snow, ice crystals, and perhaps other material, so that lichens cannot grow here although they can grow on the lee side.' (Nichols 1969:84).

The stems of trees are polished and the bark not covered by lichens on their windward sides, indicating strong erosion by sand grains and ice crystals. This feature can be frequently observed on birch bark on the sand dune areas in Sub-arctic Lapland. Wind direction and snow depth can be often noticed also from the shape and characters of vegetation. Trees are inclined downwind and they have branches only on the lee side (Fig. 2.13). The top surfaces of juniper bushes are cut by frost drying and moving ice crystals into certain shapes (Fig. 2.20). All the branches rising above the snow surface will dry in winter. Small cush-ion forming plants (e.g. *Loiseleuria, Diapensia, Dryas* etc.) are eroded on the windward side and migrate downwind (Figs. 2.26 and 2.27).

From Iceland between Hofsjkull and Vatnajökull glaciers Nielsen (1933: 249, Figs. 28 & 29) presented a very good example of the direction of the effective wind. The same surface was photographed from the south and from the north. The exposure from the south shows how the surface is covered with an apparently continuous carpet of mosses (*Grimmia*), whereas the photograph towards the south demonstrates that the carpet of mosses is not continuous but consists of a number of cushions, separated by wind-blown, stone flats. The cushions display a most asymmetric structure with a steep, undermined north side and an evenly

Fig. 9.1 Directions of effective winds in a mountain area (Abisko-Björnliden), northern Sweden, mapped from erosion scars, damages on the vegetation, snow patches etc. (Magnusson according to Rudberg 1970: Fig. 28).

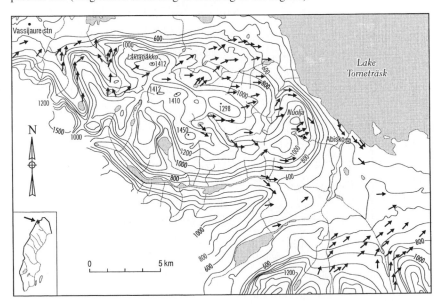

sloping plane towards the south. The moss cushions seem to be in some kind of balance with erosion. They regenerate on the lee side at such a speed that the growth there keeps pace with the erosion on the other side.

The directions of strongest winds in a mountain area in northern Sweden were mapped by means of erosion scars in soil, damage to the vegetation, and snow patch distribution (Rudberg 1968, 1970) (Fig. 9.1). Similar studies have

Fig. 9.2 Directions of surface winds in the Indian Peaks region, Colorado, U.S.A., mapped with different wind direction indicators as trees and deflation features by Holtmeier (1978 and 1996).

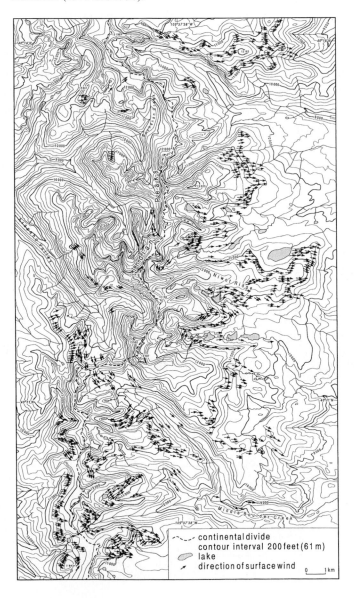

been carried out, for example, on Axel Heiberg Island, Arctic Canada (Rudberg 1968), in Finnish Lapland (Hellemaa 1991) and on the continental divide in Colorado (Holtmeier 1978, 1996) (Fig. 9.2). All these studies show the very local wind directions guided by the main topography and small relief features. In the Fjord Region, northeast Greenland, it has been concluded from the irregular distribution and orientation of the aeolian features that local topography exercises a greater control over wind directions than does the general orientation with respect to the Greenland Ice Sheet (Flint 1948: 210).

Surface winds in mountainous regions are very variable (Zotov 1940) (Fig. 3.13). At the edge of ridges surface winds converge and wind velocities generally increase, in the wake behind the ridge wind velocity generally decreases. The distance between the sites is often small. Aerial photographs have been used in Japan to determine the strong wind directions by using aerial photographs taken after strong winter monsoons. Snow surface patterns contain information of the wind directions and velocities (Nakamori 1994: 308). This technique has been used for practical purposes when planning ski lifts. Information about prevailing wind direction in Antarctica also presents itself in the orientation of sastrugi (Fig. 14.3).

Cailleux (1968) observed the prevailing easterly summer winds directly in Victoria Valley, Antarctica, from the dissymetry of the sand dunes and ripple marks, and deduced W and SW wind directions from the facts that (1) the big

Fig. 9.3 Rime, hoar ice on Fjell Njulla, northern Sweden. September, 1991.

accumulations of sand are all situated in the east and northeast parts of the valleys, and (2) basaltic and doloritic boulders on the surface have their E and NE sides coated by iron hydroxide, whereas the W and SW sides are black and scoured by wind, probably during the winter time.

Wind oriented rime is found adhering to upright objects facing a light wind. Hoar crystals are formed on the windward side of the objects. Rime may form quite large deposits of ice, for example in the Alps and in Scotland, and adds quite a lot to the mass of snow (Seligman 1980: 86–88). (Plate 1). The direction of the slight wind can be measured from the rime orientation (Fig. 9.3).

9.2   Deflation basins and furrows

Deflation basins and furrows have usually oval or longitudinal form and furrows which lead drifted sand out from the blowouts. The directions of these deflation features can be measured in the field and drawn in a diagram which indicates the direction of sand movement and wind (Fig. 6.9).

9.3   Dune orientation

Transverse dunes are perpendicular to the effective wind direction. Their windward slopes are gentle and leeward slopes steep. Longitudinal dunes are parallel to the dune forming winds and their cross-sections are quite symmetric. Coastal dunes are normally parallel to the shoreline and they are not such good indicators of regional winds as inland dunes.

Most of the sand dunes in cold environments are parabolic in their form. The axis of the parabolic dune is parallel to the dune forming wind. It is not so easy to measure the direction of axes in the field, but if we have aerial photographs at our disposal then it is easy to determine the direction and draw a wind rose-like diagram showing the directions of effective winds (Fig. 6.9). By studying several dune fields it is possible to map the directions of palaeo-winds in large areas, as is done in central Europe (Fig. 16.2) (e.g. Enquist 1932) and in northern Sweden and Finland (Fig. 16.3).

10

Ice-wedge casts and sand wedges

10.1   Frost cracking

Soil freezing causes cracking because the material contracts when it cools. Frost cracking (or thermal contraction cracking) takes place in regions with permafrost as well as seasonal frost. Other cracking processes such as desiccation and dilation can leave features similar to frost crack phenomena. Ice wedge formation starts from tension cracks that open during the winter periods of rapid and profound drop in temperature. Later, water trickles into the cracks and freezes, adding to the permafrost. The cracks open in the same places in the following winters and the ice wedges gradually thicken (e.g. Wahrhaftig 1965: 17). Thin snow cover supports the cracking.

The physical reason for frost cracking is that ice has a coefficient of linear contraction of about $45 \times 10^{-6}\,°C^{-1}$ at $-40\,°C$ (Hobbs 1974: 347–349). The ice content of frozen ground is highly variable, and the linear coefficient of thermal contraction can be much greater for frozen ground than for pure ice (Washburn 1979: 104). Frost cracking is more dependent on the rate of temperature drop than on the actual subfreezing temperature (Lachenbruch 1966: 65–66). Svensson (1977), for example, observed sudden, occasional frost cracking on grass mat in southern Sweden when the temperature dropped after a period of thaw to $-8\,°C$ or $-10\,°C$. The insulating effect of snow is an important factor. There is a high inverse correlation between snow depth and cracking frequency (Mackay 1974: 1376) and wind makes the snow cover thinner. The first time cracking occurs in a permafrost environment it starts at the surface. The crack can extend to a depth of 4 m (Mackay 1972). Frost cracks filled by ice and the growth of the ice wedge start as an ice vein which is the focus of subsequent cracking (Washburn 1979: 105). Repeated cracking at the same place year after year leads to a well-developed ice wedge. Because of the lateral growth of the wedge, adjacent sediments commonly develop an upward bend as the result of expansion during warming (Washburn 1979: 105).

Ice wedges and ice-wedge polygons are common features in the Arctic and, although less so, in Antarctica. They give the basis of formation of sand wedges and sand-wedge casts which are rarer because, for example, for the aeolian filling material transport is needed on the surface, which is often very wet and flat (e.g. Seppälä *et al.* 1991). Often the frost cracks are filled with water in summer time.

The cracking of ice wedges cannot be forecast solely from a knowledge of winter air temperature because the frequency of ice-wedge cracking is dependent, to a considerable extent, on the thickness of snow cover (Mackay 1978: 524). At Garry Island in North West Territories, Canada, when the snow cover approaches 1 m, ice wedges rarely crack (Mackay 1978: 524).

10.2   Ice-wedge casts

Upon thawing of an ice wedge there can be infilling by collapsed material and it becomes an ice-wedge cast. The filling can be aeolian in origin (e.g. Cailleux 1973a). Permafrost soil wedges formed in dry polar climates differ from ice-wedge casts in being thinner, vertically foliated, and limited to fine material which has entered narrow cracks lacking ice and only 1–2 cm wide. New cracks tend to form between older ones and with time result in the development of many vertical narrow wedges. Ice-wedge casts are broader, with more horizontal foliation and contain coarser material and even stones (Black 1969a: 229; Washburn 1979: 114–115).

A typical sand wedge cross-section has vertical fabric in the wedge and up-turned beds in enclosing material, and an ice-wedge cast cross-section shows slump structures in cast and upturned beds in enclosing material (Black 1976: Figs. 8 & 9) (Fig. 10.1). In some ice-wedge casts stones are present and show

Fig. 10.1 Sketches of the cross-sections of an ice-wedge cast and a sand wedge with vertical fabric in contraction crack fillings. Redrawn after Black (1976: Figs. 8 and 9). Reproduced with the permission of © Elsevier.

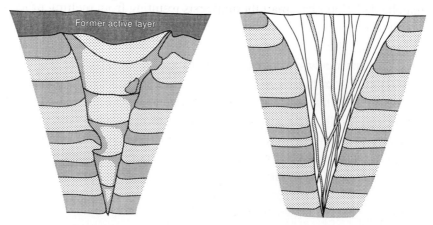

that filling was not accomplished in narrow contraction cracks, but in the full width of the wedge (Black 1976: 12).

10.3   Sand wedges

When the frost cracks are filled by sand then we call these features sand wedges. Typical for sand wedges is that they start forming at the earth surface, while the ice-wedge casts penetrate downwards from the bottom of the active layer, where the ice wedge is initiated. In the arid parts of Antarctica sand wedges are formed by repeated frost cracking and infilling with dry sand or loam (Péwé 1959; Berg & Black 1966; Ugolini *et al.* 1973). The infilling may take place by aeolian processes. The crack on the surface stays open during the dry season and sand and snow drifted on the surface by wind drop into the crack and fill it. When the frost cracking is repeated during the forthcoming seasons the infilling penetrates deeper and gets wider because of continuous deposition of aeolian sand and forms a wedge shape.

Nichols (1969: 72) found in Inglefield Land, north Greenland frost-crack polygons more than 30 m across with 1 m or more wide and up to 60 cm deep frost furrows on horizontal surfaces and also on slopes up to 15–20 degrees. Almost all of them occur on unvegetated kame terraces, valley trains, dissected deltas, alluvial fans, modern flood plains, and solifluction and beach deposits; but they were not seen on the broad expanses of till. A few were seen in tundra. Due to thermal contraction, cracking takes place more frequently in well-sorted material with high porosity, like sand and gravel, than in till, which has a lower porosity. The furrows are due to the melting of ice wedges or to the falling of surficial material into the contraction cracks. Nichols (1969: 73) described these features: 'Many of the furrows in the area around Thule have been nearly filled with wind-blown sand so that only a slight surface sag remains. Holes 4 feet long, 1 foot and more in depth, and 1 foot across are found in some of these furrows. They may be deflation pits, small thermokarsts resulting from melting of ice wedges, or holes formed by the trickling of the sand down into contraction cracks.' The description gives us a chance to interpret these holes as niveo-aeolian in origin. The furrows are filled in winter by a mixture of snow and sand and when the snow melts a hole remains. Nichols (1969: 73–74) continues that the surface of the ground adjacent to the furrows has not been significantly deformed. However, marginal levees are found along both sides of some furrows. The levees may be 60–100 cm high, and the width from the outer edge of one levee to that of the other may be 6 m. A thin veneer of wind-blown sand is commonly found in the furrows between the levees. Less commonly, miniature V-shaped trenches, 5–10 cm wide at the angle of repose, may be found in that sand. They result from the trickling of the wind-blown sand down into open contraction cracks. They prove that the furrows and levees are being formed at the present time. Nichols (1969: 74) presents as evidence of modern formation stones without lichen cover on the levees. However, the reason for barren rocks on ground surface, in these High Arctic conditions with thin snow cover, could also be eroding drift snow on the elevated levees. Nichols (1969: 74) considered these sand wedges to be similar to those described from Antarctica by Péwé (1959). Then the levees are formed because the material in the polygons between the contraction cracks, which has progressively increased in volume during the growth of either ice wedges or sand wedges, expands during the warm season and causes the deposits next to the contraction cracks to be deformed and pushed upwards. Miniature holes and tunnels, small cracks in the ground and collapsed grass, associated with the wind-blown sand found in some of the furrows, also prove that the formation of the furrows is a current process (Nichols 1969: 74).

As examples of the character of different kinds of sand wedges we shall take here some descriptions from Antarctica. From Victoria Dry Valley, South Victoria Land, Antarctica, Webb and McKelvey (1959: 126–127) reported that much of the valley floor moraine and some scree slopes are patterned with frost polygons up to 12 m in diameter. In profile each polygon is broadly concave and separated from adjacent polygons by a sand-filled trough about 60 cm wide. In coarse moraine, polygons are smaller and have deeper peripheral troughs. During the summer months the permafrost level of the valley floor retreats below the surface. The water saturated glacial drift is drained by meltwater streams and when completely dry becomes a fine wind-blown sand. From this description we may expect that sublimation dries the surface and the sand-fills of troughs are aeolian. Also elongated trough and mound forms are sculptured from moraine debris and these troughs are commonly filled with wind drifted sand.

Black and Berg (1963: 122) mapped the distribution and reviewed patterned ground in Antarctica. In the McMurdo Sound region sand wedges and ice wedges are two dominant end members between which mixtures with all proportions of sand and ice may be found locally. Sand wedges characterize the inland dry areas, whereas ice wedges are more common along the more humid coasts and islands in the Ross Sea. Overall, sand wedges and composites with less than 50% ice are most abundant. Non-sorted polygons result from annual fillings of thermal contraction cracks. All such non-sorted polygons are commonly 4- to 6-sided in plan and 5 to 30 m in diameter. In many places master polygons 20 to 40 m in diameter, as outlined by troughs 2 to 5 m wide and 0.3 to 1.5 m deep, are subdivided into halves, thirds, or quarters by narrower and shallower troughs. Beneath the troughs are wedge-shaped fillings of ice, sand, or mixtures of ice and sand with some stones, the apexes of which point downward. Wedges range from thin dikes a few millimetres wide to massive wedges up to 6 m wide and possibly 7 m high. Wedges 1 to 4 m wide and less than 5 m high seem most common.

Black and Berg (1963: 125, Fig. 4) presented an old sand-wedge trough in the bouldery moraine of Beacon Valley, Victoria Land. The sand wedge is 4 m wide and extends under the boulders on either side of the central sand zone. The sand wedge crosses a very coarse blocky surface with signs of aeolian abrasion. A paucity or lack of sand reduces or prevents sand-wedge polygons: very low humidities and low moisture content (less than 3%) of the active layer in much of Victoria Land and in the interior mountain ranges favours sand-wedge polygons over ice-wedge polygons and tends to reduce frost action that brings about sorting.

Contraction is greatest where water content of permafrost is greatest, because the coefficient of thermal expansion of ice is roughly five times that of rock (Black & Berg 1963: 126). However, growth in sand wedges is achieved by infilling of debris into winter contraction cracks. An area of free flowing sand provides optimum material for growth – even greater than the growth of ice wedges where reversal of thermal gradients in spring moves moisture from the air and active layer into the open contraction crack. Growth of hoar-frost in contraction cracks also occurs in sand wedges in dry valleys but to a lesser degree. The presence of free running sand is required for pure sand wedges: many wedges are mixtures of sand and rubble. Sand wedges may grow about 0.5 to 1 mm per year.

It has also been pointed out that patterned ground with sand wedges must also be correlated with wind erosion and deposition, and ventifacting (Black & Berg 1963: 127). Calkin & Rutford (1974: 195 and Fig. 4) described sand-wedge contraction polygons which cut the aeolian sand sheet covering the ground moraine in Victoria Valley, Antactica. Aeolian filling of an ice-wedge cast can be found in very coarse material, too (Gray & Seppälä 1991) (Figs. 10.2 and 10.3). A polygon network was found on a large glaciofluvial outwash plain in northernmost Ungava Peninsula, Québec, in the zone of continuous permafrost. The surface of the polygon field is paved mainly by lichen-covered stones of 3–10 cm diameter. Stones are well polished by strong wind action. Most of the fine sediment particles are deflated away and deposited beyond the limits of the outwash plain, or in the polygon furrows among vegetation. Furrows give some protection for denser vegetation. The aeolian filling of furrows has thin peaty layers (Fig. 10.3).

10.4   Soil and sand wedges formed by seasonal frost

Soil wedges and soil veins mean crack fillings also in a non-permafrost environment. They are formed in seasonally frozen ground and resemble active-layer soil wedges in permafrost areas. Open fissures on glaciofluvial delta surfaces with coarse gravel without aeolian filling have been found on the Norwegian Arctic coast (Svensson 1969). Their formation might be in connection with strong winds blowing off snow from raised delta surfaces and terraces, exposing them to cold air, although the environment is not extremely cold because of the Gulf Stream.

Fig. 10.2 Frost-crack furrows (tundra polygons) on a large deflated glaciofluvial outwash plain in northernmost Ungava Peninsula, Arctic Canada (Gray & Seppälä 1991).

Fig. 10.3 Cross-section of the exposure dug at the base of a furrow (Fig. 10.2), Ungava Peninsula, Arctic Canada (Gray & Seppälä 1991: Fig. 10).

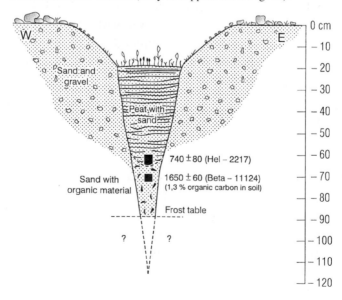

In Iceland regular polygon nets with similarities to High Arctic pattern ground have been observed in a non-permafrost environment (Friedman *et al.* 1971). The cracks are filled with material derived from the fissure walls and volcanic ash which could be dated with tephrachronology. Some cracks are filled with ash from the volcanic eruption of 1918 A.D., giving an idea that these seasonal cracks are very recent in origin. The authors (Friedman *et al.* 1971) concluded that permafrost might have occurred in Iceland still in the 17th century (the Little Ice Age) and the present seasonal cracking is probably following the old permafrost nets. In the same region palsas still exist, indicating discontinuous present-day permafrost. Iceland is a very windy place without large forests and this gives good opportunities for snow drift, which means that crests and summit areas can be snow free during the coldest time and this supports also the fast cracking of soil.

Frost cracks and frost-crack polygons in sand dunes in Finnish Lapland were first reported by Aartolahti (1972). Recent observations have been made by the author at Hietatievat (68° 27′ N, 24° 43′ E, 360 m a.s.l.) on a deflation surface lying between transverse dunes and anchored by vegetation (Seppälä 1982b; 1987b). The polygons are bounded by furrow-like depressions in the ground with a 5–15 cm deep open crack in the middle of them. The polygons are irregular in shape. The length of their sides varies between 1 and 12 m (Fig. 10.4). The largest

Fig. 10.4 Frost-crack polygons on an interdune area at Hietatievat, Finnish Lapland (Seppälä 1982b: Fig. 5).

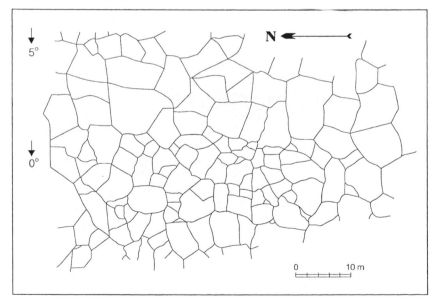

Fig. 10.5 Sand-wedge field in a blow-out on Hietatievat, Finnish Lapland. July 1965. (Seppälä 1966, 1987b).

Fig. 10.6 Cross-section of a sand wedge on Hietatievat (Fig. 10.5). July 1965. (Seppälä 1966, 1987).

polygons are to be found on the gently sloping sides of dunes (5°). At most of the junctions of the cracks the sides of three polygons meet. The depth of the frost wedges is less than 70 cm. The soil layer (podsol) extends in places to a depth of only 49 cm. Deeper down the wedge can be distinguished from its surroundings by its darker colour caused by the humus. The cracks show that the formations are still active. It is of great importance for the formation of the wedges that the ground water level lies close to the surface of the deflation basin during the spring and autumn. In this case the connection of soil wedges to aeolian features is their formation in aeolian sand. Their filling, organic debris with sand and silt, may also enter the cracks with water during the spring flood. The mean annual air temperature in the region is −1.5 °C but in the winter the air temperature often drops below −40 °C. These sands have no permafrost (King & Seppälä 1987b). These features formed in aeolian sand have no aeolian filling as do some sand wedges in the same region (Seppälä 1966, 1987b).

10.5   Sand wedges at Hietatievat

A special type of sand-wedge polygon formed by seasonal frost has been described from Hietatievat, Finnish Lapland (Seppälä 1966; 1987b). There is a wide esker which has silt deposits on the surface. After deglaciation the esker and silt were partly covered by sand dunes which are nowadays deflated, and at the bottoms of some blowouts the stratified silt is exposed. The silt surface is dissected by fissures filled with aeolian sand drifted on the surface from the edges of blowouts (Fig. 10.5). Also found are open cracks with depths ranging from 0.5 to 2.5 cm in the centre of the acolian filling. The wedges are between 35 and 50 cm deep (Fig. 10.6). Their width at the surface varies from 30 to 60 cm and tapers rapidly downwards. The most likely origin of these sand-wedges becoming polygons with diameters from 2 to 4 m is that they are connected with seasonal frost contraction cracking of the silt (Seppälä 1987b: 49). Once open, the cracks are filled with aeolian sand. During winter the silt is deeply frozen but the sand in the wedge is dry (Seppälä 1982b, 1987b).

11

Snow

11.1 General distribution of snow cover

11.1.1 Northern hemisphere

Vast snow-covered areas are uninhabited, which means that snow measurements are unavailable. The development of remote sensing techniques has given a tool to get up-to-date and synchronous data all over the world (e.g. Barry *et al.* 1993; Frei & Robinson 1993; Norton *et al.* 1993; Masson 1995) and to speak about global snow depth climatology (e.g. Foster & Davy 1989). The extent of snow cover over the northern hemisphere is greatest in January, when, on average, 46.5 million km^2 of Eurasia and North America are snow-covered. August has the least cover, averaging 3.8 million km^2, the majority found on the Greenland ice sheet (Robinson *et al.* 1993: 1692). Taking into account the southern hemisphere and sea ice, about 6% of the Earth is under a permanent snow cover and another 15% (not simultaneously) is under seasonal snow cover: this totals 115×10^6 km^2 (Kotliakov & Krenke 1982: 450). The annual average value is about 57×10^6 km^2 (Barry *et al.* 1993: 25).

Most of Eurasia north of approximately 45° N latitude is covered by snow every winter. The greatest exceptions are found in west and central Europe which have only occasional snow cover. In North America the border of the regular snow cover runs about 40° N latitude with the exception of the Pacific coast (McKay & Gray 1981: 169).

Great annual variations in snow-covered areas on a continental scale have been recently observed in North America. The winters of the early 1980s, and the period since 1987, were characterized (Frei & Robinson 1993) by low snow cover (Figs. 2.14 and 2.15).

Difficulties in snow measurements in the Arctic are described very well by Klein (1949: 108): 'Churchill – very flat barren country with practically no trees. At Churchill, completely exposed areas are unsuitable for snow measurements because of the prevalence of shifting bare patches which are produced by strong, steady winds.' According to Klein (1949: 121) only about one quarter of the snow in Churchill, on the west coast of Hudson Bay, was due to direct precipitation.

The general depth of snow cover in continental interiors north of the 50° N latitude and in Polar areas has been given as ranging between 40 and 80 cm. Snow cover depends very much on topography and geographical location. For example, in mountainous regions average depths exceed 120 cm in Norway and 180 cm in southern Alaska (McKay & Gray 1981: 170).

Climate, physiography and vegetation interact in a complex manner to govern snow cover accumulation and distribution (McKay & Gray 1981: 173), and not only on a large zonal scale but also in small detailed areas (e.g. Clark *et al.* 1985). It has been found possible to classify and map some general characteristics of snow cover according to vegetation zones.

In the tundra zone the snow cover lasts from 8 to 10 months. The thickness varies from a few centimetres to some decimetres depending on the location. The height of the shrubs often reveals the thickness of snow cover. For example, juniper and dwarf birch (e.g. *Betula nana*) are covered by snow during the coldest winter. All branches growing above the snow surface will be dessicated by frost during the winter. Some plants, for example *Diapencia*, like to be uncovered by snow and some others such as blueberry have to be covered by snow. Snow in the tundra zone is drifted and packed by wind, cold and dry, and has rather high density (>0.3 g cm^{-3}).

In the northern parts of the boreal forest zone or taiga the snow cover lasts from 6 to 8 months. The depth of snow accumulation is greater than in the tundra, ranging from 50 to 150 cm in general (McKay & Gray 1981: 179). Wind has no comparable importance in snow drifting because forest breaks the wind. Trees intercept part of the snow falling on the branches and this affects the snow depth on the ground. They also shelter the snow cover from solar radiation but the snow melts fast from branches. During very cold weather trees collect a lot of hoar snow and that increases the snow cover without precipitation (Plate 1). The average densities of snow in boreal forests range from 0.17 to 0.21 g cm^{-3} before the melting period (McKay & Gray 1981: 179).

On the northern grasslands, steppes and prairies snow cover persists for 120 to 160 days. The grassland snow cover is fairly shallow and well-drifted; the mean depths vary from 20 to 50 cm. Depressions fill with deeper layers of packed snow. The density of the prairie snow is ~ 0.2 g cm^{-3} throughout most of the winter, increasing slowly due to metamorphosis (McKay & Gray 1981: 180).

Several investigators have noticed certain relationships between snow cover properties and topography and vegetation and land use. To extrapolate the empirical relationships outside of the region for which they were developed is highly questionable (McKay & Gray 1981: 184).

Kuz'min (1963) studied the interaction of snow cover accumulation and vegetation and topography and obtained so-called *snow retention coefficients* for different landscapes. According to Kuz'min, for example, open ice is only half as effective as virgin soil or arable land in retention of snow, while river beds and edges of forests are three times more effective than virgin soil. The local conditions determine very much the snow distribution and characteristics. Although we know the local landscape parameters, the unknown variable is the wind. The aerodynamic properties of different landscapes affect snow accumulation and changes in snow cover drastically. Because winds are not constant year-to-year there are also annual variations in snow cover properties at the same sites and from site to site.

11.1.2 *Antarctica*

Antarctica has a permanent snow cover except for small areas on the Antarctic Peninsula and some dry valleys which are like deserts. In Antarctica the open ocean is the principal source of the moisture. During winter the ocean surface is frozen to a distance of 20° of latitude from the continent. However, the atmospheric currents carry moisture and heat to the coast and the continent, and the amount of transported moisture is limited by the low air temperatures that prevail. The surface temperature of the open ocean at the edge of the ice is close to $0\,°C$ and the air temperature over the water may fall to $-20\,°C$, and it is even lower at the edge of the continent (Rubin & Giovinetto 1962: 5167). Low temperatures lead to low annual precipitation in the form of snow for all of Antarctica. Cooling of moist air due to the vertical motions induced by cyclonic activity, wave motions and topographic features means that some snowing takes place (Rubin & Giovinetto 1962: 5168). There is a consistent decrease of the rate of snow accumulation from the coast toward the interior. The range from 40 to 12 cm of water in average annual snow accumulation along longitude 96° W in Antarctica has been reported (Rubin & Giovinetto 1962: Fig. 2).

11.1.3 *Snow on mountains*

There is no snow data for most of the mountainous regions. We can assume that all precipitation in mountains from late autumn to early spring is snow, but it really is difficult to measure because of the great amount of drift snow. Average winter snow depths for mountains are very inaccurate because the snow cover above the tree line is highly heterogeneous. Wind drift, snow slides and avalanches have a great influence on snow cover. Especially on the steep sloped mountains and in middle latitudes large areas are uncovered by snow even if glaciers occur in the same region. The duration of snow cover increases with altitude, as shown by Geiger (1961) in the Swiss Alps. Snow accumulation on the mountains is highly governed by wind, and this is an important factor which in turn controls other agents of slope development dependent on moisture: slushflows, mass wasting, denudation, debris flows, fluvial erosion and solute movement (e.g. Caine 1995).

A *cornice* is a deposit of snow (Plate 13) projecting over the top of a face of ice or rock. Seligman (1980: 237–270) has collected and reviewed the literature on cornice development, presenting demonstrative pictures (Fig. 11.1). Cornices are direct evidence of the vast snow drift taking place on mountain regions.

Wind-slab avalanches are formed when wind has packed snow above hard snow, soft snow or the ground (Seligman 1980: 410–438). Broken cornices and avalanches may fill the Bergschrund, which is the crevasse between the cirque glacier and its head wall. This will very effectively increase the mass of a small glacier and is very difficult to measure.

Snow cushions consist of accumulations of snow deposits on steep mountain faces in spots that are entirely calm or are exposed to very sluggish wind or eddies (Seligman 1980: 209). A snow cushion can be the initial stage of a wall glacier and, when it melts, a nivation hollow will be left.

Also in lowlands we meet troubles in snow measurements because in many locations snow cover may form and disappear several times a season. In mild climates such as in southern Finland, about 60° N latitude, this may last throughout the winter.

Fig. 11.1 Pressure and suction cornice as an example of snow drift mechanism on the crest of a mountain. Redrawn after Seligman (1980: Figs. 188 and 203).

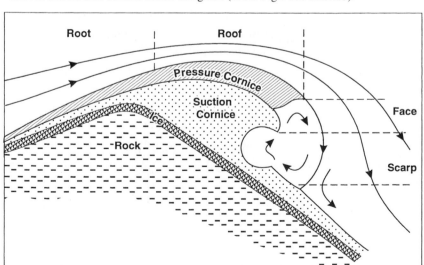

Snow depth information is important for drainage and flood forecasts or for predicting the avalanche risk. In certain cases also forecasting conditions of pastures in arid lands can be of vital importance, as in Mongolia, where strong, dry spring winds are capable of removing substantial amounts of water vapour from the thin snow pack (Norton *et al*. 1993: 387).

Snow accumulation by wind has a general trend to smooth the uneven topography of the land surface to an aerodynamic equilibrium.

11.1.4   *Snow cover and vegetation*

Landscape types reflect the character of the snow cover (Kuz'min 1963). Vegetation is the dominant control factor of the areal variation of snow cover. Adams (1976) gave a good example of this phenomenon. He measured the snow cover in an interdrumlin swale in east central Ontario and concluded that the snow cover of the vegetation zones is most varied in terms of water equivalent and the most uniform in terms of density.

The areal variability of snow cover is reflected by the landscape type and vegetation and can be used in estimations of runoff and in predicting ground temperature regimes.

The amount of snow varies from winter to winter but the relative amount in different places in the landscape is similar. Certain places collect much snow and others stay under thin snow cover or are snow free. This fact determines what kind of vegetation can grow on different places. Some plants can survive without the protecting snow cover while some others need it. This gives us a tool to make snow cover predictions from the vegetation even in summer by using, for example, remote sensing (e.g. Clark *et al*. 1985).

In Lapland it was observed that on the Alpine heath the snow depth varied from 0 to 17 cm and in the birch forest at a few hundred metres lower level it ranged from 50 to 90 cm (Clark *et al*. 1985: Table 11). On an open palsa bog snow depth was 10–20 cm and in the surrounding birch forest it varied from 40 to 120 cm, and the depth of snow depended on the density of the forest. Among scattered trees there was much less snow than on densely forested areas (Clark *et al*. 1985: Fig. 8).

Open surfaces and ridges and summits of hills are exposed to strong winds and wind drift. On vegetation covered areas the wind is less effective and drifted snow accumulates.

11.2   Physical characteristics of snow

About 25000 km^3 of water is precipitated annually on the Earth's surface in the form of snow. About one third of it melts instantaneously in the sea, while the remainder forms snow cover (Kotliakov & Krenke 1982: 449). Wind transports snow making the mantle uneven, thinning certain areas and depositing it at other sites. Wind-drifted snow means a great amount of energy transport in the form of latent heat and potential energy, which is then made free when snow slides down the slopes, flows in the metamorphosed form of glacier ice and drains in meltwaters, which all cause erosion and accumulation of mineral material and produce landforms. To understand the meaning of snow for geomorphic processes we should know the physical properties of snow.

11.2.1 Snow crystals

The simple hexagonal ice crystal in air is the initial stage in the growing of the snow particle. If the ice crystal continues to grow by sublimation a snow crystal is formed. A snowflake is an aggregation of snow crystals (Schemenauer *et al.* 1981: 136). New snow, when accumulated on the ground, is very porous and contains a lot of air. For example, if dendritic crystals form flocculated aggregates in light winds they may be deposited in a layer of extremely low density (about 10 kg m^{-3}) or, alternatively, graupel (coalesced frozen water droplets) may form a layer of high density (ca. 500 kg m^{-3}) (Langham 1981: 275). Typical fresh snow has a density of 0.1 to 0.2 g cm^{-3} which means that it is largely (90 to 80%) air-filled pore space in a fragile ice matrix (Reimer 1980: 129).

The general rule stands that the lower the temperature the smaller are the snowflakes (Seligman 1980: 41). When cold, windblown snow is first laid down, it is granular and cohesionless. Grain sizes for cold, blown snow are mainly in the range from 0.1 to 1.0 mm, and uniformity coefficients are around 2 (Mellor 1963: 529). Most of the snow settling on the ice caps and ice shelves in Greenland and Antarctica consists of fine particles less than a millimetre in diameter. Any complex crystals are soon broken down into small fragments by the persistent winds (Mellor 1963: 528). Waterhouse's grading curve for natural drift snow (Mellor 1963: 529, Fig. 1) (Fig. 11.2) indicates the size development of snow grains. Snow settling after drifting slowly grows in grain size by melting and freezing and through sublimation and refreezing.

11.2.2 Density

The density (kg m^{-3}) of snow depends of the wind speed near the surface during the deposition, settling temperature, form and size of snowflakes and if it originates from precipitation or hoar. Drifted and redeposited snow, as well as metamorphized old snow, have higher density than just deposited new snow. See Tables 11.1 and 11.2.

Yosida (1963: 485–487) presented experimental research showing the metamorphosis occurring in the texture of natural snow cover at Sapporo, Japan. The ice grains composing the snow became thicker and thicker as the days passed, at the same time joining with one another more and more firmly to make a strong network of ice. The temperature ranged from −5 to −20 °C. Wind was not present during the study.

Observations from North America indicate how the density of snow increases during the winter (McKay & Thompson 1968), see Fig. 11.3. In the Polar regions with low temperature and lacking solar radiation in winter the metamorphosis does not take place so fast, but the wind-packed snow has higher density already at the beginning. The density of surface snow on an ice cap is about 0.35 to 0.40 kg m^{-3} (Mellor 1963: 531).

The Proctor penetration test correlates very well with the density of snow and it is an easy method to get an idea of density which depends on the temperature

Fig. 11.2 Grading curve for natural drift snow. The grading was obtained by sieve analysis after disaggregation by rubbing snow on snow. Redrawn after Mellor (1963: 529, Fig. 1) obtained from Waterhouse (private communication) and CRREL internal reports. Reproduced with permission of © MIT Press.

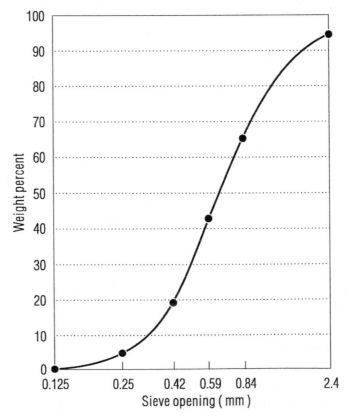

of snow, too. The colder the snow the higher is the penetration resistance and the harder it is (Mellor 1963: 544; Fig. 11.4).

11.2.3 Thermal conductivity

The thermal conductivity of snow depends on density, microstructure and temperature of the snow. The thermal conductivity of ice varies inversely with temperature by about 0.17% per °C and the same may be expected for snow (Langham 1981: 295). A temperature gradient can induce a transfer of vapour and the subsequent release of the latent heat of vaporization, and can cause a change in the thermal conductivity value (Langham 1981: 295). Yosida *et al.* (1955) suggested that the movement of water vapour contributes 37% to the

Table 11.1. *Typical densities of different types of snow (Seligman 1980: 144–145)*

New snow, immediately after falling in calm air	0.05–0.065
Damp new snow	0.1–0.2
Settled snow	0.2–0.3
Wind-packed snow	0.28–0.35
Firn	0.4–0.8
Very wet snow	0.8
Glacier ice (0.9169) with air bubbles	0.83–0.91
	(cf. Paterson 1981)
Heim (1885: 113) cited Dollfus as giving the following specific gravities for firn and glacier ice:	
Firn	0.628
White ice	0.871
Blue ice with a few air bubbles	0.897
Blue ice without air bubbles	0.909

Table 11.2. *Snow density during metamorphosis (Yosida 1963: 487, Table 1)*

Number of days after the deposition of snow	Snow density (g cm^{-3})
1	0.12
5	0.23
9	0.27
15	0.31
24	0.36
31	0.37

Fig. 11.3 Seasonal variation in average snow density in certain places in North America. Redrawn after McKay & Thompson (1968: Fig. 6).

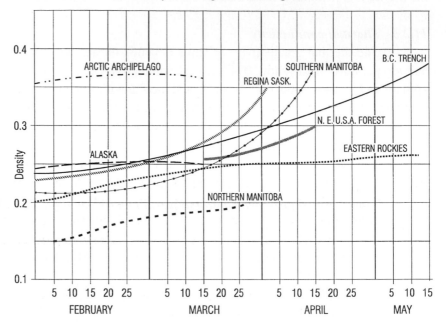

Fig. 11.4 Proctor penetration resistance (for a 1 cm² plate) of snow in relation to its density and temperature. Redrawn from Mellor (1963: 544) obtained from Agar. Reproduced with permission of © MIT Press.

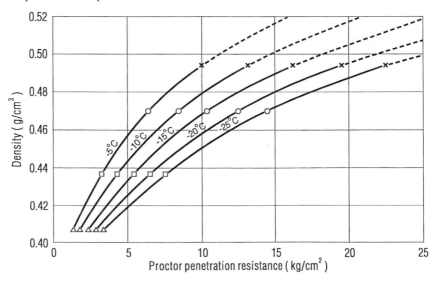

apparent thermal conductivity of snow at a density of 100 kg m^{-3}, but only 8% at a density of 500 kg m^{-3}. For low-density dry snow ($<$0.35 g cm^{-3}) Abels' formula (Mellor 1963), $k = 0.0068\rho^2$, where

$k =$ thermal conductivity in cal cm^{-1}s^{-1} °C^{-1}
$\rho =$ snow density in g cm^{-3},

appears to be satisfactory, but for higher density snows as in Greenland and Antarctica the equation of Kondrat'eva seems to be preferable (Mellor 1963: 551):

$$k = 0.0085\rho^2$$

Because of the complexity of the heat transfer processes, the thermal conductivity of snow is generally taken to be an 'apparent' or 'effective' conductivity k_e to embrace all the heat transfer processes (Langham 1981: 295). The effective thermal conductivity of snow increases when the density of snow increases (Fig. 11.5)

Fig. 11.5 Approximate relation between effective thermal conductivity of snow and ice and density. Redrawn after Langham (1981, Fig. 7.7).

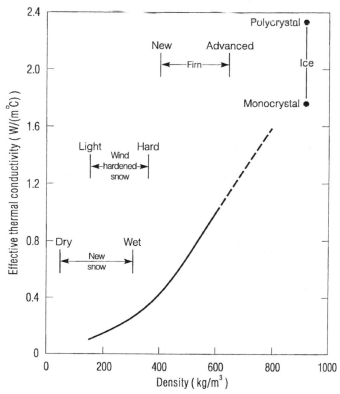

and the increase is faster the higher the temperature. We should note that also the thermal conductivity of dense snow is very low compared to that of ice or liquid water; and that is why snow is a good insulator (Mellor 1977; Reimer 1980; Langham 1981).

The thermal conductivity of snow is poor although it increases rapidly when the density of snow increases (Geiger 1961; Table 11.3).

Different investigators have measured somewhat different values of thermal conductivity, but the general trend is the same: that when density increases the thermal conductivity increases also (Fig. 11.6). When the density of snow is

Table 11.3. *Density of snow and thermal conductivity (Geiger 1961).*

Density	Thermal conductivity cal cm^{-1} s^{-1} °C^{-1}
0.1	0.00005
0.2	0.0002
0.3	0.00045
0.4	0.0008

Fig. 11.6 Temperature gradient through snow cover in central Alaska. Redrawn after Johnson (1953).

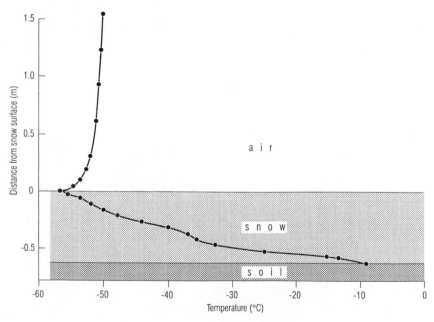

doubled the thermal conductivity will increase by roughly speaking four times. With age, morphological change in snow structure is such that surface area or surface free energy is minimized. Coarse grains grow at the expense of fine grains. Thus the thermal conductivity of snow increases with age not only because of increased density but also because of decreased path length (Reimer 1980: 131).

A knowledge of the physical properties of snow is essential background to an understanding of frost formation. The importance of the thermal conductivity of snow has to be borne in mind when considering the penetration of cold into the underlying ground. The thermal characteristics of snow are very important factors affecting the soil temperature and permafrost formation.

11.2.4 Snow as insulator

As an example of the insulating effect of snow in an old spruce forest in NE Finland (a non-permafrost locality) just south of the Arctic Circle, the freezing temperature sums (degree-days) for three months (January–March, 1985) were as follows (Havas & Sulkava 1987) (Fig. 11.7):

- Air at the height of 2 m   - 2000 °C.days
- At ground surface under a snow cover of 30–50 cm   - 1000 °C.days
- 10 cm below the ground surface   - 400 °C.days

The snow and humus layer insulates the ground effectively from the cold. The surface snow is the coldest throughout the winter. The snow near the soil is never very far below freezing (0 °C) (Johnson 1953; Reimer 1980: 133) (Fig. 11.8). The thicker the snow cover the less are the temperature fluctuations in the ground. This was well demonstrated with some temperature measurements on a palsa mire

Fig. 11.7 Snow structure, temperature and illumination in Oulu, northern Finland (Havas & Sulkava 1987).

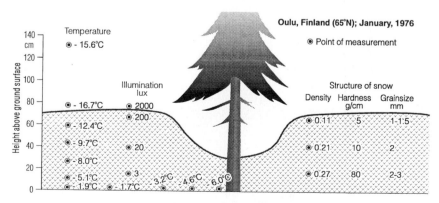

in northernmost Finland (Seppälä 1994). On the upper surface of a palsa from where the snow is blown off the minimum temperature was $-23.4\,°C$ and it stayed below $0\,°C$ over six months, while at the bottom edge of the palsa where over 1.5 m thick snow was collected the temperature under the snow stayed a constant $+0.1\,°C$ over six months (see Fig. 15.1). About 1 m thick snow cover with density of 0.3 is enough to insulate the ground almost completely from temperature fluctuations even when the air temperature drops below $-40\,°C$ (Seppälä 1983a). That is why the plants and animals in the Arctic find protection against cold under the snow (Pruitt 1984).

Mawson's (1988: 60) diary note from Antarctica tells about the practical meaning of snow: 'The snow which has banked up round the Hut lately is keeping the Hut much warmer – there is no difficulty in keeping the temp above 40 °F' (4 °C).

Thermal diffusivity, κ, is defined (Langham 1981: 297–298) by the expression

$$\kappa = k\rho C,$$
where $k = $ conductivity
$\rho = $ density
$C = $ mass heat capacity (specific heat).

Fig. 11.8 Attenuation of the amplitude of a diurnal temperature wave propagating into snow from the upper surface. Redrawn after Langham (1981, Fig. 7.8).

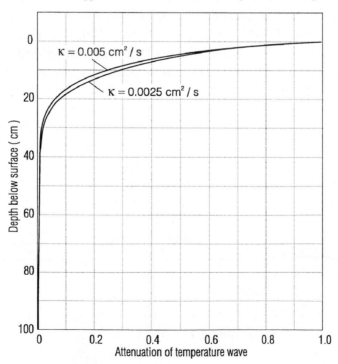

Values of $[\kappa]$ for snow lie between 0.0025 and 0.005 cm^2 s^{-1}. The thermal diffusivity is important because it governs the speed and attenuation of temperature waves propagating into snow from the surface (Fig. 11.8). If we assume that the snow cover has a constant thermal diffusivity (κ) i.e. the snow layer is homogenous, then a diurnal surface temperature wave which is assumed to follow a cosine function does not reach 1 m depth below the snow surface. Within a layered snow cover the $[\kappa]$ is not constant, which affects the attenuation of the temperature wave, too (Fig. 11.8).

In the atmosphere, strong positive vertical temperature gradients (inversions) can and do occur, while strong negative gradients cannot exist. In snow, both strong positive and negative vertical gradients occur, and negative gradients are most common. This means that forces holding air in a snow matrix are strong enough to prevent overturning (Reimer 1980: 134).

11.2.5 Gas transfer in snow

In the upper snow layer a certain amount of ventilation takes place, which may result from natural convection or from variation in atmospheric pressure and winds which force air through the snowpack (Langham 1981: 295). Wind thus affects the heat transfer in snow.

Heat flux in snow is found to be a strong function of wind speed. Strong thermal gradients in surface snow layers are observed only during light or calm winds (Reimer 1980: 129). Packing of snow, hard layers and ice in snow inhibit ventilation but will increase the effective thermal conductivity.

Snow may also contain other gases than pure air and water vapour. Snow stores and traps, for example, carbon dioxide discharged by animals, plant roots and soil micro-organisms underneath the snow layer. CO_2 is 1.5 times heavier than air and it is accumulated in the snow layer and especially in depressions in topography under the snow. There have been measured three times higher values (>1000 ppm) in a spruce forest in NE Finland (Havas & Mäenpää 1972: 11) and even four to five times higher values than in the air in black spruce-mires in Canada (Penny & Pruitt 1984). For example, rodents may cause locally 16 times higher CO_2 contents under snow (Aaltonen *et al.* 1985: 66). Relatively high temperatures at the ground surface under the snow increase the metabolism. The ventilation of snow by wind has been experimentally observed by adding CO_2 under 75 to 90 cm deep snow, and its content decreased rather rapidly (0.5–5 h) to the normal level (Aaltonen *et al.* 1985: 65). Thick and dense snow is more impermeable in this sense than thin and light snow. Despite low temperatures the accumulation of CO_2 in snow is great, lowering the pH of meltwaters, and the acidity affects the effectiveness of limestone solution in karst weathering (cf. Boyé 1952; Corbel 1959; Ford 1993). The solubility of CO_2 is inversely proportional to temperature, doubling as water is cooled from $+20\,^\circ$C to $0\,^\circ$C.

11.2.6 *Hardness of snow*

Various penetrometers have been employed to measure snow hardness in the field (e.g. Klein 1949: 111–112). Most of them apply load at a fairly rapid rate, and the resistance to penetration is usually referred to as the snow's hardness (Mellor 1963: 544). The hardness responds to changes of density, temperature and grain structure. Small grains can be packed more tightly. High density and lower temperature mean greater hardness.

The precise physical significance of snow hardness tests is not easy to interpret, but they can be usefully correlated with ultimate strength, bearing capacity and deformation resistance (Mellor 1963: 544) and this is important when expecting snow drifting by wind. Snow is usually carried along with the wind before finally settling. As a result the surface is not soft and smooth but is covered with hard sastrugi and snow dunes (Mellor 1963: 528–529).

The density expresses the proportion of ice, air and water in snow. The mobility of snow is due to how the snow crystals are compacted and sintered together. The lower the density the looser the texture. The bonding together adds considerably to the structural strength of the snow. When the snow accumulation takes place with wind then it is easily piled, forming snow patches, sastrugi and snow dunes and filling depressions as well as being deposited on the lee sides of mountain crests and fell summits.

12

Drift of snow

Deposition of snow is strongly influenced by wind. Redistribution of snow by wind is a major feature of cold environments in Polar and Alpine regions. Three different types of snow transport are: creep (ground drift), saltation and turbulent diffusion (Mellor 1965: 5). Creep means that the snow particles are moving along the snow surface. In saltation they jump up, like sand grains, and then drop back. As the wind speed increases, particles are ejected from the snow surface and carried some distance downwind before they splash back into the bed and dislodge other particles (Haehnel et al. 1993). Wind tunnel studies have shown that the saltation layer is only a few cm thick but the jumps (leaps) can be 10–100 cm high. In the field, saltating particles (Fig. 12.1) stream along near the surface in a 10 to 20 cm layer (Mellor 1965: 6). Turbulent air currents with vertical velocities can sweep the snow particles much higher into the air, up to 100 m altitude. Basic questions when speaking about snow drift are: (1) the quality of drifted snow; (2) conditions during the drift; (3) the amount of snow drifted; (4) accumulation; (5) the abrasion effects of moving snow.

12.1 Quality of drifted snow

Drifting snow is an ubiquitous phenomenon in cold regions. Although, in many respects, the process of snow drifting is similar to sand blowing mechanisms (Bagnold 1941), it should be recognized that during the movement the mass, shape and other properties of a snow crystal change significantly, whereas a sand particle is transported much further before becoming worn. Blowing snow particles are usually fragmented original crystals, except during the original snow fall. Snow can change its state from solid to gas by quick sublimation while moving and this is why snowstorms cannot reach great heights. In snow storms evaporation of snow occurs and the fine snow-dust lifted high does not fall back on the surface (Dyunin & Kotlyakov 1980: 293). Individual particles of snow lying on the surface are initially caused to move by the drag force (i.e. downwind force) exerted on them by the moving air. The drag force per unit area (plan) is the shear stress (τ) between the moving air and the snow surface. Before the snow drift can occur, the shear stress must exceed some threshold value sufficient to overcome the resisting forces. The threshold value of shear velocity depends on the size, shape and weight of each snow particle and on the

Fig. 12.1 Paths of saltating snow grains. Wind speed at 1 m height 4.8 m s^{-1}. Photographed by Daiji Kobayashi.

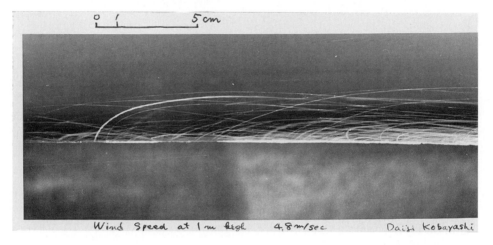

Fig. 12.2 The hardness of the snow surface at a temperature of $\sim -15\,°C$ on threshold shear velocity (based on Antarctic field data reported by Kotlyakov (1961), obtained from Kind (1981: Fig. 8.4)).

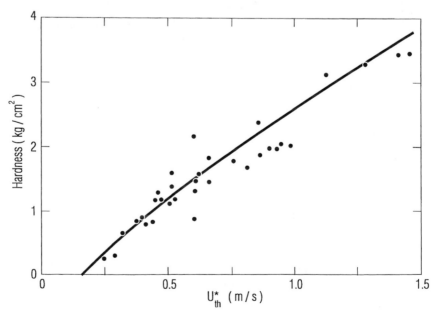

cohesive forces. The diameter of snow particles is commonly about 0.5 mm and their threshold velocities may range from about 0.1 to 0.2 m s^{-1} (Kind 1981: 343). The threshold velocities of wet and hard snow are much higher rising up to 1.5 m s^{-1} (Kind 1981) (Fig. 12.2). Ishimoto and Takeuchi (1984:104–107) applied the Snow Particle Counter (SPC) designed by Schmitt to measure the number, velocity and size of blowing snow particles. The SPC is based on two light beams. Two phototransistors detect the shadow of any particle crossing these beams. The amplifier installed in the arm of the SPC connects the two phototransistors in such a way that a particle passing through the first beam produces a positive pulse at the first window and a negative pulse at the second window. The distance between the two windows and observed time interval between positive and negative pulses provide the horizontal speed of the particle. The amplitude of the pulse depends on the size of the shadow, which corresponds to the particle size. Wendler (1987: 264) measured photoelectrically the number and sizes of the snow particles in Adélie Land, East Antarctica. The technique is very useful because it does not disturb the air flow as do mechanical devices (traps). Grain size ranged from about 150 to 300 nm.

12.2 Transport

The normal size snow particles move, mainly by saltation, close to the snow surface. The very fine snow dust can rise to great heights in the atmosphere and move suspended. Usually the largest portion of the total mass flux of the snow in movement is by saltation or ground drift and takes place in a thin layer above the surface, as reported, for example, by Oura *et al.* (1967), Kobayashi (1972) and Kobayashi *et al.* (1988) (Fig. 12.3). This is the dominant transport mechanism involved in the erosion or deposition of snow through wind action (Kind 1981: 347). The critical wind speed (threshold velocity) at which snow is picked up from the surface for transport of loose snow is from 3 to 8 m s^{-1} (measured at 10 m height) and for a dense and bonded snow winds faster than 30 m s^{-1} are needed (Mellor 1965: 1). Stronger winds pick up more particles and they are larger in size (Wendler 1987: 265). Schmidt (1980) especially analysed with theoretical calculations the wind speeds needed to transport metamorphosed snow with intergranular cohesive bonds. However, the real natural snow is so irregular that he had to make the primary simplifying assumption that snow is formed of spherical ice particles. Schmidt's (1980: 465) conclusions are the following:

(1) The initial threshold at the time of deposition will be higher at warmer temperatures, with higher humidity, and if deposition occurs with wind. Initial cohesion increases with temperature, humidity, and contact force.

(2) Threshold wind speed will increase with time since deposition. The increase slows with time and is slower at colder temperatures. This assumes that the mechanism is similar to sintering of ice spheres, although more rapid, due probably to convection.

(3) The threshold for wind transport of a snow surface will be much lower if there is a source of particles, such as precipitating snow, snow on trees, or surface hoar, that will create initial saltation, since the particle-impact forces are much higher than wind drag on surface grains or projections.

(4) The distribution of saltation trajectory heights depends on the distribution of cohesive bonds for exposed particles. The greater the impact threshold wind speed, the higher will be the particle trajectories, not only at the threshold but also at greater wind speeds. The impact threshold will be influenced by cohesion between impacting particles and surface grains so that a given wind speed will be less effective in eroding a surface at warmer temperatures and higher humidities.

Fig. 12.3 Approximate distribution of horizontal mass flux of blowing snow with height, based on data reported by Oura (1967) and Kobayashi (1973) obtained from Kind (1981: Fig. 8.5).

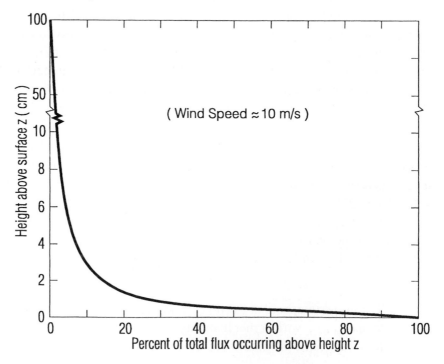

12.3   Conditions during the drift

A smooth cover of snow reduces the frictional drag and this increases the wind speed near the surface compared with a vegetation covered surface at the same site, because winds near the ground are affected by the roughness of the underlying surface (Berry 1981: 43). Mean measured snow surface roughness z_0 amounts to 1.3 mm and seems to be independent of snow surface features (Föhn 1980: 480). Several notes and descriptions of the snow drift in Arctic and Antarctic conditions can be found from the literature. In the North American Arctic 'the hard-packed wind-blown snow cover is an excellent walking surface' (Baird 1964: 57). Only after a heavy snow fall or at the late season in April is the snow soft. Very characteristic for the conditions of Antarctica are Mawson's (1988) short diary notes of 1912 which he repeated day after day:

> Blizzard heavy.
> An enormous quantity of drift has passed us.
> The air is heavily laden with snow.
> Blizzard very hard. Drifting.
> Strong blizzard – much drift.

From Franz Josef Land Luigi Amedeo of Savoy wrote (according to Nordenskjöld and Mecking 1928: 154):

> The snow never fell in large flakes, as we see at home, but was granulated, and hardened by the wind as soon as it fell, so that walking over it left no trace. It was carried by the wind like the desert sand; under a light breeze it ran along the ground, but when the wind freshened, the level of the driven snow rose to the height of several feet, and if there was a violent storm, it was impossible to know if the snow which enveloped us fell from the sky or was carried by the fury of the tempest. The snow did not lie evenly on the ground, but was piled up against every obstacle; it filled the hollow places and did not stay on flat surfaces, which made it impossible to calculate how much had fallen.

At the southern end of the Taimyr Peninsula the snow is often driven in very severe blizzards, called *purgas*. After Arved Schultz (according to Nordenskjöld & Mecking 1928):

> The purga is no snowstorm, no burgan. These snowstorms are only, as it were, preliminaries to the purga; the natives pay no attention to them and drive out on the tundra without further ado. But when the real purga sets in, i.e. when hard, dustlike snow whirls over the land and in the air, when this snow dust fills one's eyes, stops one's breathing, penetrates into the deepest folds of one's clothes, and throws down man and beast, when 'the purga thunders', when 'the dark purga has begun' – then all seek protection and safety. . . . With slight interruption the purga may continue as long as twelve days.

Wegener (1911: 345) described snow drifting in NE Greenland, on an expedition in 1906–08: When the wind velocity is about 6–7 m s^{-1} snow is drifted just some decimetres above the surface, but when the velocity reaches 10–15 m s^{-1} snow is flying in a several metres thick layer and obstructs the visibility. It means that wind with velocity of 15 m s^{-1} is counted as a storm in the Polar regions. He observed one longer lasting storm with velocity of 20 m s^{-1} and then the drift snow layer was about 15–20 m in height. Wegener noticed also that the snow drift started with wind of 6–7 m s^{-1} and it depended of the quality of the snow cover and of the air temperature. The threshold wind velocity for occurrence of drift snow at Mizuho Station, East Antarctica, was from 6 to 7 m s^{-1} measured at 1 m height above the snow surface (Takahashi *et al.* 1984a: 11). In nature the wind speeds are often so high above the threshold that the snow surface is moving fast and this is especially the case in cold Polar regions. When surface wind reaches 10 m s^{-1} it is called blizzard, and above 15 m s^{-1} heavy blizzard (Morris & Peters 1960), which means that there is a sufficiently dense and thick layer of moving snow in air to reduce visibility to near zero.

In Adélie Land, Antarctica, wind speed of 12 m s^{-1} corresponds to the appearance of blowing snow, and 20 m s^{-1} is the limit for a strong katabatic wind (Périard & Pettré 1993: 321). Wendler (1987: 265) gives lower limits for blowing snow in Adélie Land. Drifting snow was observed with wind speeds of about 8 m s^{-1}, with a speed of 14 m s^{-1} well developed blowing snow occurred, and when the wind speed reached 20 m s^{-1}, visibility went down to 20 m.

The velocity of snow particles can be greater than the horizontal wind speed (Ishimoto & Takeuchi 1984: 110). The velocity of some particles even exceeds the instantaneous maximum wind speed (Fig. 12.4). The same phenomenon, of horizontal velocities of descending particles in the saltating layer being higher than mean wind speeds, has been reported in cold wind tunnel experiments (Araoka & Maeno 1981).

Ishimoto and Takeuchi (1984: 110–111) reported field measurements showing the snow particle velocity could much exceed the wind speed. For example, the mean (10 min) wind speed was 5 m s^{-1} at the height of 80 cm above the snow surface with the instantaneous maximum speed of 8.8 m s^{-1}. The snow particle velocity was 15.9 m s^{-1} (Fig. 12.4). Generally, the heavier (larger) particles with greater inertia may be expected to have higher velocities than the lighter (smaller) particles if the descending particles maintained their large horizontal velocities. If a snow particle with higher velocity than the wind speed descended from the upper layer preserving its speed, it should have descended from 11 m above the surface, assuming a log-linear wind profile. Therefore it is unreasonable to attribute the high velocity snow particle to descending particles. The snow particles with faster velocities than the wind velocity were relatively small

(<60 mm). The weight ratio of the larger particles to the smaller ones was more than 50. The probable force to accelerate small snow particles may be the collision with other particles, even if some energy scattering occurs by destruction of snow particles. These are very basic observations when we try to understand and interpret the abrasion of rock surfaces and man-made constructions by wind-drifted snow.

The seasonal change of the snow drift flux as a function of wind velocity is explained by the air temperature. The maximum values (389–480 kg m^{-2} day^{-1}) were observed when the air temperatures were below $-50\,°$C and $-40\,°$C respectively (Takahashi *et al.* (1984a: 118). The drift flux decreased when the temperature rose above $-20\,°$C. Though we cannot fully understand this temperature dependence, the following explanations can be given. The fall velocity of snow particles depends on particle shape. The shape of drifting snow particles shows a temperature dependence, and therefore the fall velocity would depend on temperature and can cause a temperature dependence of drift flux, because the drift density is expressed as a power of the fall velocity and the drift flux is the product of drift density and wind velocity (Takahashi *et al.* 1984a: 118). Schmidt (1980) examined the temperature dependence of the threshold wind velocity for transport of snow. Air turbulence depends on temperature, seasonal variations of the roughness parameter of the surface, friction velocity

Fig. 12.4 Distribution of wind speed (dark columns) and velocity of snow particles (wide, grey columns). Redrawn after Ishimoto & Takeuchi (1984: Fig. 8). Maximum wind speed 8.8 m s^{-1}.

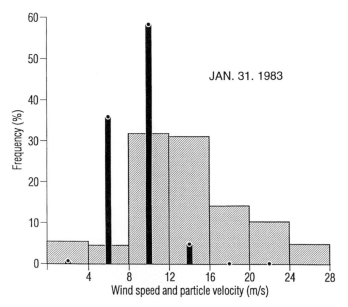

and atmospheric stability, as has been reported for Mizuho Station by Takahashi. During precipitation the drift flux increased.

12.4   Amount of snow drift

'There is one thing of special interest to the glacialist – the transportation of snow on the ice-cap by the wind. No one who has not been there can have any conception of its magnitude.' (Peary 1898: 233). Andrée (1883: 35–41) studied the amount of snow drift at Kapp Thordsen, Spitsbergen during the first Polar year in 1882–83, and reported the following results. With the average wind velocity of 6 Beaufort (about 10.7 m s^{-1}) and at an air temperature of $-10.7\,^\circ$C about 425 kg of snow per hour passed through a gate of 1 m^2. According to his observations a single snowstorm, which was 10 km broad, cleared an area of 10 km^2 with 25 cm thick snow cover in 48 hours. The density of the snow was 0.08. This was a minimum estimation because the measurements were made just up to 1 m above the surface and snow which was moving above that height was not recorded.

Snow drift transport correlates with the wind velocity. Some total transport values have been calculated and show that there is a straight regression line when transport values are on a logarithmic scale and wind speed on a linear scale (Budd *et al.* 1966: 108, Fig. 31) (Fig. 12.5).

Fig. 12.5 Snow drift transport in the layer from 1 mm to 300 m above the snow surface as a function of wind speed at a height of 10 metres (V_{10}) (Budd *et al.* 1966, Fig. 31). Dots are 5-run average transports. Reproduced with permission by © American Geophysical Union.

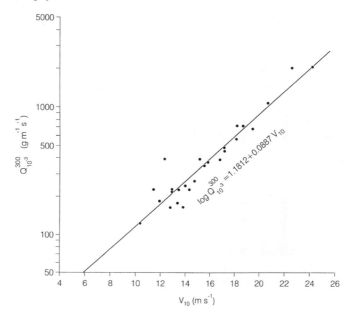

Investigations have been carried on, especially in Antarctica, with the main task of understanding the physics of the snow drift process (Mellor 1965; Budd *et al.* 1966; Loewe 1970; Dyunin & Kotlyakov 1980), the character of drifted snow (Budd 1966; Budd *et al.* 1966) and the drift quantities. Snow drift is quite difficult to measure directly by trapping because of lightness and turbulence. A rocket-type drift gauge was first constructed by Mellor (1960) and its modifications have been used by other investigators, including Budd *et al.* (1966), Föhn (1980), and Takahashi *et al.* (1984a: Fig. 2) who also used a cyclone type of collector with vanes and a slit type of collector. Takeuchi (1980: Fig. 2) introduced a snow trap with porous fabric sampling bags at different levels above each other (see also Ishimoto & Takeuchi 1984: Fig. 1). The inlet heights of the five lowest sections were at 2 cm intervals, and 5 cm for the upper four compartments, allowing measurements of the vertical profile of drift flux up to 30 cm above the snow surface (Takeuchi 1980: 484). Japanese scientists (Takahashi *et al.* 1984a) have made detailed studies of snow drift flux at Mizuho Station, East Antarctica (70° 42′ S, 44° 20′ E, 2230 m a.s.l.). They could measure snow drifted up to 30 m above the snow surface by katabatic winds. They tried to separate the drift snow values from precipitation, and there was found a good correlation between mass flux (kg m^{-2} day^{-1}) and wind velocity (m s^{-1}) at the height of 1 m on a logarithmic plot. The maximum snow drift flux (780 kg m^{-2} day^{-1}) was measured in July (the southern winter) and the lowest (20 kg m^{-2} day^{-1}) in December. This seasonal change can probably be explained by the temperature change (Takahashi *et al.* 1984a: 117). Kobayashi *et al.* (1988) estimated the amount at 10^9 kg km^{-1} yr^{-1} (10^6 kg m^{-1} yr^{-1}) at Mizuho Station, Antarctica, which is rather big in comparison with the mass transport by ice sheet flow (Takahashi *et al.* 1984a: 113). Wendler (1987: 265) had a good data base to make calculations of the snow flux in Adélie Land. The frequency was about 700 particles cm^{-2} s^{-1}, while the size of particles ranges between 150 and 250 nanometres. The frequency distribution of the wind speed year round is known. A total drifting of snow of 6.3×10^6 kg m^{-1} yr^{-1} was found which corresponds well with the values of up to 7×10^6 kg m^{-1} yr^{-1} for windy areas suggested by Loewe (1970). The flux density decreased strongly with altitude. This high snow flux has its effects on the mass balance of Antarctica, because the winds blow from inland towards the coast and large amounts of snow are lost to the oceans (Wendler 1987: 264).

Drift flux decreases with the height from the surface and it cannot be uniquely defined until the drifting snow has reached saturated drift transport (Takeuchi 1980: 487). Winds near the ground are affected by the roughness of the underlying surface. Snow cover can alter the wind speed at levels above the ground surface by changing the roughness height. The wind speed profile over relatively smooth surfaces, up to a height of about 30 m, can be represented by the logarithmic profile (Berry 1981: 43)

$$U_z = U^* \ln(z/z_0)/\kappa,$$

where

U_z = wind velocity at a height z above the surface,
U^* = friction velocity ($= \sqrt{\gamma}/\rho$ where γ is shear stress and ρ is the air density),
z_0 = roughness height,
κ = von Karman's constant (usually taken as 0.4).

Wind speed does not follow the logarithmic profile when it crosses the crest of a mountain (Föhn 1980). The mean friction velocity ($\hat{U}^*$) of snow is 0.1 m s^{-1} (Föhn 1980: 480).

12.5   Practical implications of snow drift

Snow drifting and accumulation cause serious practical problems in planning and design of buildings, roads, railways, and other constructions. In places with high precipitation, as on the Juneau Icefield, SE Alaska, the huts have a second door on the roof for the entrance in winter and spring when the camp site is totally covered by snow. Roofs have to carry heavy snow loads. Therefore on the mountain regions with heavy snow falls the houses have steep ridge roofs.

Snow drift studies are concerned especially with practical aims such as the design of snow fences (Mellor 1965) for roads and for protecting buildings, or for controlling snow characteristics in avalanche source regions. The measurements of snow drift should be focussed on terrain morphology and the turbulent boundary layer, which carries most of the suspended snow (Föhn 1980: 471).

Snow accumulations in built-up areas cause high costs in clearing works. Correct planning improves living conditions and makes life easier in snowy regions. Snow drifts around groups of buildings need field study, but studies can also be carried out with models in wind tunnels in the laboratory. Two main methods have been used to model wind action on snow: a water flume and a wind tunnel in which the snow particles are simulated by sand grains (Irwin & Williams 1985). The main task has been to find design guidelines to site the facilities in the right position to minimize snow control problems both on the ground and on roofs. Instead of snow, activated clay may be used to simulate snow in wind tunnel experiments (Tobiasson 1988: 5). The packed snow in nature tends to get hard and it freezes so that subsequent winds cannot move it, and this is difficult to simulate with other materials. Snow drifts accumulated with snow fences tend to have high density, too (Tabler 1980a: 416–417).

Attempts have been made to avoid the piling of drift snow against constructions by building houses on at least 1 m high posts, for example on the Finnish camp site at Basen, Antarctica. Construction always causes turbulence with upwind and downwind accumulation, but when the site is open the snow is drifted further than in the close vicinity of the building. The entrances of the houses are located on the sides parallel with the prevailing snow drift so that the wind channels (horse shoe vortices) keep them clear.

In the mountainous regions, for example in the Alps, Norway and the Rocky Mountains, roads have to be closed often in the winter time because of the heavy snowfall and also avalanche risks. Snow fences have been constructed to protect roads, railways and other facilities from blowing snow. Three-metre high snow fences have been rather common along the roads in northern Fennoscandia. In the United States snow fences were developed as far south as in the Rocky Mountain states, especially the flat, windswept state of Wyoming (Wangstrom 1989: 13). The basic idea with the snow fence is to cause turbulence which stops the snow drift close to the fence. The fence is porous, letting the wind blow through it. Solid snow fences also collect blowing snow, but have only one-third the collecting capacity of porous fences with a bottom gap (Wangstrom 1989: 14). The stronger the winds the further the fence should be placed from the road. Snow will be piled between the road and the fence and partly on the other side of the fence, too. 'Wind transported snow will be deposited where the wind induced shear stress is too low for transport. Because the wind induced shear stress is proportional to the velocity gradient ($\mathrm{d}u/\mathrm{d}z$), rather that the absolute wind speed, the change of wind speed with respect to elevation above ground is the important factor. Thus, the shape of the wind profile will determine where the snow is deposited and where it is eroded.' (Wangstrom 1989: 14). Tabler (1980b) studied the wind-profiles of snow fences on lake ice in Wyoming and found that 1:30 models of 1.8 and 3.8 m high snow fences can be satisfactorily used instead of full scale models.

The trapping efficiency of an empty collector snow fence is about 95%, and it slowly declines as the fence fills with snow. When the accumulation is about 80% it is in aerodynamic equilibrium with the wind, and does not create the important shear stress reducing wind profile and the trapping efficiency starts to decline to zero, and no more snow is caught by the fence (Wangstrom 1989: 14–15). This kind of mature stage of a snow fence could be avoided with higher construction and by placing several parallel snow fences at intervals.

During recent years in Fennoscandia in many difficult places with large snow drifts the road constructors have given up the building of snow fences. Instead of fences they have lifted the road some metres above the ground surface to support the wind in drifting of the snow, and it keeps the road surface rather well uncovered (Seppälä 1999).

13

Snow accumulation

Drifted snow forms similar accumulation features to sand drift. Cornish (1914) gave a detailed description of his observations of surface forms of snow in middle latitudes. Among others are snow ripples, asymmetric snow-waves, ripples in granular snow, snow barchans, erosion waves in compact snow, longitudinal erosional structures and accumulation features at obstacles and wind channels around them.

Snow patches are in many ways involved in different slope processes. They can form a passive platform for debris rolling down the slope, and when big amounts of rock material deposits and forms a ridge on the distal edge of the perennial snow patch it is called a protalus rampart (Washburn 1979: 234).

13.1   Snow patches and nivation

Snow has a significant role in geomorphic development of landforms. It is temporally and spatially at least as important as glacial ice (Thorn 1978). Wind-drifted snow is deposited on the lee sides of crests without any depression and in sheltered basins which form snow traps. The deeper the hollows on the slope the greater amounts of snow can be deposited and the larger the glaciers which can develop. Snow is collected also into rock fractures, faults and all kind of lineaments with depressions. This can be easily seen from aerial photographs in autumn and early summer time on the Subarctic and Arctic rock surfaces which appear as lace figures since snow is still filling the fractures (Fig. 13.1). Melting snow concentrates water in the weak zones and gives more support to the chemical weathering than is expected from the amount of the mean annual precipitation. Solifluction is also activated below the snow patches (Thorn 1988: 15–17).

Especially in carbonate rock (limestone and dolomite) regions we have to take into account that under snow the concentration of carbon dioxide increases because snow traps the CO_2 produced by organisms and soil. This makes the meltwaters acid which causes chemical dissolution (Williams 1949; Boyé 1952) forming solution alveoles on rocks at the snow patches. Organic debris, often dry leaves and short branches, accumulates on snow banks by wind drift, and in thawing periods accumulates at the base of nivation hollows. Its decomposition increases the CO_2 content in snow. Concentrated chemical weathering occurs on carbonate rock surfaces beneath melting snow. The etched, hackly limestone surfaces are widespread in the Arctic (Bird 1974: 716–717). Karst relief is also possible in cold environments, for example, the Lac Glacé depression in the central Pyrénées (Boyé 1952). Even in Antarctica bedrock surfaces may reach temperatures suitable for chemical weathering (Miotke 1983). Chemical weathering may be much more important at snow patch sites than has traditionally been recognized (Thorn & Hall 1980; Thorn 1988).

Snow collects in cracks, gullies and hollows along sand, silt and clay terraced river banks. As these concentrations of snow melt, small streams wash out the fine sediments, causing slumping and circular hollows to develop on the terrace side (Bird 1974: 716).

The processes involved in snow patch erosion fall into a larger category of nivation phenomena, and the hollow in which the snow lies is referred to as the nivation hollow (Lewis 1939: 153). In this context we do not want to go into the details of the weathering and erosion deepening of the nivation hollows, corries and cirques. It is a complex process accomplished by the alternate freeze

Fig. 13.1 Aerial view of snow filling of rock fractures and depressions in Baffin Island, NWT, Canada, July 1984.

and thaw of the meltwater at the receding margins of a snow patch in summer (Matthes 1900). The frost action comminutes the surface material, which is then carried away by the abundant water set free from the melting snow. This process Matthes (1900) termed nivation, which is used to designate all aspects of weathering and transport which are accelerated or intensified by the presence of late-lying snow (Thorn 1979b: 41). Nivation means weathering and erosion related to a snow patch. Thorn (1978: 424; 1988: 27) does not like the term and finds nivation confusing and an 'inadequate and obsolescent term' but he did not finally replace it with any alternative terminology.

Lewis (1939) called nivation 'snow patch erosion' and according to him a nivation hollow is a periglacial phenomenon, a 'snow niche' or 'amphitheatre' where snow patches tend to dig themselves into the basin. Rapp (1983) defined the term nivation as the geomorphological impact of late-lying snow patches which are accumulations of wind-drifted snow. Many authors have used the term nivation as a synonym for snow-patch erosion. Nivation hollows or nivation cirques are the results of late-lying snow patches. Nivation hollows are formed by snow patches and they are not only developed on rock surfaces but also on slopes of moraines, glaciofluvial and fluvial terraces. A snow patch keeps the slope underneath wet and the soil saturated during the thawing season, and this increases erosion (Rudberg 1974; Thorn 1976; Hall 1980; Rapp 1986; Nyberg 1991). Snowpatches accelerate solution and transport.

In the traditional model nivation hollows are considered as the initial stage of glacial cirque formation (Embleton & King 1975a: 218; Thorn 1988: 22; Rapp 1986; Rapp *et al.* 1986), but even so the nivation alone cannot form cirque size features. A glacial action is needed, which causes much faster erosion and effective material removement (Evans 1997). According to Thorn (1988) 'nivation is so complex that it is operationally unmanageable'. There is freezing of meltwater at night which makes frost action possible also underneath the snow patch, and material is carried away by the trickling water (Lewis 1939). Frost creep, solifluction (gelifluction), small-scale mudflows and overland flow are considered potential transport mechanisms in a nivation hollow (Thorn 1979b: 42). The weathering aspect is irrelevant or problematic according to Thorn (1979b: 42). Instead hydration (at the molecular level) might be one of the basic processes in the formation of the nivation hollows and terraces (White 1976; Thorn 1988: 12–13). Melting snow produces saturation and a shallow frozen subsurface produces a perched water table. Caine (1995) has pointed out that the indirect influences of snow patterns on erosion are much greater than the mechanical effects of avalanches.

Our task is to concentrate on the formation of the snow accumulations, which then cause the nivation hollows and small glaciers on the slopes. Where, how and why is snow packed? The orientation and location of snow patches tell us about the effects of wind. An even snow cover is rare and most snowfall is rapidly redistributed. Thorn (1978: 415) gives an example from Colorado where normal winter snowpack is discontinuous and oriented and according to him 'Discontinuity reflects the occupation of topographic lows; orientation is produced by prevailing winds'.

Russell (1896–97: 391) wrote about Mount Rainier, Washington State: 'The eastern slope of the mountain is more heavily snow covered than any other portion, mainly for the reason that the prevailing westerly winds cause the snow to be deposited there in greatest abundance. The great peak rising in the path of the moist winds from the Pacific produces something like an eddy in the air currents on its eastern side, and thus favours deep snow accumulation.' Drifting snow accumulation takes place in gullies and canyons. Thick snow deposits on the steep slopes (>20°) cause avalanches (e.g. Rapp 1959, 1960; Luckman 1977) which shape slopes and valleys. Avalanches can remove slope debris, erode gullies, deposit fans and boulder tongues. Often there are several parallel gullies, avalanche tracks, on steep slopes indicating the continuous avalanche activity.

In Hornsund, SW Spitsbergen, winds blow mainly from the eastern sector which cleans the eastern slopes of hills, and they are frequently free of snow and there are large snow drifts on their lee sides. That in turn causes a differentiation of the morphogenetic processes on both sides and an asymmetric development of the form (Czeppe 1966: 123).

Nichols (1969: 77–78) described miniature nivation cirques which are formed by small snow patches on the lee sides of river channels. In Inglefield Land, which is below the climatic snowline, perennial snow patches are in general less than 30–60 m long and are inset about one metre into the topography of the till and glaciofluvial deposits on which they rest. Meltwater from the snow patches forms small sand flows with channels, levees, small fans, etc. The volume of material moved in present-day conditions is insignificant and the distance it is moved is generally only some ten metres. The ground immediately below the snow patches is soft and soupy during the summer months. Here the meltwater from the snow patches facilitates solifluction. Nichols' description shows how the nivation takes place and how the wind has an indirect effect on solifluction and mass wasting on slopes. The snow layers deposited on the slopes keep the surficial material wet and promote down wasting.

13.2   Accumulation and redistribution of snow in the mountains

Every obstacle reduces the wind speed and produces secondary turbulence affecting the snow drift and deposition. Close to the obstacle the drifting force increases and causes erosion on the snow surface. According to Föhn

(1980: 477) wind-speed profiles do not follow the usual logarithmic form when wind crosses the crest. They show a wind speed maximum in the lowest 1 to 4 metres, increasing with higher wind speeds. The drift-density profiles have similar humps which may be caused by the flow acceleration over the top of the ridge, leading to reduced pressure and therefore increased particle lift. Another explanation for a uniform snow-filled ground layer of 1–1.5 m depth above the snow surface could be the presence of a separation layer, starting at the edge of the ridge (Föhn 1980: 477). Dyunin and Kotlyakov (1980: 289), using the different motions of snow particles, distinguished four types of snowstorms: (1) upper snowstorms, when precipitation occurs and snowflakes do not move on the surface; (2) deflation snowstorms without precipitation, already settled snow grains are removed by rolling or saltation; (3) suspension snowstorms without precipitation, elevated small snow particles rise by turbulent diffusion to the upper zones of the atmospheric boundary layer and move above the surface in a suspended state; (4) general snowstorms, when all three types of snowstorms act simultaneously. According to their observations snowstorms are especially surface phenomena and deflation of surface snow is the main factor influencing snow drifts (Dyunin & Kotlyakov 1980: 290). The upper snowstorms have difficulties in developing snow cover in the mountains because the surface deflation powerfully relocates the snow deposits and makes the snow distribution non-uniform (Fig. 13.2). Diffusion is present in snow transport, but the smallest snow particles

Fig. 13.2 Mean real snow-cover water equivalent in the upper part of Gaudergrat ridge, Swiss Alps, on two consecutive days. Redrawn from Föhn (1980: Fig. 7).

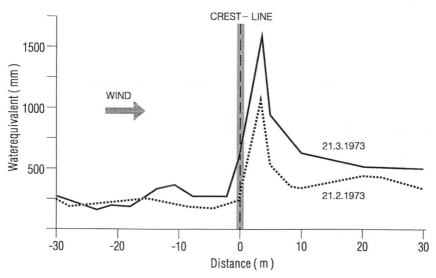

lifted to the upper atmosphere are unstable because they evaporate quickly in the turbulent air (Dyunin & Kotlyakov 1980: 291). In the mountains the roles of upper and deflation snowstorms are different. Due to the negligible rise from the surface to great heights and due to unsaturated snow flow, snow is transported for relatively small distances by deflation snowstorms and its redistribution occurs only within small basins. Due to blowing snow accumulating irregularly on the mountains, often excessive snow deposits cause snow cornices, avalanches and small corrie glaciers (Dyunin & Kotlyakov 1980: 294). On the other hand, strong upper snowstorms cause considerable macro-redistribution of snow

Fig. 13.3 Glaciers on the east side of the Continental Divide on the Rocky Mountains, Wyoming, after Enquist (1916: Fig. 13). Altitudes in feet.

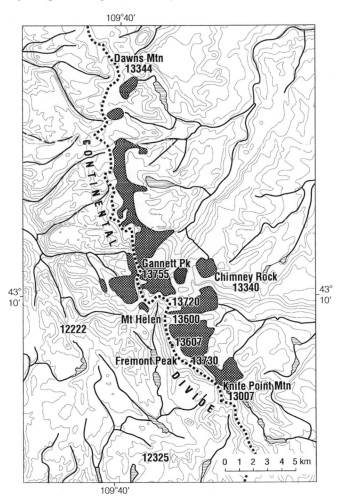

in the mountains. The topography causes non-uniform precipitation and deforms the wind flow, and as a result snow accumulates mainly on the leeward slopes of the mountain range (Dyunin & Kotlyakov 1980: 294) (Fig. 13.2). The distribution of snow patches in relation to the directions of strong winds in Spitsbergen was presented by Samuelsson (1926).

On the Front Range of the Colorado Rockies snow blown off the summit peneplain feeds the east-facing cirque glaciers (Enquist 1916). The importance of an 'orientation gradient' has been shown in both the temporal and spatial variation of accumulation and ablation in the Alpine terrain. It is a good predictor of snow distribution patterns also in the Snowy Range in southern Wyoming (Fig. 13.3). In the much less windy Bridger Range, in southwestern Montana, both orientation and elevation appear to contribute to the distribution of snow (Alford 1980). Snow accumulations and small glaciers on the mountains are rather often located on lower elevations, not on the highest peaks where it is too windy to accumulate enough snow for ice formation. Snow drift on glaciers has a great importance not only for the snow accumulation but also for snow deflation, which has a negative influence on the mass balance of glaciers and the temperature of the surroundings. Mawson (1915: 124) presented it in the following way: 'In regard to the drift, a point which struck me was the enormous amount of cold communicated to the sea by billions of tons of low-temperature snow thrown upon its surface'. There is no doubt that snow is drifted from the windward slope to the lee slope over the crest and the summit area has very little snow deposit. The areal mass balance measurements made in the upper part of Gaudergrat ridge, Swiss Alps (Föhn 1980: Fig. 7) (Fig. 13.2) demonstrate this character of snow accumulation already in moderate wind speeds (5–15 m s^{-1}). The total mass of snow on the windward slope was 61 kg m^{-2} and on the leeward slope 236 kg m^{-2}. A horizontal distance of about 200 m on both sides of the crest-line seems to delimit snow drift processes caused by the ridge, though the wind speeds were not maximal (Föhn 1980: 479, Fig. 1). In the wind shadow of the ridge a strong reverse eddy flow of the order of 50 to 150 m long results in a large depression or cavity in the snow pack. Where snow pack depth is substantially decreased, zones of increased stresses develop, often giving rise to snow-slab failure (Föhn 1980: 479).

13.3   Orientation of cirques

Cirque aspect is defined as the direction the cirque headwall is facing, i.e. the cirque headwall aspect (Evans 1977: 155). In the case of an elongated shape, the median axis might point to that direction, too. Enquist (1916: 22) studied the orientation of valley glaciers in general and came to the following conclusion:

glaciers and snow patches are frequently located on the lee side of the mountain crests in relation to the dominating snow drifting winds (Fig. 13.4). Especially important are the strong winds in the winter. They pack and consolidate snow on the lee sides and basins so that it is difficult to transport further. Topography has an important role for the winds and snow drift, too. Other factors with impact on glacier location are the amount of precipitation, temperature, elevation and

Fig. 13.4 Orientation of present-day glaciers, after Enquist (1916).

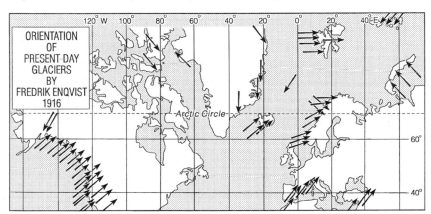

Fig. 13.5 Aspect of snow patches and distribution of winds in the Black Forest, Germany. Redrawn from Enquist (1916: Figs. 7 and 8).

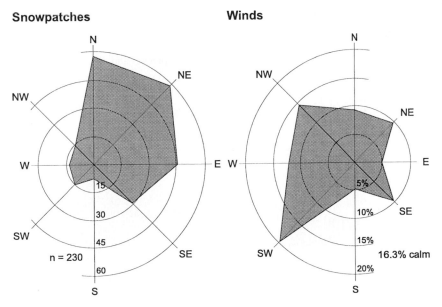

inclination of slopes. Enquist (1916: 43–44; Fig. 13) gives several good examples of glacier orientation. In the Wind River Mountains, Wyoming, there are a dozen or more small glaciers on the east side of the continental divide (Fig. 13.3). The reason for this distribution appears to lie in the better protection of cirques from the sun and also in the westerly winds drifting much snow over the crest of the range to the lee side. On the Colorado Front Range snow is blown off the summit peneplain and feeds all the east-facing cirque glaciers. Similar distribution of snow patches on the north-southeast sector was observed in the Black Forest (Schwarzwald), Germany, and the main wind directions were from southwest to northwest (Enquist 1916: Figs. 7 and 8) (Fig. 13.5).

Aleshkoff (1933) reported eastward-facing cirques in the northern Ural Mountains with predominantly westerly winds drifting the snow. He also pointed out that the snow in the cirques is protected from solar radiation by high southern and southwestern cirque walls.

Snow patches and cirques are predominantly located on the lee sides and somewhat also on the fore sides of the mountains, for example in SE Alaska (Miller 1961, 1967) and in Scandinavia (Vilborg 1977). For example, in northern Sweden this preferred east orientation means that the glacial cirques are nourished by snow blown over the mountain crest by winds predominantly from the westerly directions (SW-NW quadrant) (Rapp 1983: 99; Vilborg 1977) (Fig. 13.6). The concept of cirque orientation in Scandinavia controlled by snow drifting from westerly directions has been advocated by many authors (e.g. Enquist 1916; Mannerfelt 1945; Schytt 1959; Vilborg 1977; Rapp 1983; Rudberg 1994). The east-facing cirques of Kärkevagge, Swedish Lapland, are large and well

Fig. 13.6 The orientation of perennial snow patches and glacial cirques in Swedish Lapland. Redrawn after Vilborg (1977: Figs. 35 and 29b).

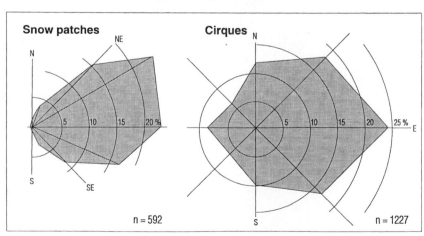

developed, the west-facing are small and vague (Rapp 1983: 100). SE is the strongest aspect of cirques of central Sweden according to Vilborg (1984: 67). He assumed that the orientation of the mountain crest has its effects on cirque formation, too, because N and S aspects are rather strong. Topography and rock structures are the basic elements to give a suitable position for drifting snow to accumulate. This seems to be the case in northern England, too. The glacial cirques in the Lake District, Cumbria, are oriented to the east and northeast, where the mountain crest is in a N-S direction and perpendicular to the main wind directions. When the crest is in a W-E direction and about parallel to the winds then the glacial cirques face to the N and S (Fig. 13.7).

Evans (1977) concluded his world-wide study of the direction and concentration of cirque and glacier aspects in the following way:

> Topoclimatic effects on glacier balance encourage shade, lee and east-facing glaciers in varying combination in different regional climates and at different altitudes. Present-day glaciers usually face slightly east of poleward on average: in Scandinavia and the Urals, consistent westerly winds strengthen the lee tendency, especially where gentle summits favour drifting of snow. Eastward cirque aspects are found in the Falkland Islands, New Hampshire and Central Spain, east-northeast aspects in the American Rockies, the

Fig. 13.7 Glacial cirques (black) in the Lake District, Cumbria, north England, according to personal information from Ian S. Evans (1995). Lakes indicated in white.

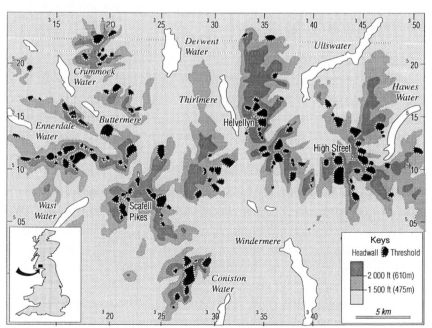

Faeroes, Central Europe, Japan and Tasmania. In Scandinavia, the British Isles and the Carpathians, northeastward resultants are more common, showing a lee effect tempered with a shade effect (in the Canadian Arctic, Alaska, British Columbia, western U.S.A., central Chile, the Alps, southwest and central Asia, the Altai and even Papua-New Guinea), both cirque and glacier aspect resultants are poleward or within 30° eastwards. This situation reflects differential ablation from direct solar radiation as a more important factor than wind.

However, most of these places belong to the zone of westerlies and if there is not enough snow for ice formation on the slope the solar radiation will melt it anyway. Direct measurements have recorded much more snow on the lee side of the mountain crests in general than on the fore-side. Heavy snow accumulation together with smaller solar radiation tends to cause glaciers. If the lack of melting by radiation was the dominant reason for cirque glaciation then the Polar aspects would be favoured. Lewis (1939: 153) studied the snow-patch erosion in Iceland and presented his observations as follows: 'The snow patches were more frequent on slopes facing N and W, but protection from the sun's rays seemed to be only one of the factors tending to preserve them, the chief factor being their bulk'.

McCabe (1939: 465) concluded: 'During the amelioration of climate, shade and the direction of moisture-bearing winds have become of increasing importance in the location of the glaciers. As the glaciers diminish, shade has become the most important factor in deciding the final refuge of the shrinking ice mass.' He does not think the original locations of snow accumulation are so much determined by the aspect.

In Scandinavia cirque formation depends on altitude, relative relief, and slope angles. 'Slope angle in this context has a lower limit below which there is no question of wind separation, and a higher level above which wind-driven snow cannot rest, except in minor parts in chutes of different kinds or perhaps local wet snow. Cirque formation depends, furthermore, on snow precipitation, and not least the annual percentage of dry snow, available for snow drifting. Other important factors are wind stability, wind strength, and temperature limits that decide the glacier movement' (Rudberg 1994: 179). McCabe (1939: 465) noticed that in West Spitsbergen all the studied corries were intimately related to the structure of the region. The floors of the corries were determined by a platform of more resistant strata and were clearly eroded along lines of weakness due to faults. The structure of the underlying rock should be considered as an important factor in deciding the form, orientation and location of snow patches, cirques, and corries (Seppälä 1975a).

It seems obvious that snow drifting decides the aspect of cirques (Rudberg 1994: 187). Wind directions may change somewhat in the same region during the development of cirques, but probably they also guide the further development and partly the old orientation of the cirque persists in later phases with a slightly different wind direction (Rudberg 1994: 187). The wind direction can be deflected locally with a further development of cirques, in both number and size.

In the northern Scandinavian mountains the NE aspect of the cirques is more pronounced, while there is a maximum in the E direction in SE Norway. Mainly the cirques are located on the N-E facing sector of slopes (Rudberg 1994: 187). Some large specimens of cirques occur with a westerly aspect.

In Victoria Dry Valley, South Victoria Land, Antarctica, some high ranges are dissected by several ice free cirques with accordant floors about 800 m above the valley floor. They are conspicuous along the southern side of the valley. The cirques have gently sloping backwalls and are mostly about 800 m long and some 400 m wide. Most cirques contain small lakes and moraine charged masses of ice lie nearer the backwall. The shape of the cirque walls is controlled by the rocks in which they are cut (Webb & McKelvey 1959: 124–125). These cirques apparently are transitional forms between the nivation hollows and mature cirques with vertical headwalls.

13.4   Snow drift and glacier mass balance

Snow drift has a great importance on glacier mass balance in general. The measured precipitation does not give accurate information for glacier ice deposition. Drift snow disturbs measurements and in windy conditions rain gauges do not record all the precipitation because of turbulence. Heavy deposition takes place in certain sheltered places and also on the firn areas of glaciers. A good example of the concentrated deposition can be found from British Columbia where snow ridges exist on the lee sides of the nunataks of the firn (personal communication from Ian Evans) (Fig. 13.8). On the edges or fore-sides of these nunataks we can see deflation lakes on snow (see section 14.2) and bare blue ice surfaces which indicate strong local erosion. When the mass balance of a glacier is to be explained we must think of the wind conditions. A calm winter with reasonable snow fall might cause less snow deposition on the firn than low precipitation with stormy weather.

Both in Antarctica and Greenland and on the mountain areas with ice caps and ice fields as in Alaska, nunataks rising above the glacier surface are free from ice and often snow-free during the summer (Fig. 13.9). The reason is strong wind drifting of snow to the lower levels.

On the lateral slopes of glaciated valleys wind deposits steep snow piles as cornices (Plate 13) which then can collapse on the glacier as an avalanche and make the glacier mass balance measurements very difficult.

Fig. 13.8 Snow accumulations on the lee sides of nunataks on Pacific Ranges, Bridge River Icefields, Stanley Smith Glacier, British Columbia (map 92J13). Interpreted from an aerial photograph.

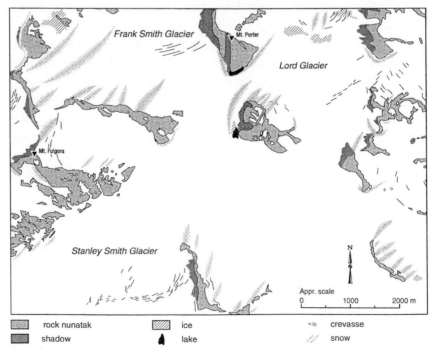

High mountains can also cause orographic obstacles for snow drift, such as in Antarctica near the coast, preventing wind-snow currents and glaciers from reaching potential open land (Solopov 1967: 4). This is thought to be one of the reasons for the dry oasis valleys in Antarctica.

13.5 Asymmetric valleys

The asymmetric shape of valleys in the Arctic is a well known phenomenon. It has been decribed, for example, from east Greenland and West Spitsbergen by Poser (1932; 1936), who explained the asymmetry with permafrost and snow accumulation differences on the slopes. Stäblein (1983) realized in Jameson Land at Scoresby Sound, east Greenland that the east and southeast facing slopes collected much snow because they are the lee sides of the NE wind. The opposite slope is worn by much stronger 'Abluation' (German) which means meltwater erosion.

There are several common interpretations of the processes involved in the formation of periglacial asymmetric valleys: differential insolation and freeze-thaw, differential solifluction, wind and snow, lateral stream erosion, with decline of N and E facing slopes and steepening of S and W facing slopes. French

Fig. 13.9 Taku Towers arête ridge formed by several cirque glaciers on both sides of the ridge by the large Taku Glacier in the foreground, Juneau Icefield, Alaska. Photograph by Matti Seppälä (1972).

(1976: 254–255) has summarized these factors from different studies. There seems to be a substantial agreement that the periglacial environment tends to give valley and slope asymmetry. Controversy has risen when old asymmetric valleys have been interpreted as signs of former permafrost environment.

Karrasch (1983) supported a model of periglacial asymmetry of relief. Wind has great impact on the asymmetry and especially snow drift which modifies the thermal conditions of the slopes. There has been some discussion and controversy concerning slope asymmetry in geomorphological text books (Washburn 1973, 1979; Embleton & King 1975a; French 1976) over what may be described as the insolation hypothesis versus the wind hypothesis. Black (1969a: 231) presented his reservations in the following way: 'Asymmetrical valleys are found throughout the world and obviously are the result of many processes and factors. Distinguishing processes inherent only to periglacial climates seems most difficult in light of our present state of knowledge.' According to Karrasch (1983: 325) the controversy can, however, be overcome by a more subtle investigation of the phenomenon which also takes into account different climatic variation factors. His conclusion is that the valley asymmetries of the Polar zone seem to be niveal-fluvial asymmetries where the wind influence with unilateral or unilaterally enlarged snow accumulations is decisive.

Relief asymmetry has a more common meaning. It refers not only to valley asymmetry but also other types of slope asymmetry, for instance a combination of corries with rock glaciers with NE aspect and smoothly curved slopes with SW aspect (Karrasch 1983: 325). Cirques and nivation hollows in fact are also examples of relief asymmetry even though they are often controlled by the rock structures (e.g. McCabe 1939; Seppälä 1975a).

In the summer months, wind action results in significant evaporation and latent heat loss from exposed slopes, which in turn influences the depth of the active layer and the dampness of the slope and this may affect the slope and valley asymmetry (French 1976: 202).

Snow accumulation in gullies activates gully erosion because the melt waters pursue the erosion. The original reason for the gully formation in a Polar climate can be melting of an ice wedge or piping along it (Seppälä 1997c).

13.6   Slushflows

Slushflows or slush avalanches are defined as rapid, channelled wet-snow movement downslope (Washburn 1979: 193–194). Slushflows characteristically take place in first-order channels traversing valley slopes during rapid snowmelt in spring (Nyberg 1985). They are common in mountainous regions with steep gorge-like valleys which collect large amounts of wind-drifted snow (Fig. 13.10). Factors affecting the formation of slushflows (Nyberg 1985) are:

(1) great depth of snow,
(2) an ice surface beneath the snow,
(3) topographical features which facilitate damming of meltwater,
(4) rapid thawing of the snow.

Slushflows take place when the air temperature rises rapidly and probably heavy rain occurs at the same time (Nyberg 1985; Clark & Seppälä 1988; Hardy 1993). Snow in the gorge becomes saturated by water and starts to move down the valley. It brings all available debris with large boulders, and on coming to rest, leaves a characteristic depositional feature with a whale-backed fan of predominantly non-sorted debris at the mouth of the gorge (Washburn 1979; Nyberg 1985). After snowmelt several stones on the fan can be piled on each other in a very unusual way (Clark & Seppälä 1988).

Drift snow can totally fill deep creek furrows which have waterfalls during the summer time. Those places are especially favourable for slushflows (Nyberg 1985, 1989; Clark & Seppälä 1988; Hardy 1993) which start to move down along the channel after heavy warm rains and/or sudden rise in air temperature. Due to

Fig. 13.10 Deep snow filled furrow of Skirhasjohka creek in Kilpisjärvi, Finnish Lapland. A typical site for slushflows (Clark & Seppälä 1988).

high radiative input in Polar regions, the transition from winter to summer occurs abruptly, often without spring. This gives hydrological processes enormous strength for a very short period when the snow melts. Slushflows and floods transport significant amounts of debris and have a marked effect on denudation in the mountain areas. Slushflows take place repeatedly at the same sites and it seems that they have continuously fresh rock debris to transport, and this gives the impression that the snow patch margin is particularly susceptible to freeze-thaw weathering (Gardner 1969; Thorn 1979a, 1979b; Thorn & Hall 1980). This may partly also explain cirque development (Gardner 1987, 1992).

14

Deflation of snow cover

14.1 Erosional features of snow

Strong winds without new snow fall wear snow deposits into holes and bumps (Fig. 14.1). Because wind-drifted snow is very compact its uneven microrelief can have very steep slopes.

One way to observe the amount of erosion of snow cover is to observe footprints or sledge tracks. Seligman (1980: Fig. 179) gives a nice example of fox spoors in snow which indicate the different erodibility of soft and slightly hardened snow. The present author has seen similar features in Lapland on reindeer hoofprints which were about 20 cm above the eroded snow surface. When the reindeer ran on the soft snow its cloven hooves pressed the snow surface down at least 20 cm and the new position of these footprints indicated that at least 40 cm thick layer of snow had blown away from the path (Fig. 14.2).

Sastrugi (see section 11.2.6) is a scalloped snow surface caused by strong wind. Over Low Dome, Antarctica, an ice cap 200 km in diameter, the constant strong winds provide a constant sastrugi direction which is often an aid to navigation in bad weather (Potter 1987: 10) (Fig. 14.3).

If the snow layer is hardened and stratified by ice layers, then the erosion of the snow surface becomes very pronounced and leaves continuous erosion ridges (Seligman 1980: 231) and it is called *skavler* (a Norwegian word meaning waves of the sea). These features have no direct geomorphic significance as such, but they indicate the drift of snow and thinning of snow cover which means less insolation for the underlying soil.

After settling, snow crystals become bonded together so rapidly that the wind has little chance to move these larger particles (Schmidt 1980: 453). When wind commences again to move an old snow surface, protrusions at the surface are usually much less fragile than just after a calm snow fall when they are merely interlocking of fresh snow crystals. Dry and cold snow is light material and the crystals are small. It is very easily transported by weak winds. Wind packs snow and increases its density. Heavy snow is hard to erode and drift is less active.

Fig. 14.1 Eroded snow surface with 'snow waves'. Palsas in the background.

Fig. 14.2 Reindeer hoofprints on eroded snow surface in Lapland. March 1985.

Temperature affects the drifting capacity of wind. High temperatures sinter the snow crystals together in bigger crystal clumps and drifting is restrained. Higher wind speeds are needed to deflate sintered snow. At snow temperatures greater than about −2 °C the cohesive forces are negligible in newly-fallen snow, and after a little movement the large snow particles are broken into smaller ones (Kind 1981: 343). If snow is deposited by wind drift, the particles are typically small ice grains, and cohesive bonds between them must be broken to start movement again. If the bond strength has increased so that inter-particle bonds cannot be broken by the initial wind, then a higher threshold wind speed for further transport is needed (Schmidt 1980: 453–454). This means also that snow particles deposited after wind drift on the surface cannot be used to determine the velocities of the drifted winds in the same manner as we can do with sand grains. Another difference of snow compared with sand accumulation is that less time is required to attain equilibrium. Snow moves quicker and easier and is then rapidly compacted into certain forms.

A pass between two mountains forms a wind funnel where the speed of wind increases when penetrating through it, and this prevents snow accumulation and clears the pass of snow (Seligman 1980: 218).

Fig. 14.3 Sastrugi surface in Queen Maud Land, Antarctica. Two different wind directions can be seen from the patterns. The dark spot on the left is the shadow of a helicopter. January 1989.

14.2   Deflation lakes on glaciers

Alongside nunataks and moraine ridges on glaciers there have been found minor marginal lakes which, according to earlier explanations, may have developed owing to reflected heat from the rock (summarized by Embleton & King 1968: 424; 1975a: 537).

This type of lake has been briefly described, for example, from Iceland on Vatnajökull (Koch & Wegener 1930: 395–399). Three of them were located along a moraine close to Esjufjoll. They had drained partly through the ice and it had left silt and sand deposits on the edges and coloured the ice blocks at the bottom. One more lake was reported at the lee side of Snjofjall (Prestfjall). It was about 40 m deep in an empty hollow below the glacier surface. The depth of the lake had been 5–10 m but in one place it touched the soil below the glacier and there the lake depth had been up to 20 m.

These small 'moat lakes', sometimes empty, sometimes filled with water (Fig. 14.4), are also encountered on the edges of glacial firn areas of the Juneau Icefield in southeastern Alaska (Seppälä 1973b). One of the oval-shaped lakes

Fig. 14.4 A basin formed by deflation, Lake Linda on the Lemon Creek Glacier, SE Alaska. The long axis of the basin is parallel to the effective wind. There is some water at the bottom (Seppälä 1973b).

was bordered for the most part by snow and ice, its long axis about 300 m long runs in SE-NW direction parallel to the prevailing wind. It is 60 m broad and 25 m deep (Fig. 14.5) with a maximum dip of slopes of 60°. Another lake facing south under the steep, snow covered nunatak slope (Fig. 14.6) was about 550 m long, had a maximum width of about 120 m and was about 12 m deep. The maximum dip was about 45° (Seppälä 1973b: 267–268). The glacier flow tried to fill this basin.

The origin of these lakes is explained as deflation basins. In the first case the drift takes place along the side of the moraine and rock surface and in the second case the steep slope perpendicular to the strong winds causes eddying currents and turbulences and wind channels with less snow accumulation (Fig. 14.6). The depression of the glacier surface then collects meltwater from the surface of the glacier. Such a pond enlarges its basin by melting the walls and floor by the heat of the water. Melting lumps of ice and snow rise from the bottom and float in the water. Meltwater draining from the nunatak or moraine carries with it considerable quantities of sediment, which make the water muddy and at the same time dirty the floor of the lake and the ice and snow floating in the water. The lake may be drained catastrophically. A distinct dirty area is then left on the

Fig. 14.5 An empty deflation lake (Salla Lake, Fig. 14.6) on the Ice Basin of the Taku Glacier, SE Alaska. View looking down the slope of the nunatak. (Seppälä 1973b).

lake basin (Fig. 14.7). The sediments decrease the albedo and ablation increases. All these events together deepen the lake basin originally caused by deflation of snow. Seppälä (1973b: 273) concluded that 'too much importance has been hitherto attached to the influence exerted by heat reflected from nunataks'. In the study area of Evans (1990) on the glaciers of the Southern Coast Mountains, British Columbia, Canada, many similar lakes are found at the feet of nunataks (Fig. 13.8).

By the steep or almost vertical edges of nunataks of Vestfjella mountains, Dronning Maud Land, Antarctica, we have recognized several depressions with frozen lakes (cf. Jonsson 1988). They are partly formed by melt waters running down the slopes and caused by strong radiation, but it seems to be obvious that deflation on the snow surface plays an important role in their formation, too. Firn snow depth in Antarctica is much bigger (up to 100 m) than in Alaska.

In Antarctica the formation process of lake basins is partly abrasion of the ice bed, too. In Mawson's (1988: 62) diary we can find an interesting observation:

Fig. 14.6 Position of a deflation lake (Salla Lake, Fig. 14.5) on the Juneau Icefield, SE Alaska. Profile line A-B approximately in the direction of the effective wind. Redrawn from Seppälä (1973b, Figs. 4 and 6).

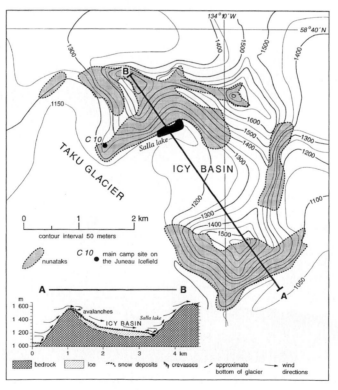

'Abrasion has undercut sastrugi of accretion and erosion. 18 ins (45 cm) of ice went from the glacier in late summer by semi-melting, i.e. granulation and blowing away.'

Seligman (1980: 214–217) already in 1936 described the mechanism of the formation of the basins by the steep cliffs though he did not mention any lakes filling them. According to Seligman, wind stream funnelling by the obstruction causes an upward spiral eddy and snow is carried back into the main air stream and removed. When wind blows against a steep or vertical cliff, the snow is hollowed out from the foot of the cliff. The hollow is called an eddy-pool. Allen (1965: 336–337) studied small scour marks in snow and pointed out that since the eddy occurs in a zone of generally compressed stream lines, vortices in the eddy could be sufficient to prevent snow deposition and cause actual erosion, thus leading to growth of the furrow containing the eddy. Wind erodes a crescent-shaped furrow to the front and sides of the resistant mass, and deposits to the lee of the mass a ridge of snow. The exact same process takes place on a large scale in the accumulation areas of glaciers at the nunataks. Nunataks with steep slopes function in the same way as snow fences (cf. Anno 1984: Fig. 11).

Very spectacular wind-formed snow scoops have been found at the edges of nunataks, for example in Vestfjella, Antarctica (Plate 14). They are formed by the

Fig. 14.7 Debris at the bottom of an emptied deflation lake (Salla Lake, Figs. 14.5 and 14.6) on the Juneau Icefield, SE Alaska (Seppälä 1973b).

echo effect (cf. Tsoar 1983). Sand and snow accumulations are found in front of or behind topographic obstacles such as cliffs, shrubs, boulders, etc. Obstacles deform the wind direction and velocity around them and a wind channel is formed. We should separate the effect of the obstacles such as cliffs parallel to the wind direction from those which are transverse to the wind direction. It depends also on whether the cliff is vertical or inclined. In both cases marginal basins can be formed (Seppälä 1973b).

15

Snow and frost formation

Wind drifts snow from place to place, eroding and accumulating (Seppälä 1990). Because snow is a very good insulator it can protect the underlying ground from cold in winter and from warmth in summer. Under a thick layer of snow the temperature is almost constant, as has been observed on a palsa bog (often called a mire) in Finnish Lapland. Under about one and a half metres of snow by the edge of a palsa the temperature remained constant (+0.1 °C) for a period of six and a half months (Seppälä 1994: 286–287) (Fig. 15.1). The snow was piled by winds at the edges of the palsa, making the relief more even (Fig. 15.2). In that kind of place the frost depth in the peat is only some 20 cm (Seppälä 1990).

Thin snow cover means more chances for cold to penetrate into the soil and rock. The palsa surface itself was almost uncovered by snow caused by deflation of the snow surface (Fig. 15.2). The same feature can be observed on a smaller scale on the fields of earth hummocks called pounus (Seppälä 1998; van Vliet-Lanoë & Seppälä 2002) (Fig. 15.3). At the beginning of winter snow cover has very much the same thickness all over the ground but then the wind starts to transport it. This has been clearly demonstrated with the observations on a string bog in northeastern Finland (Seppälä & Koutaniemi 1985: Fig. 17; Koutaniemi & Seppälä 1986) (Fig. 15.4). Freezing in the early winter tends to proceed more rapidly in the pools than it does in the strings with greater accumulation of snow. This situation evens itself out in the course of winter as the snow smooths over the small-scale relief of the mire surface (Eurola 1968, 1975; Koutaniemi & Seppälä 1986: 66). Wind smooths the snow surface. At the end of January the strings could hardly be identified under the snow cover. The freezing of the ground did not continue in this stage any more and the ice cover on pools between the strings under the snow could thaw because of the ground water flow through the strings (Fig. 15.4).

Wind strongly affects the snow cover and its physical characteristics. The density of snow determines the frost penetration in soil. A hard and dense snow layer has greater thermal conductivity than soft and light snow. In this way wind affects the depth of freezing of the ground.

Wind-drifted snow accumulates also on certain parts of riverbanks and in gullies and affects ground freezing there. When ice break-up and flooding

Fig. 15.1 Temperature fluctuations at a palsa: A. on the surface, B. at the edge, in Finnish Lapland. The profile at the bottom indicates the positions of the instruments. (Seppälä 1994). © John Wiley & Sons Limited. Reproduced with permission.

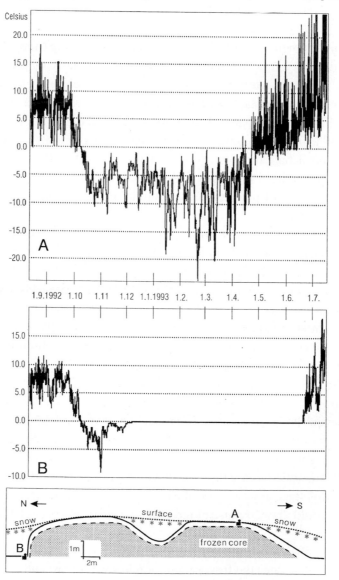

Fig. 15.2 Palsa surface in March, 1985.

Fig. 15.3 Snow conditions on peat hummocks (pounus) with height about 65 cm.

Fig. 15.4 Development of snow surface (grey) and frost (black) on a string mire in northern Finland (Seppälä & Koutaniemi 1985: 17). Reproduced by permission of © Taylor & Francis AS.

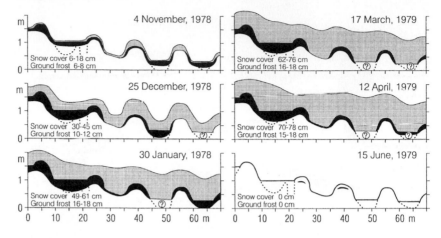

Fig. 15.5 Fell summit at Paistunturit, Finnish Lapland.

starts the frost-free banks erode easier than the deeply frozen sites (Koutaniemi 1984).

15.1 Effects on permafrost formation

Different materials have different thermal conductivity. Mineral soils freeze easily and thaw deeply in a short time, if they are not dry. Peat and other organic materials behave differently in the cold. Dry peat has a very low thermal conductivity because it contains much air. Wet saturated peat can have five times higher thermal conductivity and frozen peat even 28 times higher than dry peat (Washburn 1979: 177). Permafrost and palsa formation is based on this character of peat. Wind blows snow off freezing mire surfaces and frost penetrates deep into the ground. When thawing starts in late winter then the sublimation and wind dry the surface of the peat very fast, thus slowing down thawing and so permafrost can be formed.

It has been found that in Lapland new permafrost was formed in small peat hummocks (pounus), less than one metre in height and two metres in diameter (Seppälä 1998). The reason might be strong winter winds which kept the hummocks snow-free.

Small bushes, birches and willows cause turbulence and uneven snow cover is formed by wind. This may be the initial reason for peat hummock formation by frost heave.

Reindeer dig up lichens and make holes in snow and this may have effects on microtopography because holes collect cold air which can freeze the ground deeply. Upheaval of the ground in these limited areas may also start the hummock formation which is then carried further by wind blowing snow off the hump surface. So far we have not found any reports of this feature in the literature.

King (1984; 1986) has shown very well how the fells (low mountains in Scandinavia called fjäll; fjell) have in their summit areas permafrost in the bedrock. The same feature has been recognized in Finnish Lapland: there is permafrost in bedrock above the tree line (King & Seppälä 1987, 1988; Seppälä 1997a) due to the strong winds blowing snow off the summits of fells (Fig. 15.5). However, the summits of fells get a snow cover before the valleys (Fig. 15.6), it stays thin through the winter and does not prevent deep frost penetration. In spite of the height of fells the mean temperatures in winter can be higher than in the valleys. Cold air drains down in the valleys and the inversion between the summits and valley bottoms some 250 m below can often be 20 °C (Tabuchi & Hara 1992; Barry 1992: 180, 218–219).

Even very limited small openings in the snow cover mean deeper frost penetration (Fig. 11.7). Permafrost has been noted underneath spruces in Alaska and northern Sweden (Viereck 1965; Kullman 1989). Spruce branches collect much snow and therefore the ground surface on the roots is almost uncovered. Wind can also form a wind channel in the snow around the tree (Plate 1) and that supports the frost penetration under the tree.

15.2   Palsa experiment

Experimental geomorphological studies at a natural scale are often impracticable. The time scale, for example, in weathering processes should be changed as is done in frost shattering studies in the laboratory, when one freeze-thaw cycle takes 24 hours (e.g. Lautridou & Seppälä 1986). Deflation measurements in the field are rather easy to arrange, as explained earlier. The problem is that we are not able to control the process.

The aim of the experimental palsa study carried out in Finnish Lapland (Seppälä 1982a, 1995b) was to test the hypothesis (Fries & Bergström 1910) that palsa formation is triggered when wind turbulence is responsible for thinning the snow cover on certain parts of a mire, so that in places frost could penetrate particularly deeply into the peat. This caused initial upheaval of the

Fig. 15.6 Early snow on the fell by Kofjord, Norway.

surface, and during subsequent winters the hump would have a greater tendency
to become snow-free, and the thickness of the frozen layer would be increased.

An experimental area on a palsa mire was cleared of snow several times during
the winter while control plots were left untouched in their natural condition. At
the experimental site, frost penetrated the peat to a depth of about 80 cm, but
reached to only about 40 cm in the control plots (Fig. 15.7). During the melting
period, the seasonal frozen soil thawed completely from the control plots by
mid-August. However, at the centre of the experimental site about 35 cm of
frost still persisted when freezing from the ground surface began again in early
October.

The experiment was continued for a further two years, after which the small
artificial palsa in the experimental plot (25 m^2 in area) had reached some 30 cm
in height. Its core remained frozen for almost eight years. However, the summer
of 1984 was very wet and the whole mire was severely flooded even at the end of

Fig. 15.7 Depth of snow cover and frost layer on the experimental site (points 1 and 2),
two control points close to the experimental site (points 3 and 4) and in a thermokarst
hollow with deep snow (point 5). The arrows and vertical lines indicate the position of
the snow clearing operations. (Seppälä 1982a, 1995a).

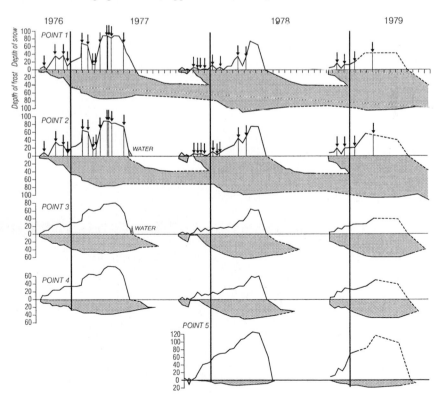

September, after which no ground frost could be detected in the artificial palsa (Seppälä 1986, 1988, 1995b).

This experiment confirms that variation in snow thickness is the main controlling factor for frost and palsa formation under Subarctic conditions (Seppälä 1990) and wind is the main agent modifying the snow depth on open mires.

16

Aeolian landforms indicating palaeowind conditions

16.1   Environmental conditions during the Pleistocene

What is the difference between present cold environments located on the high latitudes or mountain regions and ancient cold regions in connection with the ice sheets regarding their aeolian activities?

At present, globally, landscapes are more in balance with the climatic conditions, mainly covered by vegetation and humid. During the ice age and deglaciation there were enormous vast uncovered fine sediment deposits and the vegetation succession had not yet reached its climax, even where the conditions were favourable for certain vascular plants, lichens and mosses.

Meltwaters from thawing ice sheets every summer produced fresh uncovered sediments to be drifted by winds. Probably also the wide air temperature difference between cold ice masses and barren outwash surfaces caused strong winds which prevented the development of anchoring vegetation. The moving sand and silt surfaces were unfavourable for the colonizing plants. Because of the lower latitudes there was strong solar radiation during the summers, which was fatal for many plants and dried the surface layers quickly after the snowmelt. In the High Arctic the radiation is much less effective and soils stay moist most of the time even though the precipitation is low, but also evapotranspiration is less effective. Sublimation takes place especially together with frost in autumn but is seldom enough to give loose particles to form sand dunes. In the about 7000–8000 years old landscapes in Arctic and Subarctic conditions in the present cold environments, the vegetation succession has reached a relative balance with the present climatic conditions. There are only occasional deflation and major aeolian events when the balance is disturbed or rivers or meltwaters produce fresh sand and silt surfaces, frequently near to glaciers.

A fully open question is why there have been ice ages on the Earth. Several hypotheses have been presented (Wilson *et al.* 2000). We know a lot about the conditions especially around the ice sheets during and after deglaciation, but the initial stage is unknown. From the morphological characteristics of the ancient periglacial regions we can conclude that during the Pleistocene winds were much stronger and frequent than today and formed large sand dune fields and thick layers of cover sands and loess. The strong winds during the Pleistocene might have caused large polynyas on the high latitude oceans, as they do today around Antarctica and in Arctic waters. From the polynyas the heat flux can be 10–100 times that of ice-covered ocean (Cotton & Michael 1994). This increased the formation of sea ice, cooled down the sea water and supported the glaciation by increasing the air humidity. This type of atmosphere–ice–ocean interaction has just recently been understood and could be one of the main processes of formation of ice sheets. Why the winds got stronger during the glaciation is still an open question. Probably the thermal gradients at the Polar fronts became steeper and then the regime worked cybernetically.

Cold climate regions change continuously and have changed in the past (Fairbridge 1972). Continental ice sheets covered continent-size areas (Flint 1971) and outside of them periglacial processes dominated, which formed many cold climate landforms which do not belong to present climato-morphological conditions in the environment where we find them today.

Aeolian deposits, loess and cover sand, are widespread and form relict layers which indicate former wind conditions and different vegetation during their formation (Koster 1988; Schwan 1988; Feng *et al.* 1994) (Fig. 16.1). During the

Fig. 16.1 Pleistocene periglacial zone in North America, redrawn after Jahn (1975: Fig. 3) and a map by Brunnschweiler (1962).

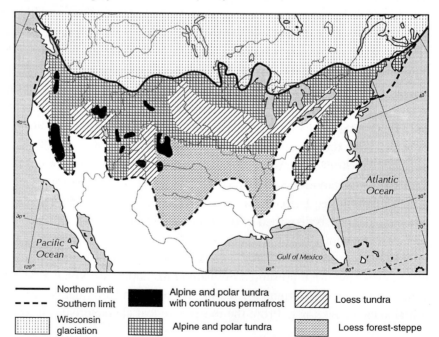

deglaciation huge amounts of loose sediments were uncovered by vegetation and subject to wind transport. Ice sheets formed special aeolian regimes. Ice sheets have continuous high pressure conditions and form anticyclones (Thorson & Schile 1995) around which the winds circulated along the ice margin. Strong katabatic winds blew down from the ice sheet but their effects did not reach far from the ice margin.

Brunnschweiler (1962) separated three different periglacial zones in North America during the Wisconsin glaciation: (1) Polar and Alpine tundra with continuous and discontinuous permafrost, (2) loess tundra, (3) loess steppe with scattered forests (Fig. 16.1) which also reveal the wind conditions during the Wisconsin glaciation. The southern limit of loess forest-steppe was up to ca. 1500 km from the ice margin and the tundra zone was some 1000 km in width.

Büdel (1951) compiled a basic map of the vegetation zones in the Würm (Weichelian, Wisconcin glaciation) period in Europe. Outside the glaciated regions were the zones of periglacial desert (*Frostschuttzone*), loess tundra, loess steppe, forest tundra, steppe and galerie forest, undifferentiated loess steppe areas, etc. Tricart (1967) used Büdel's information when limiting the periglacial regions of Europe, and all of them had loess deposits indicating also strong aeolian activity. (1) Periglacial desert with strong aeolian processes. A belt from Holland through central Europe to the Pechora River. Traces found also in uplands of west and south Europe. (2) Loess accumulation zone which was, prior to loess deposition, a periglacial desert. Covering northern France, Germany and Poland. (3) Loess forest and brush steppe. Rather continental climatic conditions existed in east Europe and in the Upper Don and Volga river basins after the glacial maximum. (4) Cold loess steppe. Major aeolian accumulation in the Pannonian and Walachian lowlands, Ukraine and Caucasus area. (5) Cold loess steppe and park forests with local frost activity in central Iberia, Yugoslavia and Bulgaria.

There is considerable evidence that wind action played a dominant role in periglacial conditions during the Pleistocene in central Europe and in North America (e.g. Tricart 1970: 143–150). Large areas with sand dunes with frosted sand grains, loess and wind-worn blocks and pebbles, and stone pavements all testify to intense aeolian activity (Troll 1948). Ice sheets produced at their margins vast deposits of fine sediments which laid uncovered by vegetation.

16.2   Past wind directions

The palaeo-wind directions of former dune-building winds can be inferred from (1) dune morphology, (2) dune sand bedding and/or (3) the disposition of dunes or dune fields with respect to restricted sources of sand (Smith 1949b: 1487).

The axes of parabolic dunes in central Europe (Fig. 16.2) and in northern Fennoscandia (Fig. 16.3) indicate that winds which formed the dunes blew along the ice margin and around the high pressure. Many investigators have made this kind of interpretation of palaeo-wind conditions, for example from central Europe (Poser 1948, 1950), northern Fennoscandia (Klemsdal 1969; Seppälä 1971a, 1972a, 1973a; Aartolahti 1976), Alaska (Lea 1990; Lea & Waythomas 1990; Galloway & Carter 1994) and New England (Thorson & Schile 1995).

The ancient periglacial sand dunes found in central Europe and North America have much greater dimensions than active sand dunes at present in cold environments in the Arctic and in Antarctica. During the Pleistocene glaciation the periglacial zone or underground glaciation (permafrost) dominated much wider areas than today.

16.3   Pleistocene abrasion

Maybe the first observations of wind abrasion and ventifact formation were described by Blake (1855). Along the eastern base of the Sierra Nevada in California, in situations where wind abrasion is now inhibited and probably ceased long ago, are found blocks with surfaces that squarely confronted the

Fig. 16.2 Late glacial summer wind directions and air pressure in central Europe. Redrawn after Poser (1950).

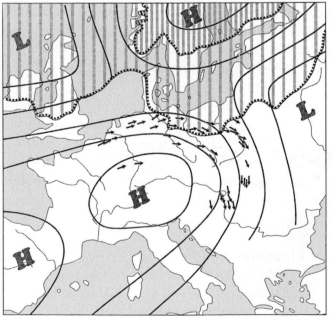

wind and are covered with pits or wells 5–10 mm wide and 10–25 mm deep. Where the wind was nearly tangential to the surface, grooves (5–15 mm) were carved out (Blackwelder 1929b: 259). The origin of the abrading winds could be the Pleistocene valley glaciers.

Woodworth (1894 – according to Thorson & Schile, 1995) considered post-glacial aeolian action in New England, Thiesmeyer (1942) in the Middle West, and Needham (1937) in New Mexico. Wentworth & Dickey (1935) collected data on ventifacts with a questionnaire, and ventifacts were well represented, for example in Wyoming, Colorado, Massachusetts, Michigan, New Jersey and New York State.

Wind-worn pebbles are reported from many localities in the British Isles (Bather 1900). Ventifacts are reported from vast areas in central Europe (Nitz 1965; Schönhage 1969; Maarleveld 1983). In southern Sweden ventifacts are common features (Johnsson 1958; Svensson 1981, 1991; Schlyter 1991, 1992,

Fig. 16.3 Directions of dune-forming winds in northern Fennoscandia observed from the axis directions of parabolic sand dunes compiled from Seppälä (1972a: Fig. 2, 1973a), Aartolahti (1976) and Sollid *et al.* (1973) (see Fig. 8.12).

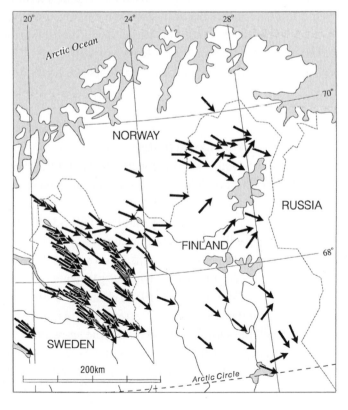

1994, 1995). Schlyter tried to reconstruct the palaeo-wind directions using the abrasion marks of big ventifacts which have not been removed during the wearing processes or after them. He found effective winds from the east, which is at present an uncommon direction in this climatic zone of westerly winds.

From present-day Polar glacial and periglacial conditions it is hardly possible to find similar intensive aeolian environments such as prevailed around the ice sheets during the Ice Age. Attempts have been made to find analogues from Alaska and Greenland (Koster & Dijkmans 1988; Dijkmans 1990) but only miniature features have been found compared with the ancient features. Probably in dry valleys in Antarctica there could be somewhat similar aeolian conditions, but there the deglaciation does not affect the sedimentation in so large a scale as during the late glacial time in Europe or North America.

An odd thing is that relict ventifacts are relatively rare in the Canadian Arctic (Pissart 1966), and Bird (1967) even talks about an absence of wind erosional features in northern Canada. Some wind eroded landforms have been reported from sandstone and silt regions on Ellef Ringnes Island (St-Onge 1965). In Alaska ventifacts of late glacial origin are not rare features (Péwé & Reger 1983: 62). A strong argument against the silt abrasion of ventifacts is that Vierhuff (1967) found that more wind-affected stones occur in stone pavements underlying sand than in pavements underlying sand containing loess. Modern observations from Arctic Canada (personal communication with J. Ross Mackay) (Fig. 5.7) confirm the idea that many of the ventifacts in cold environments are formed by wind-blown snow.

16.4   Other evidence of past aeolian activities

Aeolian filling in sand wedges and ice wedge casts (Berg 1969; Black 1976, 1983) indicates that in frost conditions open sand surfaces were also available during the wedge formation. Cailleux (1942, 1952) made studies of Quaternary periglacial wind-worn sand grains in Europe and the U.S.S.R. In the fillings of wedges it is easy to identify wind-worn quartz grains.

Cirques are clear evidence of the past wind drift of snow and there are many studies of the orientation or aspect of cirques without glacier ice (e.g. Enquist 1916; Miller 1961; Evans 1977; Vilborg 1977, 1984; Hassinen 1998). A general conclusion of these studies could be that wind directions during the ice formation in the cirques were very much the same as nowadays but the snow limit was often at much lower levels. The cirque formation of course takes thousands of years and therefore they are much rougher indications of wind conditions than, for example, sand dunes.

16.5   How to make progress in palaeoenvironmental studies

Without knowing the present-day geomorphological processes we may come to very risky conclusions when interpreting relict features and trying to reconstruct palaeoenvironmental conditions. We can present here in a flow diagram (Fig. 16.4) a general scheme of how to make, for example, sand dune studies (Seppälä 1975b; Koster 1988).

Fig. 16.4 Flow diagram of the progress of sand dune studies in the cold environment (Seppälä 1975b; Koster 1988).

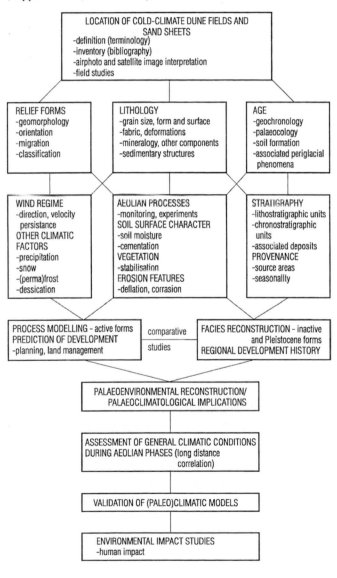

The idea is that we start from cold-climate dune fields and learn to know their geomorphological characteristics, grain composition and age. We need detailed observations of the conditions: wind regime and other climatic factors should be measured and we must avoid transferring observations which are made far from the observation site. These data should then be combined with observed aeolian processes and the stabilisation of aeolian features. The inner structures of the aeolian deposits have to be carefully observed.

The data can then be used in comparisons of relict features formed in ancient conditions and we can reach some progress in palaeoenvironmental implications.

This may be the way to avoid some mistakes in land utilization, especially in Subarctic regions where human activated aeolian processes can destroy, for example, the pasture areas of reindeer (cf. Käyhkö & Pellikka 1994; Käyhkö *et al.* 1999).

Bibliography

Aaltonen, Katri; Pasanen, Seppo & Aaltonen, Heikki 1985. Measuring system for analysing of the CO_2 concentration under snow. *Aquilo, Ser. Botanica* **23**: 65–68.

Aartolahti, Toive 1966. Über die Einwanderung und die Verhäufigung der Fichte in Finnland. *Annales Botanici Fennici* **3**: 368–379.

— 1970. Fossil ice-wedge, tundra polygons and recent frost cracks in southern Finland. *Annales Academiae Scientiarum Fennicae, Ser. A III. Geologica – Geographica* **107**: 1–26.

— 1972. Dyynien routahalkeamista ja routahalkeamapolygoneista. (English summary: Frost cracks and frost crack polygons in Finland.) *Terra* **84(3)**: 124–131.

— 1976. Lentohiekka Suomessa (Aeolian sand in Finland). In Finnish only. *Suomalainen Tiedeakatemia. Esitelmät ja pöytäkirjat* 1976: 83–95.

Abrahams, Athol D. & Parsons, Anthony J. (eds.) 1994. *Geomorphology of Desert Environments*. Chapman & Hall, London. 674 pp.

Adams, W. P. 1976. Areal differentiation of snowcover in East-Central Ontario. *Water Resources Research* **12(6)**: 1226–1234.

Ahlbrandt, Thomas S. & Andrews, Sarah 1978. Distinctive sedimentary features of cold-climate eolian deposits, North Park, Colorado. *Palaeogeography, Palaeoclimatology, Palaeoecology* **25**: 327–351.

Åhman, Richard 1976. The structure and morphology of minerogenic palsas in northern Norway. *Biuletyn Peryglacjalny* **26**: 25–31.

— 1977. *Palsar i Nordnorge* (Summary: *Palsas in Northern Norway*). Meddelanden från Lunds Universitets Geografiska Institutionen, Avhandlingar **78**: 1–156.

Åkerman, Jonas 1980. *Studies on Periglacial Geomorphology in West Spitsbergen*. Meddelanden från Lunds Universitets Geografiska Institution, Avhandlingar **89**: xii and 297 pp. and maps.

— 1982. Observations of palsas within the continuous permafrost zone in eastern Siberia and in Svalbard. *Geografisk Tidsskrift* **82**: 45–51.

Aleksandrova, V. D. 1980. *The Arctic and Antarctic: their Division into Geobotanical Areas*. Cambridge University Press, Cambridge. 247 pp.

— 1988. *Vegetation of the Soviet Polar Deserts*. Cambridge University Press, Cambridge. 288 pp.

Aleshkoff, A. N. 1933. The glaciers of the northern Urals. *Scottish Geographical Magazine* **49**: 359–362.

Alestalo, Jouko & Häikiö, Jukka 1976. Ice features and ice-thrust shore forms at Luodonselkä, Gulf of Bothnia, in winter 1972/73. *Fennia* **144**: 1–24.

Alford, Donald 1980. Spatial patterns of snow accumulation in the alpine terrain. *Journal of Glaciology* **26**: 517.

Allard, M.; Seguin, M.-K. & Levesque, R. 1986. Palsas and mineral permafrost mounds in northern Quebec. In: V. Gardiner (ed.): *International Geomorphology* Vol. 2, John Wiley & Sons, Chichester: 285–309.

Allen, J. R. L. 1965. Scour marks in snow. *Journal of Sedimentary Petrology* **35**: 331–338.

— 1985. *Principles of Physical Sedimentology*. George Allen & Unwin, London. 272 pp.

Alley, R. B.; Meese, D. A.; Shuman, C. A.; Gow, A. J.; Taylor, K. C.; Grootes, P. M.; White, J. W. C.; Ram, M.; Waddington, E. D.; Mayewski, P. A. & Zielinski, G. A. 1993. Abrupt increase in Greenland snow accumulation at the end of the Younger Dryas event. *Nature* **362**: 527–529.

Anderson, Robert S. 1987. A theoretical model for aeolian impact ripples. *Sedimentology* **34**: 943–956.

Anderson, Robert S. & Hallet, Bernard 1986. Sediment transport by wind: toward a general model. *Geological Society of America Bulletin* **97**: 523–535.

Andreas, Edgar L. 1995. A physically based model of the form drag associated with sastrugi. *CRREL Report* **95–16**: 1–12.

Andrée, S. A. 1883. Om yrsnön i de arktiska trakterna. *Öfversigt af Kongliga Vetenskaps-Akademiens Förhandlingar, Stockholm* **40(9)**: 33–41.

Anno, Yutaka 1984. Requirements for modeling of a snowdrift. *Cold Regions Science and Technology* **8**: 241–252.

— 1985a. Froude number paradoxes in the modeling of a snowdrift. *Cold Regions Science and Technology* **10**: 191–192.

— 1985b. Supplement to Anno's modeling conditions for a snowdrift. *Cold Regions Science and Technology* **10**: 193–195.

Anno, Yutaka & Tomabechi, Tsukasa 1985. Development of a snowdrift wind tunnel. *Cold Regions Science and Technology* **10**: 153–161.

Araoka, K. & Maeno, N. (1981). Dynamical behaviors of snow particles in saltation layer. *Mem. National Institute of Polar Research*, Special Issue **19**: 253–263.

Arnalds, Olafur; Aradóttir, Ása L. & Thorsteinsson, Ingvi 1987. The nature and restoration of denuded area in Iceland. *Arctic and Alpine Research* **19**: 518–525.

Ashwell, Ian Y. 1966. Glacial control of wind and of soil erosion in Iceland. *Annals of the Association of American Geographers* **56(3)**: 529–540.

Aufrère, L. 1931. Le cycle morphologique des dunes. *Annals Géographique* **40**: 362–385.

Bagnold, R. A. 1938. The measurement of sand storms. *Proceedings of the Royal Society of London*, Ser. A **167**: 282–291.

— 1941 (reprinted 1965). *Physics of Blown Sand and Desert Dunes*. Methuen, London. 265 pp.

Baird, Patrick D. 1964. *The Polar World*. Longmans, Green & Co., London. 328 pp.

Bakker, Th. W., Jungerius, P. D. & Klijn, J. A. (eds.) 1990. Dunes of the European coasts: Geomorphology – Hydrology – Soils. *Catena Supplement* **19**: 1–223.

Barry, Roger G. 1992. *Mountain Weather and Climate*. 2nd edition. Routledge, London. 402 pp.

Barry, R. G. & Hare, F. K. 1974. Arctic climate. In: Ives, J. D. & Barry, R. G. (eds.): *Arctic and Alpine Environments*. Methuen, London, 17–54.

Barry, R. G.; Armstrong, R. L. & Krenke, A. N. 1993. An approach to assessing changes in snow cover. An example for the former Soviet Union. *50th Eastern Snow Conference, 61st Western Snow Conference, Quebec City*. 25–33.

Bather, F. A. 1900. Wind-worn pebbles in the British Isles. *Geological Association of London, Proceedings* **16**: 396–420.

Beaty, Chester B. 1975. Sublimation or melting: observations from the White Mountains, California and Nevada, U.S.A. *Journal of Glaciology* **14(71)**: 275–286.

Begét James E. & Hawkins, Daniel B. 1989. Influence of orbital parameters on Pleistocene loess deposition in central Alaska. *Nature* **377**: 151–153.

Begét James; Edwards, Mary; Hupkins, David; Keskinen, Mary & Kukla, George 1991. Old Crow tephra found at the Palisades of the Yukon, Alaska. *Quaternary Research* **35**: 291–297.

Bégin, Christian; Michaud, Yves & Filion, Louise 1995. Dynamics of a Holocene cliff-top dune along Mountain River, Northwest Territories, Canada. *Quaternary Research* **44**: 392–404.

Belly, Pierre-Yves 1962. Sand movement by wind. *University of California, Berkley, Hydraulic Engineering Laboratory, Institute of Engineering Research, Technical Report* **72(7)**: 1–91.

— 1964. Sand movement by wind. *U.S. Army Coastal Engineering Research Center, Washington, D.C., Technical Memorandum* **1**, Addendum III.

Benoit, R. E.; Campbell, W. B. & Harris, R. W. 1972. Decomposition of organic matter in the wet meadow tundra, Barrow; a revised word model. *Proceedings of 1972 Tundra Biome Symposium, University of Washington (Seattle)*: 111–115.

Berg, T. E. 1969. Fossil sand wedges at Edmonton, Alberta, Canada. *Biuletyn Peryglacjalny* **19**: 325–333.

Berg, Thomas E. & Black, Robert F. 1966. Preliminary measurements of growth of nonsorted polygons, Victoria Land, Antarctica. In: Tedrow, J. F. C. (ed.): *Antarctic Soils and Soil-Forming Processes*. American Geophysical Union Antarctic Research Series 8: 61–108.

Bergqvist, Erik 1981. Svenska inlandsdyner. Översikt och förslag till dynreservat. (Ancient inland dunes of Sweden. Survey and proposals for dune reserves). Statens naturvårdsverk (The national Swedish Environmental Protection Board). *Naturvårdsverket Rapport* PM **1412**: 1–109.

Bernes, Claes 1996. *The Nordic Arctic Environment – unspoiled. Exploited, Polluted?* Nord 1996: 26, The Nordic Council of Ministers, Copenhagen. 240 pp.

Berry, M. O. 1981. Snow and climate. In: Gray, D. M. & Male, D. H. (eds.): *Handbook of Snow. Principles, Processes, Management & Use*. Pergamon Press, Toronto. 32–59.

Billings, W. D. & Peterson, K. M. 1980. Vegetational change and ice-wedge polygons through the thaw-lake cycle in Arctic Alaska. *Arctic and Alpine Research* **12**: 413–432.

Bird, J. Brian 1967. *The Physiography of Arctic Canada. With Special Reference to the Area South of Parry Channel*. The Johns Hopkins Press, Baltimore. 336 pp.

— 1974. Geomorphic processes in the Arctic. In: Ives, Jack D. & Barry, Roger G. (eds.): *Arctic and Alpine Environments*. Methuen, London. 703–720.

Black, Robert F. 1951. Eolian deposits of Alaska. *Arctic* **4**: 88–111.

— 1969a. Climatically significant fossil periglacial phenomena in northcentral United States. *Biuletyn Peryglacjalny* **20**: 225–238.

— 1969b. Thaw depressions and thaw lakes. A review. *Biuletyn Peryglacjalny* **19**: 131–150.

— 1976. Periglacial features indicative of permafrost: ice and soil wedges. *Quaternary Research* **6**: 3–26.

— 1978. Historical wind fluting and ventifacting. *Geological Society of America, Abstracts* **2**: 33.

—— 1983. Pseudo-ice-wedge casts of Connecticut, northeastern United States. *Quaternary Research* **20**: 74–89.

Black, Robert F. & Barksdale, William L. 1949. Oriented lakes of northern Alaska. *Journal of Geology* **57**: 105–118.

Black, Robert F. & Berg, Thomas E. 1963. Patterned ground in Antarctica. *Proceedings Permafrost International Conference*. National Academy of Sciences – National Research Council, Washington, D.C. 121–128.

Blackwelder, Eliot 1929a. Cavernous rock surfaces of the desert. *American Journal of Science* **5(17)**: 393–399.

—— 1929b. Sandblast action in relation to the glaciers of the Sierra Nevada. *Journal of Geology* **37**: 256–260.

—— 1940. The hardness of ice. *American Journal of Science* **238**: 61–62.

Blake, W. P. 1855. On the grooving and polishing of hard rocks and minerals by dry sand. *American Journal of Science* **20**: 178–179.

Böcher, Tyge W. 1949. Climate, soil, and lakes in continental West Greenland. *Meddelelser om Grønland* **147(2)**: 1–63.

Bogacki, M. 1970. Eolian processes on the forefield of the Skeidarárjökull (Iceland). *Bulletin de l'Akadémie Polonaise des Sciences, Série des Sciences géologiques et géographiques* **18**: 279–287.

Borówka, Ryszard K.; Gonera, Przemyslaw; Kostrzewski, Andrzej & Zwolinski, Zbigniew 1982. Origin, age and paleogeographic significance of cover sands in the Wolin end moraine area, North-West Poland. *Questiones Geographicae* **8**: 19–36.

Boulton, G. S. 1978. Boulder shapes and grain-size distributions of debris as indicators of transport paths through a glacier and till genesis. *Sedimentology* **25**: 773–799.

Bout, P. 1953. Etudes de géomorphologie dynamique en Islande. *Expeditions Polaires Françaises* **3**: 218 pp.

Bovis, Michael J. & Barry, Roger G. 1974. A climatological analysis of north polar desert areas. In: Smiley, Terah L. & Zumberge, James H. (eds.): *Polar Deserts and Modern Man*. The University of Arizona Press, Tucson. 23–31.

Boyd, L. A. 1948. The coast of Northeast Greenland. *American Geographical Society, Special Publication* **30**: 1–339.

Boyé, M. 1952. Névés et erosion glaciaire. *Revue de Geomorphologie Dynamique* **3**: 20–36. Translation: Névés (firn) and glacial erosion, in: Evans, David J. A. (ed.) 1994: *Cold Climate Landforms*. John Wiley & Sons, Chichester. 205–220.

Brewer, M. C.; Carter, L. David & Glenn, R. 1993. Sudden drainage of a thaw lake on the Alaskan Arctic Coastal Plain. Permafrost. *Sixth International Conference, Proceedings, Beijing*. Vol. 1: 48–53.

Britton, Max E. 1957. Vegetation of the arctic tundra. In: Hansen, Henry P. (ed.): *Arctic Biology*. Oregon State University Press, Corvallis. 2nd edition 1967. 67–130.

Brockie, W. J. 1973. Experimental frost-shattering. *Proceedings 7th New Zealand Geographical Conference, Hamilton 1972*, Conf. Ser. 7: 177–186.

Brown, R. J. E. 1970. *Permafrost in Canada. Its Influence on Northern Development*. University of Toronto Press, Toronto. 234 pp.

Brunnschweiler, Dieter 1962. The periglacial realm in North America during the Wisconsin glaciation. *Biuletyn Peryglacjalny* **11**: 15–27.

Bruun, P. 1953. Forms of equilibrium coasts with a littoral drift. *University of California Institute of Engineering Research* **347**: 1–7.

Bryant, Ian D. 1982. Loess deposits in Lower Adventdalen, Spitsbergen. *Polar Research* **2**: 93–103.

Buckley, J. R. & Pond, S. 1976. Wind and the surface circulation of a fjord. *Journal of the Fisheries Research Board of Canada* **33**: 2265–2271.

For lake drainage

Budd, W. F. 1966. The drifting of nonuniform snow particles. Studies in Antarctic Meteorology. *American Geophysical Union, Antarctic Research Series* **9**: 59–70.

Budd, W. F.; Dingle, W. R. J. & Radok, U. 1966. The Byrd snow drift project: Outline and basic results. Studies in Antarctic Meteorology. *American Geophysical Union, Antarctic Research Series* **9**: 71–134.

Büdel, Julius 1951. Die Klimazonen des Eiszeitalters. *Eiszeitalter und Gegenwart* **1**: 16–26.

— 1977. *Klima-Geomorphologie*. Gebrüder Borntraeger. Berlin-Stuttgart. 304 pp.

— 1982 *Climatic Geomorphology*. Princeton University Press. Princeton. 443 pp.

Buk, Enrique & Trombotto, Dario 1995. Monitoring mountain permafrost in the Central Andes, Argentina: El Salto – Morenas Coloradas. *Frozen Ground Workshop: Our current understanding of processes and ability to detect change. 9–11 December 1995, Hanover, New Hampshire, U.S.A. Abstracts:* 5.

Bull, Colin 1966. Climatological observations in ice-free areas of Southern Victoria Land, Antarctica. Studies in Antarctic Meteorology. *American Geophysical Union, Antarctic Research Series* **9**: 177–194.

Businger, Steven 1991. Arctic hurricanes. *American Scientist* **79(1)**: 18–33.

Butterfield, G. R. 1971. The instrumentation and measurement of wind erosion. *Proceedings Sixth Geography Conference, New Zealand Geographical Society, Christchurch*, 125–130.

Butzer, K. W. 1965. *Environment and Archeology*. Methuen. London. 524 pp.

Cabot, E. C. 1947. Northern Alaska coastal plain interpreted from aerial photographs. *Geographical Review* **37**: 639–648.

Cailleux, André 1939. Action du vent sur les formations volcaniques en Islande. *Bulletin Volcanologique* **5**: 19–64 and 8 plates.

— 1942. Les actions périglaciaires en Europe. *Mémoires de la Societé Géologique de France, Nouvelle série, Mémoire* **46**. 176 pp.

— 1952. Morphoskopische Analyse der Geschiebe und Sandkörner und ihre Bedeutung für die Paläoklimatologie. *Geologische Rundschau* **40**: 11–19.

— 1963. *Géologie de l'Antarctique*. Sedes, Paris. 201 pp.

— 1968. Periglacial of McMurdo Strait (Antarctica). *Biuletyn Peryglacjalny* **17**: 57–90.

— 1969. Quaternary periglacial wind-worn sand grains in USSR. In: Troy L. Péwé (ed.): *The Periglacial Environment. Past and Present*. McGill-Queen's University Press. 285–301.

— 1972. Les formes et dépôts nivéo-éoliens actuels en Antarctique et au Nouveau-Québec. *Cahiers de Géographie de Québec* **16(39)**: 377–409.

— 1973a. Éolisations périglaciaires quaternaires au Canada. *Biuletyn Peryglacjalny* **22**: 81–115.

— 1973b. Répartition et signification des différents critères d'éolisation périglaciaire. *Biuletyn Peryglacjalny* **23**: 50–63.

— 1974. Formes précoces et albédos du nivéo-éolien. *Zeitschrift für Geomorphologie, N.F.* **18**: 437–459.

— 1976. Formes et dépôts nivéo-éoliens sur le pied de glace, à Poste-de-la-Baleine, Québec subarctique. *Revue de Géographie de Montréal* **30**: 213–219.

Cailleux, A. & Calkin, P. 1963. Orientation of hollows in cavernously weathered boulders in Antarctica. *Biuletyn Peryglacjalny* **12**: 147–150.

Caine, Nel 1992. Sediment transfer on the floor of Martinelli Snowpatch, Colorado Front Range, U.S.A. *Geografiska Annaler* **74A**: 133–144.

— 1995. Snowpack influences on geomorphic processes in Green Lakes Valley, Colorado Front Range. *Geographical Journal* **161**: 55–68.

Calkin, P. & Cailleux, A. 1962. A quantitative study of cavernous weathering (taffonis) and its application to glacial chronology in Victoria Valley, Antarctica. *Zeitschrift für Geomorphologie* **6**: 317–324.

Calkin, P. E. & Rutford, R. H. 1974. The sand dunes of Victoria Valley, Antarctica. *Geographical Review* **64(2)**: 189–216.

Carlson, P. R.; Roy, C. J.; Hussey, K. M.; Davidson, D. T. & Handy, R. L. 1959. Geology and mechanical stabilization of cenozoic sediments near Point Barrow. *Iowa Engineering Expedition Station Bulletin* **186**: 101–128.

Carson, Charles E. 1968. Radiocarbon dating of lacustrine strands in arctic Alaska. *Arctic* **21**: 12–26.

Carson, Charles E. & Hussey, Keith M. 1959. The multiple working hypothesis as applied to Alaska's oriented lakes. *Iowa Academy of Science, Proceedings* **66**: 334–349.

— 1960. Hydrodynamics in three arctic lakes. *Journal of Geology* **68**: 585–600.

— 1962. The oriented lakes of Arctic Alaska. *Journal of Geology* **70**: 417–439.

Carson, M. A. & MacLean, P. A. 1986. Development of hybrid aeolian dunes: the William River dune field, northwest Saskatchewan, Canada. *Canadian Journal of Earth Sciences* **23 (12)**: 1974–1990.

Carter, L. David 1981. A Pleistocene sand sea on the Alaskan Arctic Coastal Plain. *Science* **211**: 381–383.

— 1987. Oriented lakes. *Geological Society of America, Centennial Special Volume* 2: 615–619.

— 1988. Loess and deep thermokarst basins in Arctic Alaska. *Permafrost Fifth International Conference, Trondheim, Proceedings* Vol. 1: 706–711.

— 1993. Late Pleistocene stabilization and reactivation of eolian sand in northern Alaska: Implications for the effects of future climatic warming on an eolian landscape in continuous permafrost. *Permafrost. Proceedings of the Sixth International Conference on Permafrost, Beijing*, Volume 1: 78–83.

Carter, L. David; Heginbottom, J. Alan & Woo, Ming-ko 1987. Arctic lowlands. *Geological Society of America, Centennial Special Volume 2*: 583–628.

Carter, R. W. G.; Curtis, T. G. & Sheehy-Skeffington, M. J. (eds.) 1992. *Coastal Dunes. Geomorphology, Ecology and Management for Conservation*. A. A. Balkema, Rotterdam. 533 pp.

Catto, N. R. 1983. Loess in the Cypress Hills, Alberta, Canada. *Canadian Journal of Earth Sciences* **20**: 1159–1167.

Chapman, V. J. 1964. *Coastal Vegetation*. MacMillan, New York. 245 pp.

Chepil, W. S. 1945–46. Dynamics of wind erosion I–VI. *Soil Science* **60**: 305–320, 397–411, 475–480. *Soil Science* **61**: 167–178, 257–263, 331–340.

— 1953. Factors that influence clod structure and erodibility of soil by wind. *Soil Science* **75**: 473–483.

— 1956. Influence of soil moisture on erodibility of soil by wind. *Soil Science Society of America Proceedings* **20**: 288–292.

— 1957. Sedimentary characteristics of dust storms: I. Sorting of wind-eroded soil material. *American Journal of Science* **255**: 12–22.

— 1965. Transport of soil and snow by wind. *Meteorological Monographs* **6(28)**: 123–132.

Chepil, W. S. & Woodruff, N. P. 1963. The physics of wind erosion and its control. *Advances in Agronomy* **15**: 211–302.

Church, M. 1972. Baffin Island sandurs: a study of arctic fluvial processes. *Geological Survey of Canada Bulletin* **216**. 208 pp.

Clark, Michael J. & Seppälä, Matti 1988. Slushflows in a subarctic environment, Kilpisjärvi, Finnish Lapland. *Arctic and Alpine Research* **20**: 97–105.

Clark, Michael J.; Gurnell, Angela M.; Milton, Edward J.; Seppälä, Matti & Kyöstilä, Maarit 1985. Remotely-sensed vegetation classification as a snow depth indicator for hydrological analysis in subarctic Finland. *Fennia* **163**: 195–216.

Clark, Richard & Wilson, Peter 1992. Occurrence and significance of ventifacts in the Falkland Islands, South Atlantic. *Geografiska Annaler* **74A**: 35–46.

Clement, Pierre & Vandour, Jean 1968. Observations on the pH of melting snow in the Southern French Alps. In: Wright, H. E. Jr. & Osburn, W. H. (eds.): *Arctic and Alpine Environments 10*. Indiana University Press, Bloomington. 205–213.

Cline, D. W. 1991. Modeling alpine snow distribution using geographic information processing techniques. *EOS Transactions. American Geophysical Union* **72** (**44**, Supplement): 165.

COHMAP 1988. Climatic changes of the last 18,000 years; Observations and model simulations. *Science* **241**: 1043–1052.

Collins, F. 1985. *Map Showing a Vegetated Dune Field in Central Alaska*. United States Geological Survey Miscellaneous Field Studies Map I-50.

Committee on Sedimentation 1965. Sediment transportation mechanics: Wind erosion and transportation. *Journal of the Hydraulics Division, Proceedings of the American Society of Civil Engineers* **91 HY 2**, Paper 4261: 267–287.

Cooke, C. W. 1934. Discussion of the origin of the supposed meteorite scars of South Carolina. *Journal of Geology* **42**: 88–96.

Cooke, Ronald U. & Warren, Andrew 1975. *Geomorphology of Deserts*. B. T. Batsford, London. 394 pp.

Cooke, Ron; Warren, Andrew & Goudie, Andrew 1993. *Desert Geomorphology*. UCL Press, London. 526 pp.

Cooper, William S. 1935. The history of the upper Mississippi River in late Wisconsin and postglacial time. University of Minnesota, *Minnesota Geological Survey Bulletin* **26**: 1–111.

Corbel, Jean 1959. Érosion en terrain calcaire (vitesse d'érosion et morphologie). *Annales de géographie* **68**: 97–120.

Corbet, Philip S. 1972. The microclimate of arctic plants and animals, on land and in fresh water. *Acta Arctica* **18**: 1–43.

Cornish, Vaughan 1914. *Waves of Sand and Snow and the Eddies which Make Them*. T Fisher Unwin, London. 383 pp.

Cotton, Jeremy H. & Michael, Kelvin J. 1994. The monitoring of katabatic wind-coastal polynya interaction using AVHRR imagery. *Antarctic Science* **6**: 537–540.

Craig, B. G. 1965. Glacial Lake McConnell, and the surficial geology of parts of Slave River and Redstone River map-areas, District of Mackenzie. *Geological Survey of Canada, Bulletin* **122**: 1–33.

Cummings, Craig E. & Pollard, Wayne H. 1990. Cryogenetic categorization of peat and mineral-cored palsas in Schefferville area, Quebec. *Collection Nordicana* **54**: 95–102.

Czeppe, Zdzislaw 1966. Przebieg głównych procesów morfogenetycznych w poludniowa-zachodnim Spitsbergenie. (Summary: The course of the main morphogenetic processes in South-west Spitsbergen). *Prace Geograficzne, Kraków*, **13**: 1–124.

Czudek, Tadéás & Demek, Jaromír 1970. Thermokarst in Siberia and its influence on the development of lowland relief. *Quaternary Research* **1**: 103–120.

Dahl, Ragnar 1966. Block fields, weathering pits and tor-like forms in the Narvik mountains, Nordland, Norway. *Geografiska Annaler* **48A**: 55–85.

Dalrymple, Paul C. 1966. Physical climatology of Antarctic plateau. Studies in Antarctic Meteorology. *American Geophysical Union. Antarctic Research Series* **9**: 70–134.

Dansereau, Pierre 1955. Biogeography of the land and the inland waters. In: Kimble, George H. T. & Good, Dorothy (eds.): *Geography of the Northlands*. American Geographical Society Special Publication 32: 84–118.

David, Peter P. 1977. Sand dune occurences of Canada. A theme and resource inventory study of eolian landforms of Canada. *Indian and Northern Affairs. National Parks Branch Contract* No 74–230. 183 pp.

— 1979. Sand dunes in Canada. *Geos* Spring 1979: 12–14.

— 1981. Stabilized dune ridges in northern Saskatchewan. *Canadian Journal of Earth Sciences* **18**: 286–310.

Davies, William E. 1974. Geological and limnological factors of cold deserts. In: Smiley, Terah L. & Zumberge, James H. (eds.): *Polar Deserts and Modern Man*. The University of Arizona Press, Tucson. 53–61.

Debenham, F. 1921. Recent and local deposits of McMurdo Sound region. *British Antarctic ("Terra Nova") Expedition, 1910, Natural History Reports, Geology* **I**: 63–100.

Defant, Friedrich 1951. Local winds. In: Malone, Thomas F. (ed.): *Compendium of Meteorology*. American Meteorological Society, Boston. 655–672.

Denton, G. H.; Prentice, M. L.; Kellogg, D. E. & Kellogg, T. B. 1984. Late Tertiary history of the Antarctic ice sheet: Evidence from the Dry Valleys. *Geology* **12**: 263–267.

Derbyshire, Edward & Evans, Ian S. 1976. The climate factor in cirque variation. In: E. Derbyshire (ed.): *Geomorphology and Climate*. John Wiley & Sons, Chichester. 447–494.

Derbyshire, E. & Owen, L. A. 1996. Glacioaeolian processes, sediments and landforms. In: John Menzies (ed.): *Past Glacial Environments. Sediments, Forms and Techniques*. Butterworth-Heinemann Ltd, Oxford. *Glacial Environments*: Volume 2: 213–237.

Dewdney, John C. 1979. *A Geography of the Soviet Union*. Third edition. Pergamon Press, Oxford. 175 pp.

Dietrich, R. V. 1977a. Impact abrasion of harder by softer materials. *Journal of Geology* **85**: 242–246.

— 1977b. Wind erosion by snow. *Journal of Glaciology* **18**: 148–149.

Dijkmans, J. W. A. 1990. Niveo-aeolian sedimentation and resulting sedimentary structures; Søndre Strømfjord area, western Greenland. *Permafrost and Periglacial Processes* **1**: 83–96.

Dijkmans, Jos W. A.; Koster, Eduard A.; Galloway, John P. & Mook, Willem G. 1986. Characteristics and origin of calcretes in a subarctic environment, Great Kobuk Sand Dunes, northwestern Alaska, USA. *Arctic and Alpine Research* **18**: 377–387.

Dionne, Jean-Claude 1978. Dunes et dépôts éoliens en Jamésie et Hudsonie, Québec subarctique. *Environnement-Canada, Rapport d'information, Québec*. 35 pp.

— 1994. Les cordons de blocs frangeants (boulder barricades). Mise au point avec exemples du Québec. *Revue de Géomorphologie Dynamique* **43**: 49–70.

Dionne, Jean-Claude & Quessy, Sylvain 1995. Évaluation quantitative de l'affouillement glaciel de la plate-forme intertidale à Neuville, Québec. *Canadian Coastal Conference 1995 – Conférence Canadienne sur le Littoral 1995*. 243–257.

Dorn, Ronald I. 1986. Rock varnish as an indicator of aeolian environmental change. In: Nickling, William G. (ed.): *Aeolian Geomorphology*. Allen & Unwin, Boston. 291–307.

Dorn, Ronald I. & Oberlander, Theodore M. 1982. Rock varnish. *Progress in Physical Geography* **6**: 317–367.

Dorsey, Herbert G. 1951. Arctic meteorology. In: Malone, Thomas F. (ed.): *Compendium of Meteorology*. American Meteorological Society, Boston. 942–951.

Dylik, Jan 1956. Coup d'oeil sur la Pologne periglaciaire. *Biuletyn Peryglacjalny* **4**: 195–238. Translation: A general view of periglacial Poland. In: Evans, David J. A. (ed.) 1994: *Cold Climate Landforms*. John Wiley, Chichester. 45–81.

Dylikowa, Anna 1958. Próba wyróznienia faz rozwoju wydm wokolicach Lodzi (Résumé: Phases du developpement des dunes aux environs de Lódz). *Acta Geographica Universitatis Lodziensis* **8**: 233–270.

Dyunin, A. K. & Kotlyakov, V. M. 1980. Redistribution of snow in the mountains under the effect of heavy snow-storms. *Cold Regions Science and Technology* **3**: 287–294.

Easterbrook, Donald J. 1993. *Surface Processes and Landforms*. Macmillan Publishing Company, New York. 520 pp.

Edlund, S. A. 1993. The distribution of plant communities on Melville Island, Arctic Canada. *Geological Survey of Canada, Bulletin* **450**: 247–255.

Edwards, John S. 1987. Arthropods of alpine aeolian ecosystems. *Annual Review of Entomology* **32**: 163–179.

Einarsson, Thorleifur 1973. Geology of Iceland. In: Pitcher, Max G. (ed.): *Arctic Geology*. Tulsa, Oklahoma. 171–175.

Embleton, Clifford & King, Cuclaine A. M. 1968. *Glacial and Periglacial Geomorphology*. Edward Arnold, London. 608 pp.

— 1975a. *Glacial Geomorphology*. Edward Arnold, London. 573 pp.

— 1975b. *Periglacial Geomorphology*. Edward Arnold, London. 203 pp.

Enquist, Fredrik 1916. Der Einfluss des Windes auf die Verteilung der Gletscher. *Bulletin of the Geological Institute of the University of Uppsala* **14**: 1–108.

— 1932. The relation between dune-form and wind-direction. *Geologiska Föreningens i Stockholm Förhandlingar* **54**: 19–59.

Eurola, Seppo 1968. Über die Ökologie der nordfinnischen Moorvegetation im Herbst, Winter und Frühling. *Annales Botanici Fennici* **5**: 83–97.

— 1975. Snow and ground frost conditions of some Finnish mire types. *Annales Botanici Fennici* **12**: 1–16.

Evans, Ian S. 1977. World-wide variations in the direction and concentration of cirque and glacier aspects. *Geografiska Annaler* **59A**: 151–175.

Evans, Ian S. 1990. Climatic effects on glacier distribution across the southern Coast Mountains, B.C., Canada. *Annals of Glaciology* **14**: 58–64.

— 1997. Process and form in the erosion of glaciated mountains. In: Stoddart, D. R. (ed.): *Process and Form in Geomorphology. Festschrift for Richard J. Chorley*. Routledge, London. 145–174.

Evans, Ian S. & Cox, Nicholas J. 1995. The form of glacial cirques in the English Lake District, Cumbria. *Zeitschrift für Geomorphologie N.F.* **39**: 175–202.

Evans, James R. 1962. Falling and climbing sand dunes in the Cronese ("Cat") Mountain area, San Bernardino County, California. *Journal of Geology* **70**: 107–113.

Evans, J. W. 1911. Dreikanter. *Geological Magazine* **8**: 334–335.

Everett, K. R. 1971. Soils of the Meserve Glacier Area, Wright Valley, South Victoria Land, Antarctica. *Soil Science* **112**: 425–438.

— 1979. Evolution of the soil landscape in the sand region of the Arctic Coastal Plain as exemplified at Atkasoon, Alaska. *Arctic* **32**: 207–223.

Everett, K. R. & Parkinson, R.J. 1977. Soil and landform associations, Prudhoe Bay area, Alaska. *Arctic and Alpine Research* **9**: 1–19.

check —   Evteev, S. A. 1962. At what speed does wind "erode" stones in Antarctica? *Soviet Antarctic Expedition Information Bulletin* **2(17)**: 211.

Fairbridge, Rhodes W. 1972. Climatology of a glacial cycle. *Quaternary Research* **2**: 283–302.

Fedorovich 1969. The closed Quaternary lakes of Middle and Central Asia. *VIIIe Congres INQUA Paris 1969, Résumés des Communications*, Section **1**: 17.

Feng, Zhao-Dong; Johnson, William C. & Diffendal, Robert F., Jr. 1994. Environments of aeolian deposition in south-central Nebraska during the last glacial maximum. *Physical Geography* **15**: 249–261.

Fernald, Arthur T. 1964. Surficial geology of the central Kobuk River valley, northwestern Alaska. *United States Geological Survey Bulletin* **1181-K**: K1-31.

— 1965. Recent history of the upper Tanana River valley lowland, Alaska. *United States Geological Survey Professional Paper* 525C: C124–127.

Filion, Louise 1984. A relationship between dunes, fire and climate recorded in the Holocene deposits of Quebec. *Nature* **309** (5968): 543–546.

— 1987. Holocene development of parabolic dunes in the central St. Lawrence Lowland, Quebec. *Quaternary Research* **28**: 196–209.

Filion, Louise & Morisset, Pierre 1983. Eolian landforms along the eastern coast of Hudson Bay, Northern Québec. *Nordicana* **47**: 73–94.

Filion, Louise; Saint-Laurent, Diane; Desponts, Mireille & Payette, Serge 1991. The late Holocene record of aeolian and fire activity in northern Québec, Canada. *The Holocene* **1**: 201–208.

Flint, Richard Foster 1948. Glacial geology and geomorphology. In: Boyd, Louise A.: *The Coast of Northeast Greenland*. American Geographical Society, Special Publication 30: 91–210.

— 1971. *Glacial and Quaternary Geology*. John Wiley & Sons, New York. 892 pp.

Flohn, H. 1969. Local wind systems. In Flohn, H. (ed.): *World Survey of Climatology*, Volume 2, *General Climatology 2*. Elsevier, Amsterdam 139–171.

Fogelberg, Paul & Seppälä, Matti 1979. Geomorphological map of Finland. *Atlas of Finland*. Fifth edition. Folio 121–122: 3–4.

Föhn, Paul M. B. 1980. Snow transport over mountain crests. *Journal of Glaciology* **26**: 469–480.

Föhn, Paul M. B. & Meister, Roland 1983. Distribution of snow drifts on ridge slopes: measurements and theoretical approximations. *Annals of Glaciology* **4**: 52–57.

Folk, R. L. 1966. A review of grain-size parameters. *Sedimentology* **6**: 73–93.

Folk, R. L. & Ward, W. C. 1957. Brazos River bar: A study in the significance of grain size parameters. *Journal of Sedimentary Petrology* **27**: 3–26.

Ford, Derek C. 1993. Karst in cold environments. In: French, Hugh M. & Slaymaker, Olav (eds.): *Canada's Cold Environments*. McGill-Queen's University Press, Montreal & Kingston. 199–222.

Forman, Steven L.; Goetz, Alexander F.H. & Yuhas, Roberta H. 1992. Large-scale stabilized dunes on the high plains of Colorado: Understanding the landscape response to Holocene climates with the aid of images from space. *Geology* **20**: 145–148.

Foster, Dudley J. (Jr.) & Davy, Robert D. 1989. Global snow depth climatology. *Sixth Conference on Applied Climatology, March 1989, Charleston, South Carolina*. American Meteorological Society. 145–148.

Fox, Adrian J. & Cooper, A. Paul R. 1994. Measured properties of the Antarctic ice sheet derived from the SCAR Antarctic digital database. *Polar Record* **30**: 201–206.

Frei, A. & Robinson, D. A. 1993. North American snow cover variability from satellite data (1972–1993) and comparison with model output. *50th Eastern Snow Conference, 61st Western Snow Conference, Quebec City*. 43–50.

French, H. M. 1976. *The Periglacial Environment*. Longman, London. 309 pp.

French, Hugh M. & Slaymaker, Olav (eds.) 1993. *Canada's Cold Environments*. McGill-Queen's Press, Montreal. 340 pp.

Frich, Povl 1988. *En Analyse af Klimatiske og Geomorfologiske Betingelser for Vindsblibning med Ispartikler i Vestgroenland*. Master thesis. Institute of Geography, University of Copenhagen.

Friedman, J. D.; Johansson, C. E.; Oskarsson, N.; Svensson, H.; Thorarinsson, S.; Williams, Jr., R. S. 1971. Observations on Icelandic polygon surfaces and palsa areas. Photo interpretation and field studies. *Geografiska Annaler* **53A**: 115–145.

Fries, Thore C. E. 1913. *Botanische Untersuchungen im nördlichsten Schweden. Ein Beitrag zur Kenntnis der alpinen und subalpinen Vegetation in Torne Lappmark*. Akademische Abhandlung. Vetenskapliga och praktiska undersökningar i Lappland anordnade af Luossavaara-Kiirunavaara Aktiebolag. Uppsala & Stockholm. 361 pp.

Fries, T. & Bergström, E. 1910. Några iakttagelser öfver palsar och deras förekomst i nordligaste Sverige. *Geologiska Föreningen i Stockholm Förhandlingar* **32**: 195–205.

Fristrup, Børge 1952a. Danish expedition to Peary Land, 1947–50. *Geographical Review* **42**: 87–97.

— 1952b. Physical geography of Peary Land. I. Meteorological observations for Jørgen Brønlunds Fjord. *Meddelelser om Grønland* **127(4)**: 1–143.

— 1952–53. Winderosion within the Arctic deserts. *Geografisk Tidsskrift* **52**: 51–65.

— 1953. High arctic deserts. *Congrés Géologique International, Alger 1952, Comptes Rendus de la Dix-neuviéme session*: 91–99.

Fürbringer, Werner & Haydn, Rupert 1974. Zur Frage der Orientierung Nordalaskischer Seen mit Hilfe des Satellitenbildes. *Polarforschung* **44**: 47–53.

Gallet, Sylvian; Jahn, Bor-ming; Van Vliet-Lanoë, Brigitte; Dia, Aline & Rossello, Eduardo 1998. Loess geochemistry and its implications for particle origin and composition of the upper continental crust. *Earth and Planetary Science Letters* **156**: 157–172.

Galloway, John P. & Carter, L. David 1993a. Dune activity on the western Arctic Coastal Plain of Alaska coincident with neoglacial cirque-glacier expansion in the Brooks Range. *23rd Annual Arctic Workshop Program and Abstracts*, Byrd Polar Research Center, Ohio State University, Columbus, BPRC Miscellaneous Series M-322: 35–36.

— 1993b. Late Holocene longitudinal and parabolic dunes in northern Alaska: Preliminary interpretations of age and paleoclimatic significance. *U.S. Geological Survey Bulletin* **2068**: 3–11.

— 1994. Paleowind directions for late Holocene dunes on the Western Arctic Coastal Plain, Northern Alaska. *U.S. Geological Survey Bulletin* **2107**: 27–30.

Galloway, John P.; Huebner, Mark; Lipkin, Robert & Dijkmans, Jos W. A. 1992. Early Holocene calcretes from the subarctic active Nogahabara Sand Dune field, northern Alaska. *U.S. Geological Survey Bulletin* **1999**: 100–111.

Galon, Rajmund 1958. Z problematyki wydm sródladowych w Polsce (Résumé: Sur les dunes continentales en Pologne). *Wydmy Sródladowe Polski* **I**: 13–31.

— 1959. New investigations of inland dunes in Poland. *Przeglad Geograficzny* **31** (Supplement): 93–110.

Gardner, James S. 1987. Evidence for headwall weathering zones, Boundary Glacier, Canadian Rocky Mountains. *Journal of Glaciology* **33**: 60–67.

— 1992. The zonation of freeze-thaw temperatures at a glacier headwall, Dome Glacier, Canadian Rockies. In: Dixon, J. C. & Abrahams, A. D. (eds.): *Periglacial Geomorphology. Proceedings of the 22nd Annual Binghamton Symposium in Geomorphology*. John Wiley & Sons, Chichester. 89–102.

— 1969. Snowpatches: their influence on mountain wall temperatures and the geomorphic implications. *Geografiska Annaler* **51A**: 114–120.

Gary, Margaret; McAfee, Jr., Robert & Wolf, Carol (eds.) 1972. *Glossary of Geology*. American Geological Institute, Washington, D.C. 805 and appendix 52 pp.

Geiger, Rudolf 1961. *Das Klima der bodennahen Luftschicht*. Vierte Auflage. Friedr. Vieweg & Sohn, Braunschweig. 646 pp.

Giaever, John 1955. *The White Desert. The Official Account of the Norwegian–British–Swedish Antarctic Expedition*. E. P. Dutton & Company, Inc., New York. 256 pp.

Gilbert, Robert 1983. Sedimentary processes of Canadian Arctic fjords, *Sedimentary Geology* **36**, 147–175.

Glasby, G. P.; McPherson, J. G.; Kohn, B. P.; Johnson, J. H.; Keys, J. R.; Freeman, A. G. & Tricker, M. J. 1981. Desert varnish in southern Victoria Land, Antarctica. *New Zealand Journal of Geology and Geophysics* **24**: 389–397.

Good, Dorothy 1955. Introduction. In: Kimble, George H. T. & Good, Dorothy (eds): *Geography of the Northlands*. American Geographical Society, Special Publication 32: 1–10.

Good, Timothy R. & Bryant, Ian D. 1985. Fluvio-aeolian sedimentation – An example from Banks Island, N.W.T., Canada. *Geografiska Annaler* **67A**: 33–46.

Gorshkov, S. G. (ed.) 1980. *Atlas Okeanov: Severnyy Ledovityy Okean*. Ministerstvo Oborony SSSR.

Goudie, Andrew S. 1997. Weathering processes. In: Thomas, David S. G. (ed.): *Arid Zone Geomorphology. Process, Form and Change in Drylands*. John Wiley & Sons, Chichester. 25–39.

Graf, William L. 1976. Cirques as glacier locations. *Arctic and Alpine Research* **8**: 79–90.

Gray, James T. & Seppälä, Matti 1991. Deeply dissected tundra polygons on a glacio-fluvial outwash plain, Northern Ungava Peninsula, Québec. *Géographie physique et Quaternaire* **45**: 111–117.

Greeley, Ronald & Iversen, James D. 1985. *Wind as a Geological Process on Earth, Mars, Venus and Titan*. Cambridge University Press, Cambridge. 333 pp.

Gripp, K. 1929. *Glaziologische und geologische Ergebnisse der Hamburgischen Spitzbergen-Expedition 1927*. Abhandlungen aus dem Gebiete der Naturwissenschaften, Naturwissenschaftlichen Verein in Hamburg 22 (3–4): 145–249 and 32 tables.

Gudmundsson, Gils 1956. *Öldin sem Leið – Minnisverð Tiðiðndi 1860–1900*. Idunn, Reykjavik.

Gunn, B. M. & Warren, Guyon 1962. Geology of Victoria Land between the Mawson and Mulock Glaciers, Antarctica. *New Zealand Geological Survey, Bulletin* **71**: 1–157.

Haeberli, W.; Schotterer, U.; Wagenbach, D.; Haeberli-Schwitter, H. & Bortenschlager, S. 1983. Accumulation characteristics on a cold, high-alpine firn saddle from a snow-pit study on Colle Gnifetti, Monte Rosa, Swiss Alps. *Journal of Glaciology* **29**: 260–271.

Haehnel, R. B.; Wilkinson, Cpt. J. H. & Lever, J. H. 1993. Snowdrift modeling in the CRREL wind tunnel. *50th Eastern Snow Conference, 61st Western Snow Conference, Quebec City*. 139–147.

Haesaerts, Paul 1985. Les loess du Pléistocène Supérieur en Belgique; comparaisons avec les séquences d'Europe Centrale. *Bulletin de l'Association Française pour l'Etude du Quaternaire* **2–3**: 105–115.

Hall, Kevin 1980. Freeze-thaw activity at a nivation site in Northern Norway. *Arctic and Alpine Research* **12**: 183–194.

— 1989. Wind-blown particles as weathering agent? An Antarctic example. *Geomorphology* **2**: 405–410.

Hamberg, Axel 1907. Die Eigenschaften der Schneedecke in den lappländischen Gebirgen. *Naturwissenschaftliche Untersuchungen des Sarekgebirges in Schwedisch-Lappland*, geleitet von Dr. Axel Hamberg Bd. I, Abt. III, Lief. I. Stockholm: 1–68.

Hamelin, L.-E. & Cook, F. A. 1967. *Illustrated Glossary of Periglacial Phenomena*. Les Presses de l'Université Laval, Québec. 237 pp.

Hardy, D. R. 1993. Snowmelt-induced slushflows, Ellesmere Island, N.W.T., Canada. *50th Eastern Snow Conference, 61st Western Snow Conference, Quebec City*. 423–431.

Hare, F. Kenneth 1955. Weather and climate. In: Kimble, George H. T. & Good, Dorothy (eds.): *Geography of the Northlands*. American Geographical Society Special Publication 32: 58–83.

— 1959. The circumpolar atmospheric circulation in high latitudes. *Cahiers de Géographie de Québec* **6**: 163–178.

— 1968. The Arctic. *Quarterly Journal of the Royal Meteorological Society* **94**: 439–459.

Hare, F. Kenneth & Hay, John E. 1974. The climate of Canada and Alaska. In: Bryson, Reid A. & Hare, F. Kenneth (eds.): *Climates of North America*. World Survey of Climatology vol. 11. Elsevier, Amsterdam. 49–129.

Hare, F. Kenneth & Ritchie, J. C. 1972. The boreal bioclimates. *Geographical Review* **62**: 333–365.

Harlin, Ben W. 1958. IGY meteorology in Antarctica. Inside Antarctica – No. 1 Little America. *Weatherwise* **13**: 116–123.

Harry, D. G. & French, H. M. 1983. The orientation and evolution of thaw lakes, southwest Banks Island, Canadian Arctic. *Permafrost, Fourth International Conference on Permafrost, Fairbanks, Trondheim*, Vol. 1: 456–461.

Hassinen, S. 1998. A morpho-statistical study of cirques and cirque glaciers in the Senja-Kilpisjärvi area, northern Scandinavia. *Norsk Geografisk Tidsskrift* **52**: 27–36.

Hastings, Andrew D. 1961. *Atlas of Arctic Environment*. Research Study Report RER-33. Headquarters Quatermaster Research & Engineering Command, US Army, Environmental Protection Research Division, Natick, Massachusetts.

Havas, Paavo & Mäenpää, Erkki 1972. Evolution of carbon dioxide at the floor of a Hylocomium-Myrtillus type spruce forest. *Aquilo, Ser. Botanica* **11**: 4–22.

Havas, Paavo & Sulkava, Seppo 1987. *Suomen Luonnon Talvi* (The Winter of Nature in Finland. In Finnish only). Kirjayhtymä, Helsinki. 222 pp.

Heginbottom, J.A. 1993. Generalised distribution of permafrost in the Northern Hemisphere. *Permafrost VI International Conference, Proceedings* Volume 2, map. Beijing.

Heim, Albert 1885. *Handbuch der Gletscherkunde*. Verlag von J. Engelhorn, Stuttgart. 560 pp.

Hellemaa, Pirjo 1991. Tuulen vaikutus deflaatioon ja lumen kinostumiseen Kilpisjärven alueella. (Winds causing deflation and snow-drifting in Kilpisjärvi area). *Terra* **103(4)**: 309–316.

Hétu, Bernard 1992. Coarse cliff-top aeolian sedimentation in northern Gaspésie, Québec (Canada). *Earth Surface Processes and Landforms* **17**: 95–108.

Higgins, Charles G. 1956. Formation of small ventifacts. *Journal of Geology* **64**: 506–516.

Hjulström, Filip 1935. Studies of the morphological activity of rivers as illustrated by the River Fyris. *Bulletin of the Geological Institution of the University of Uppsala* **25**: 221–527.

— 1955. The problem of the geographic location of wind-blown silt. An attempt of explanation. *Geografiska Annaler* **37**: 86–93.

Hobbs, Peter V. 1974. *Ice Physics*. Clarendon Press, Oxford. 837 pp.

Hobbs, William Herbert 1926. *The Glacial Anticyclones. The Poles of the Atmospheric Circulation*. Macmillan, London. 198 pp.

— 1931. Loess, pebble bands, and boulders from glacial outwash of the Greenland Continental Glacier. *Journal of Geology* **39**: 381–385.

— 1942. Wind: The dominant transportation agent within extramarginal zones to continental glaciers. *Journal of Geology* **50(5)**: 556–559.

Hodgson, D. A. 1982. Surficial materials and geomorphological processes, Western Sverdrup and adjacent Islands, District of Franklin. *Geological Survey of Canada*, Paper **81–9**: 1–34.

Högbom, Bertil 1912. Wüstenerscheinungen auf Spitzbergen. *Bulletin of the Geological Institution of the University of Uppsala* 11: 242–251.

Högbom, Ivar 1923. Ancient inland dunes of northern and middle Europe. *Geografiska Annaler* **5**: 113–242.

Holtmeier, F.-K. 1978. Die bodennahen Winde in den Hochlagen der Indian Peaks Section, Colorado Front Range. *Münstersche Geographische Arbeiten* **3**: 3–47.

Holtmeier, Friedrich-Karl 1996. Die Wirkungen des Windes in der subalpinen und alpinen Stufe der Front Range, Colorado, U.S.A. *Arbeiten aus dem Institut für Landschaftsökölogie Westfälische Wilhelms-Universität* **1**: 19–45.

Hopkins, David M. 1949. Thaw lakes and thaw sinks in the Imaruk Lake area, Seward Peninsula, Alaska. *Journal of Geology* **57**: 119–131.

— 1959. Some characteristics of the climate in forest and tundra regions in Alaska. *Arctic* **12**: 215–220.

Hopkins, D. M. & Kidd, J. G. 1988. Thaw lake sediments and sedimentary environments. *Permafrost, Permafrost Fifth International Conference, Trondheim. Proceedings*, Vol. 1: 790–795.

Horikawa, Kiyoshi & Shen, H. W. 1960. *Sand Movement by Wind Action – On the Characteristics of Sand Traps*. Department of the Army, Corps of Engineers, Beach Erosion Board, Technical Memorandum **119**: 1–51 and 6 appendix.

Hörner, N. G. 1926. Brattforsheden. Ett värmländskt randdeltakomplex och dess dyner. *Sveriges Geologiska Undersökning*, Serie C. **342**: 1–192.

Huang, Scott L. & Aughenbaugh, Nolan B. 1987. Sublimation of pore ice in frozen silt. *Journal of Cold Regions Engineering* **1**: 171–181.

Hunt, Charles B. 1954. Desert varnish. *Science* **120**: 183–184.

Huntley, B. & Birks, H. J. B. 1983. *An Atlas of Past and Present Pollen Maps for Europe: 0–13 000 Years Ago*. Cambridge University Press, Cambridge. 667 pp and maps.

Hustich, Ilmari 1948. The Scotch pine in northernmost Finland and its dependence on the climate in the last decades. *Acta Botanica Fennica* **42**: 1–75.

— 1951. The lichen woodlands in Labrador and their importance as winter pastures for domesticated reindeer. *Acta Geographica* **12(1)**: 1–48.

— 1953. The boreal limits of conifers. *Arctic* **6**: 149–162.

— 1966. On the forest-tundra and the northern tree-lines. *Reports from the Kevo Subarctic Research Station* **3**: 1–41.

Irwin, Peter A. & Williams, Colin J. 1985. Snowdrift models. *The Northern Engineer* **17(3)**: 4–11.

Ishimoto, Keishi & Takeuchi, Masao 1984. Mass flux and visibility observed by snow particle counter. *Memoirs of National Institute of Polar Research*, Special Issue **34**: 104–112.

Iversen, J. D. & White, B. R. 1982. Saltation threshold on Earth, Mars and Venus. *Sedimentology* **29**: 111–119.

Izmaiłow, Bogdana 1984. Eolian processes in Alpine belts of the High Tatra Mountains, Poland. *Earth Surface Processes and Landforms* **9**: 143–151.

Jahn, Alfred 1972. Niveo-eolian processes in the Sudetes Mountains. *Geographia Polonica* **23**: 93–110.

— 1975. *Problems of the Periglacial Zone*. PWN – Polish Scientific Publishers, Warszawa. 223 pp. and 116 photographs.

Jauhiainen, Erkki 1969. On soils in the boreal coniferous region. Central Finland-Lapland-Northern Poland. *Fennia* **98(5)**: 1–123.

— 1980. Über den Boden fossiler Dünen in Finnland. *Fennia* **100(3)**:1–32.

Jensen, J. A. D. 1879. Expeditionen til Syd-Grönland i 1878. *Meddelelser om Grønland* **1(2)**: 17–76.

— 1889. Undersøkelse af Grønlands Vestkyst fra 64° til 67° NB, 1884 0g 1885. *Meddelelser om Grønland* **8**: 33–121.

Johnson, H. M. 1953. Preliminary ecological studies of microclimates inhabited by the smaller Arctic and Subarctic mammals. *Proceedings Second Alaska Science Conference 1951*: 125–131.

Johnsson, Gunnar 1958. Periglacial wind and frost erosion at Klageruo, S.W. Scania. *Geografiska Annaler* **40**: 232–243.

Jones, H. G.; Pomeroy, J. W.; Walker, D. A. & Wharton, R. 1994. Snow ecology: A report on a new initiative. *Information North* **20(2)**: 5–6.

Jonsson, Stig 1988. *Observations on the Physical Geography and Glacial History of the Vestfjella Nunataks in Western Dronning Maud Land, Antarctica*. Naturgeografiska Institutionen vid Stockholms universitet, Forskningsrapporter STOU-NG **68**. 57 pp.

Kádár, László 1938. Die periglazialen Binnendünen des Norddeutschen und Polnischen Flachlandes. *Comptes Rendus du Congrès International de Géographie, Amsterdam, Actes du Congrès* **1**: 167–183.

— 1966. Az eolikus felszíni formák természetes rendszere. (Summary: Natural system of eolian landforms). *Földrajzi Értesitö* **15(4)**: 413–448.

Kaldi, J.; Krinsley, D. H. & Lawson, Dan 1978. Experimentally produced aeolian surface textures on quartz sand grains from various environments. In: Whalley, W. Brian (ed.): *Scanning Electron Microscopy in the Study of Sediments*. Geo Abstracts, Norwich. 261–274.

Kalinske, A. A. 1943. Turbulence and the transport of sand and silt by wind. *Annals of the New York Academy of Sciences* **44**: 41–54.

Kallio, Paavo; Laine, Unto & Mäkinen, Yrjö 1971. Vascular flora of Inari Lapland 2. Pinaceae and Cupressaceae. *Reports from the Kevo Subarctic Research Station* **8**: 73–100.

Karlov, N. N. 1969. A new classification of eologlyptolitus. *Biuletyn Peryglacjalny* **19**: 221–229.

Karlstrom, Thor N. V.; Coulter, Henry W.; Fernald, Arthur T.; Williams, John R.; Hopkins, David M.; Péwé, Troy L.; Drewes, Harald; Muller, Ernest H. & Condon, William H. 1964. *Surficial Geology of Alaska*. Maps in scale 1:1,584,000. Interior-geological Survey, Washington D.C.

Karpan, Robin & Karpan, Arlene 1991. Northern dunes. The strange, ice-age desert of Lake Athabasca. *Canadian Geographic* **111(3)**: 43–50.

Karrasch, Heinz 1983. *Die periglaziale Tal- und Reliefasymmetrie*. Abhandlungen der Akademie der Wissenschaften in Göttingen, Dritte Folge **35**: 310–327.

Kawamura, Ryuma 1964. *Study of Sand Movement by Wind*. University of California, Hydraulic Engineering Laboratory, Wave Research Projects, Institute of Engineering Research, Technical Report HEL-2-8: 1–58.

Käyhkö, Jukka 1991. *Eoliset Prosessit Hietatievoilla Enontekiössä*. (Aeolian Processes at Hietatievat, Enontekiö, Finnish Lapland). M.Sc. thesis in Finnish. Department of Geography, University of Helsinki. 166 pp.

— 1997. *Aeolian Activity in Subarctic Fennoscandia – Distribution, History and Modern Processes*. Unpublished Ph.D. thesis. Postgraduate Research Institute for Sedimentology, The University of Reading, UK. 314 pp and lxxxviii.

Käyhkö, J. & Pellikka, P. 1994. Remote sensing of the impact of reindeer grazing on vegetation in northern Fennoscandia using SPOT XS data. *Polar Research* **13**: 115–124.

Käyhkö, J.; Vuorela, A.; Pye, K. & Worsley, P. 1999. Landsat TM mapping of evidence for current wind activity in northern Fennoscandia. *International Journal of Remote Sensing* **20**: 421–443.

Kihlman, A. O. 1890. Pflanzenbiologische Studien aus Russisch Lappland. Ein Beitrag zur Kenntnis der regionalen Gliederung in der polaren Waldgrenze. *Acta Societatis pro Fauna et Flora Fennica* **6(3)**: 1–263.

Kind, R. J. 1981. Snow drift. In: Gray, D. M. & Male, D. H. (eds.): *Handbook of Snow*. Pergamon Press, Toronto. 338–359.

King, L. C. 1936. Wind-faceted stones from the Marlborough strand-plain, New Zealand. *Journal of Geology* **44**: 201–213.

King, Lorenz 1984. Permafrost in Skandinavien – Untersuchungsergebnisse aus Lappland, Jotunheimen und Dovre/Rondane. *Heidelberger geographische Arbeiten* **76**: 1–174.

— 1986. Zonation and ecology of high mountain permafrost in Scandinavia. *Geografiska Annaler* **68A**: 131–139.

King, Lorenz & Seppälä, Matti 1987. Permafrost thickness and distribution in Finnish Lapland – Results of geoelectrical soundings. *Polarforschung* **57(3)**: 127–147.

— 1988. Permafrost sites in Finnish Lapland and their environment. Occurrences de pergelisol en Lapponie Finlandaise. *V International Conference on Permafrost, Proceedings*. Tapir Publishers, Trondheim, Norway. 183–188.

Klein, G. J. 1949. Canadian survey of physical characteristics of snow-covers. *Geografiska Annaler* **31**: 106–124.

Klemsdal, Tormod 1969. Eolian forms in parts of Norway. *Norsk Geografisk Tidsskrift* **23**: 49–66.

Klute, F. & Krasser, L. M. 1940. Über Wüstenlackbildung im Hochgebirge. *Petermanns geographische Mitteilungen* **86**: 21–22.

Knott, P. & Warren, A. 1981. Aeolian processes. In: Goudie, A. S. (ed.): *Techniques in Geomorphology*. Allen and Unwin, London. 226–246.

Kobayashi, Daiji 1972. Studies of snow transport in low-level drifting snow. *Contributions from the Institute of Low Temperature Science, Sapporo*, Series **A24**: 1–58.

Kobayashi, Daiji; Ishikawa, Nobuyoshi & Nishio, Fumihiko 1988. Formation process and direction distribution of snow cornices. *Cold Regions Science and Technology* **15**: 131–136.

Kobayashi, S. 1978. Snow transport by katabatic winds in Mizuho Camp area, East Antarctica. *Journal of the Meteorological Society of Japan* **56**: 130–139.

Kobendza, Jadwiga & Kobendza, Roman 1958. Rozwiewane wydmy Puszczy Kampinoskiej. Résumé: Les dunes éparpillées de la Forêt de Kampinos. *Wydmy Sródladowe Polski* **I**: 95–170.

Koch, J. P. & Wegener, A. 1930. Wissenschaftliche Ergebnisse der dänischen Expedition nach Dronning Louises-Land und Quer über das Inlandeis von Nordgrönland 1912–13 unter Leitung von Hauptman J. P. Koch. *Meddelelser om Grønland* **75(1)**: 1–676.

Komárková, Vêra & Webber, P. J. 1980. Two low arctic vegetation maps near Atkasook, Alaska. *Arctic and Alpine Research* **12**: 447–472.

Konishchev, V. N. 1987. Origin of loess-like silt in Northern Jakutia, USSR. *GeoJournal* **15(2)**: 135–139.

Kopanev, I. D. 1964. Blowing snow in Antarctica. *Soviet Antarctic Expedition Information Bulletin* **II**: 88–90.

Koster, Eduard A. 1988. Ancient and modern cold-climate aeolian sand deposition: a review. *Journal of Quaternary Science* **3(1)**: 69–83.

— 1995. Progress in cold-climate aeolian research. *Quaestiones Geographicae*, *Special Issue* **4**: 155–163.

Koster, Eduard A. & Dijkmans, Jos W. A. 1988. Niveo-aeolian deposits and denivation forms, with special reference to the Great Kobuk sand dunes, northwestern Alaska. *Earth Surface Processes and Landforms* **13**: 153–170.

Koster, E. A.; Galloway, J. P. & Pronk, T. 1984. *Photo-interpretation Map of Surficial Deposits and Landforms of Nogahabara Sand Dunes and Part of the Koyukuk Lowland, Alaska*. U.S. Geological Survey Open-File Report 84-10, 1 sheet in scale 1:60,390.

Köster, Erhard 1960. *Mechanische Gestein- und Bodenanalyse*. Leitfaden der Granulometrie und Morphometrie. München. 171 pp.

— 1964. *Granulometrische und morphometrische Messmethoden an Mineralkörnern, Steinen und sonstigen Stoffen*. Stuttgart. 336 pp.

Kotilainen, Mia 1991. Aavikkopaholaisen jäljillä – dyynikenttien uudelleenaktivoituminen Pohjois-Lapissa. (Summary: Some reactivated dunefields in Northern Lapland, Finland). *Geological Survey of Finland – Report of Investigation* **105**: 105–113.

Kotliakov, V. M. 1959. Snow cover accumulation in a 50-kilometer-wide coastal belt in Antarctica. *Information Bulletin of the Soviet Antarctic Expedition* **13**: 73–75.

Kotliakov, V. M. & Krenke, A. N. 1982. Data on snow cover and glaciers for the global climatic models. In: Eagleson, P. S. (ed): *Land Surface Processes in Atmospheric General Circulation Models*. Cambridge University Press, Cambridge. 449–461.

Koutaniemi, Leo 1984. The role of ground frost, snow cover, ice break-up and flooding in the fluvial processes in the Oulanka river, Finland. *Fennia* **162**: 127–161.

Koutaniemi, Leo & Seppälä, Matti 1986. Aapasuon routaa säätelevät tekijät – esimerkkitapaus Kuusamosta. (Summary: Factors controlling ground frost on an aapa mire in Kuusamo, Finland). *Terra* **98**: 60–67.

Kozarski, Stefan (ed.) 1991. Late Vistulian (Weichselian) and Holocene aeolian phenomena in Central and Northern Europe. *Zeitschrift für Geomorphologie*, Supplementband **90**: 1–207.

Krinsley, D. H. & Doornkamp, J. 1973. *Atlas of Quartz Sand Grain Textures*. Cambridge University Press, Cambridge. 91 pp.

Krinsley, D. & Margolis, S. 1971. Quartz sand grain surface textures. In: Carver, R. E. (ed.): *Procedures in Sedimentary Petrology*. John Wiley & Sons, New York. 151–180.

Krinsley, David H. & McCoy, Floyd 1978. Aeolian quartz sand and silt. In: Whalley, W. Brian (ed.): *Scanning Electron Microscopy in the Study of Sediments*. Geo Abstracts, Norwich. 249–260.

Krumbein, W. C. & Pettijohn, F. J. 1938. *Manual of Sedimentary Petrography*. Appleton – Century – Crofts, New York. 549 pp.

Krygowski, Bogumil 1965. Some Pleistocene and Holocene sedimentary environments in the light of the mechanical graniformametry. *Geografia Polonica* **6**: 117–126.

Krygowski, Bogumil & Krygowski, Tabeusz M. 1965. Mechanical method of estimating the abrasion grade of sand grains (Mechanical graniformametry). *Journal of Sedimentary Petrology* **35**: 496–499.

Kuenen, Ph. H. 1960. Experimental abrasion 4: Eolian action. *Journal of Geology* **68**: 427–449.

Kuenen, Ph. H. & Perdok, W. G. 1962. Experimental abrasion 5. Frosting and defrosting of quartz grains. *Journal of Geology* **70**: 648–658.

Kuhlman, Hans 1958. Quantitative measurements of aeolian sand transport. *Geografisk Tidsskrift* **57**: 51–74.

— 1960. Microenvironments in a Danish dune area, Råbjerg Mile. *Meddelelser fra Dansk Geologisk Forening* **14**: 253–258.

Kuhry-Helmens, K. F., Koster, E. A. & Galloway, J. P. 1985. *Photo-interpretation Map of Surficial Deposits and Landforms of the Kobuk Sand Dunes and Part of the Kobuk Lowland, Alaska*. U.S. Geological Survey, Open-File Report 85-242, 1 sheet in scale 1:63,360.

Kullman, L. 1989. Excursion site 3:3. Tuolluvuoma. In: Sollid, J. L. (ed.): *Northern Finland – Norway – Sweden. Excursion Guide*. Meddelelser fra Geografisk Institutt Universitet i Oslo, Naturgeografisk Serie Rapport 12: 34–38.

— 1991. Ground frost restriction of subarctic *Picea abies* forest in northern Sweden. A dendrochronological analysis. *Geografiska Annaler* **73A**: 167–178.

— 1996. Recent cooling and recession of Norway spruce (*Picea abies* (L.) Karst.) in the forest-alpine tundra ecotone of the Swedish Scandes. *Journal of Biogeography* **23**: 843–854.

Kutzbach, John E. & Guetter, Peter J. 1986. The influence of changing orbital parameters and surface boundary conditions on climate simulations for the past 18,000 years. *Journal of the Atmospheric Sciences* **43**: 1726–1759.

Kutzbach, J. E. & Wright, H. E., Jr. 1985. Simulation of the climate of 18,000 years BP: Results for the North American/North Atlantic/European sector and comparison with the geological record of North America. *Quaternary Science Reviews* **4**: 147–187.

Kuz'min, P. P. 1963. *Snow Cover and Snow Reserves*. Israel Program for Scientific Translation, Translation 828. 140 pp.

Kuznetsov, M. A. 1960. Sculptured forms of the ice surface in Antarctica and their origin. *Information Bulletin of the Soviet Antarctic Expedition* **17**: 189–192.

Lachenbruch, Arthur H. 1966. Contraction theory of ice wedge polygons: A qualitative discussion. *Permafrost International Conference 1963, Proceedings*. National Academy of Sciences – National Research Council, Washington, D.C., Publication 1287: 63–71.

Laity, Julia E. 1992. Ventifact evidence for Holocene wind patterns with east-central Mojave Desert. *Zeitschrift für Geomorphologie N.F.* Supplementband **84**: 73–88.

Lamb, H. H. & Woodroffe, A. 1970. Atmospheric circulation during the last ice age. *Quaternary Research* **1**: 29–58.

Landsberg, S. Y. 1956. The orientation of dunes in Britain and Denmark in relation to wind. *Geographical Journal* **72**: 176–189.

Langham, E. J. 1981. Physics and properties of snowcover. In: Gray, D. M. & Male, D. H. (eds.): *Handbook of Snow*. Pergamon Press, Toronto. 275–337.

Lautridou, J.-P. & Ozouf, J. C. 1982. Experimental frost shattering: 15 years of research at the Centre de Géomorphologie du CNRS. *Progress in Physical Geography* **6**: 215–232.

Lautridou, Jean-Pierre & Seppälä, Matti 1986. Experimental frost shattering of some Precambrian rocks, Finland. *Geografiska Annaler* **68A**: 89–100.

Law, Jane & van Dijk, Deanna 1994. Sublimation as a geomorphic process: a review. *Permafrost and Periglacial Processes* **5**: 237–249.

Lea, Peter D. 1990. Pleistocene periglacial eolian deposits on southwestern Alaska: sedimentary facies and depositional processes. *Journal of Sedimentary Petrology* **60**: 582–591.

Lea, Peter D. & Waythomas, Christopher F. 1990. Late-Pleistocene eolian sand sheets in Alaska. *Quaternary Research* **34**: 269–281.

Leatherman, S. P. 1978. A new sand trap design. *Sedimentology* **25**: 303–306.

Lee, J. A. 1987. A field experiment on the role of small scale wind gustiness in aeolian sand transport. *Earth Surface Processes and Landforms* **12**: 331–335.

Leffingwell, Ernest de K. 1919. *The Canning River Region, Northern Alaska*. U.S. Geological Survey, Professional Paper **109**: 1–251 and 10 plates.

Lehman, Scott J. & Keigwin, Lloyd D. 1992. Sudden changes in North Atlantic circulation during the last deglaciation. *Nature* **356**: 757–762.

Le Ribault, Loic 1978. The exoscopy of quartz sand grains. In: Whalley, Brian (ed.): *Scanning Electron Microscopy in the Study of Sediments*. GeoAbstracts, Norwich. 319–328.

Lewis, W. V. 1939. Snow-patch erosion in Iceland. *Geographical Journal* **94**: 153–161.

— 1940. The function of meltwater in cirque formation. *Geographical Review* **30**: 64–83.

— 1949. The function of meltwater in cirque formation: a reply. *Geographical Review* **39**: 110–128.

Lewkowicz, Antoni G. & Young, Kathy L. 1991. Observations of aeolian transport and niveo-aeolian deposition at three lowland sites, Canadian Arctic Archipelago. *Permafrost and Periglacial Processes* **2**: 197–210.

Li, Zhanqing 1996. Satellite detection of forest fires. Remote Sensing in Canada. *Canadian Centre for Remote Sensing Newsletter* **24(1)**: 5.

Lindé, Krister 1984. Scanning electron microscope studies of different sands and silts. Doctoral dissertation. *Acta Universitatis Upsaliensis* **748**: 1–41 and appendices.

Lindé, Krister & Mycielska-Dowgiallo, Elzbieta 1980. Some experimentally produced microtextures on grain surfaces of quartz sand. *Geografiska Annaler* **62A**: 171–184.

Lindsay, John F. 1973a. Reversing barchan dunes in Lower Victoria Valley, Antarctica. *Geological Society of America Bulletin* **84(5)**: 1799–1805.

— 1973b. Ventifact evolution in Wright Valley, Antarctica. *Geological Society of America Bulletin* **84**: 1791–1797.

Linell, Kenneth A. & Tedrow, John C. F. 1981. *Soil and Permafrost Surveys in the Arctic*. Clarendon Press, Oxford. 279 pp.

Livingstone, Daniel A. 1954. On the orientation of lake basins. *American Journal of Science* **252**: 547–554.

Logie, M. 1982. Influence of roughness elements and soil moisture on the resistance of sand to wind erosion. *Catena Supplement* **1**: 161–173.

Loewe, Fritz 1954. Beiträge zur Kenntnis der Antarktis. *Erdkunde* **8**: 1–15.

— 1961. Fortschritte in der physikalisch-geographischen Kenntnis der Antarktis. *Erdkunde* **15**: 81–92.

— 1962. On the mass economy of the interior of the Antarctic Ice Cap. *Journal of Geophysical Research* **67(13)**: 5171–5177.

— 1970. *The Transport of Snow on Ice Sheets by Wind*. University of Melbourne, Meteorology Department Publication 13: II: 1–69.

Loope, David B.; Swinehart, James B. & Mason, Jon P. 1995. Dune-dammed paleovalleys of the Nebraska Sand Hills: Intrinsic versus climatic controls on the accumulation of lake and marsh sediments. *Geological Society of America Bulletin* **107**: 396–406.

Luckman, B. H. 1977. The geomorphic activity of snow avalanches. *Geografiska Annaler* **59A(1–2)**: 31–48.

— 1978. Geomorphic work of snow avalanches in the Canadian Rockies. *Arctic and Alpine Research* **10**: 261–276.

— 1992. Debris flows and snow avalanche landforms in the Lairig Ghru, Cairngorm Mountains, Scotland. *Geografiska Annaler* **74A**: 109–121.

Luoto, Miska & Seppälä, Matti 2000. Summit peats ('peat cakes') on the fells of Finnish Lapland: continental fragments of blanket mires? *The Holocene* **10**: 229–241.

Maarleveld, G. C. 1965. Frost mounds, a summary of the literature of the past decade. *Medelingen van de Geologische Stichting, Nieuwe Serie* **17**: 1–16.

— 1983. A note on ventifacts and the shape, angularity and surface polish of lydites in fluviatile deposits and in stine pavements. *Geologie en Mijnbouw* **62**: 451–453.

Mabbutt, J. A. 1977. *Desert Landforms*. MIT Press, Cambridge, Massachusetts. 340 pp.

MacCarthy, G. R. & Huddle, J. W. 1938. Shape-sorting of sand grains by wind action. *American Journal of Science* **35**: 64–73.

Mackay, J. Ross 1956. Notes on oriented lakes of the Liverpool Bay area, Northwest Territories. *Revue Canadienne de Géographie* **10(4)**: 169–173.

— 1957. Les lacs orientés de la région de la Baie de Liverpool. Discussion. *Revue Canadienne de Géographie* **11(2–3)**: 175–178.

— 1958. The Anderson River map-area N.W.T. *Geographical Branch, Mines and Technical Surveys, Ottawa*, Memoir **5**: 1–137.

— 1963. The Mackenzie Delta area, N.W.T. *Canadian Department of Mines and Technical Surveys, Geographical Branch*, Memorandum **8**: 1–202.

— 1972. Some observations on ice-wedges, Garry Island, NWT. In: Kerfoot, D. E. (ed.): *Mackenzie Delta Area Monograph*. International Geographical Congress, Montreal. 131–139.

— 1974. Ice-wedge cracks, Garry Island, Northwest Territories. *Canadian Journal of Earth Sciences* **11**: 1336–1383.

— 1978. The use of snow fences to reduce ice-wedge cracking, Garry Island, Northwest Territories. *Geological Survey of Canada*, Paper **78-1A**: 523–524.

— 1988. Ice wedge growth in newly aggrading permafrost, western Arctic coast, Canada. *Permafrost Fifth International Conference, Trondheim, Proceedings* Vol. **1**: 809–814.

Mackay, J. Ross & MacKay, D. K. 1974. Snow cover and ground temperatures at Garry Island, N.W.T. *Arctic* **27**: 287–296.

Maeno, N. 1981. *Science of Ice*. (In Japanese). Hokkaido University Press, Sapporo.

Malin, Michael C. 1984. Preliminary abrasion rate observations in Victoria Valley, Antarctica. *Antarctic Journal of the United States* **18(5)**: 25–26.

— 1985. Abrasion rate observation in Victoria Valley, Antarctica: 340-day experiment. *Antarctic Journal of the United States* **19(5)**: 14–16.

— 1986. Rates of geomorphic modification in ice-free areas southern Victoria Land, Antarctica. *Antarctic Journal of the United States* **20(5)**: 18–21.

— 1988. Abrasion in ice-free areas of southern Victoria Land, Antarctica. *Antarctic Journal of the United States* **22(5)**: 38–39.

— 1992. Short-term variations in the rate of eolian processes, southern Victoria Land, Antarctica. *Antarctic Journal of the United States* **26(5)**: 27–29.

Malin, Michael C. & Eppler, Dean B. 1981. *Eolian Processes in Iceland's Cold Deserts. Reports of Planetary Geology Program – 1981*. NASA Technical Memorandum 84211. NASA Office of Space Science, Washington, D.C. 247–248.

Mannerfelt, C. M. 1945. Några glacialmorfologiska formelement och deras vittnesbörd om inlandsisens avsmältningsmekanik i svensk och norsk fjällterräng. *Geografiska Annaler* **27**: 1–239.

Mansikkaniemi, Hannu & Laitinen, Tuija 1990. Pattern of local wind changes in a fjell region, northern Finland. *Reports from Kevo Subarctic Research Station* **21**: 11 20.

Margolis, Stanley V. & Krinsley, David H. 1971. Submicroscopic frosting on eolian and subaqueous quartz and sand grains. *Bulletin of the Geological Society of America* **82**: 3395–3406.

Marin, Pierre & Filion, Louise 1992. Recent dynamics of subarctic dunes as determined by tree-ring analysis of white spruce, Hudson Bay, Québec. *Quaternary Research* **38**: 316–330.

Markov, K. K.; Bardin, V. I.; Lebedev, V. L.; Orlov, A. I. & Suetova, I. A. 1970. *The Geography of Antarctica*. Israel Program for Scientific Translations, Jerusalem. 370 pp.

Marrs, Ronald W. & Gaylord, David R. 1982. Techniques for interpretation of windflow characteristics from eolian landforms. *Geological Society of America*, Special Paper **192**: 3–17.

Martinelli, M. Jr. 1973. Snow-fence experiments in alpine areas. *Journal of Glaciology* **12**: 291–303.

Martinelli, M. Jr. & Ozment, Arnold 1985. Some strength features of natural snow surfaces that affect snow drifting. *Cold Regions Science and Technology* **11**: 267–283.

Mason, Owen Kenneth (1990). *Beach Ridge Geomorphology of Kotzebue Sound: Implications for Paleoclimatology and Archeology*. Ph.D. Thesis. University of Alaska, Fairbanks. xxii and 262 pp.

Masson, Robert 1995. Satellite remote sensing of polar snow and ice: present status and future directions. *Polar Record* **31**: 99–114.

Mather, K. B. 1960. Katabatic winds south of Mawson, Antarctica. In: *Antarctic Meteorology. Proceedings of the Symposium held in Melbourne, February 1959.* Pergamon Press, London. 317–320.

— 1962. Further observations on sastrugi, snow dunes and the pattern of surface winds in Antarctica. *Polar Record* **11**: 158–171.

Mather, K. B. & Miller, G. S. 1966. The problem of the katabatic winds on the coast of Terre Adélie. *Polar Record* **13 (85)**: 425–432.

— 1967. *Notes on Topographic Factors Affecting the Surface Wind in Antarctica, with Special Reference to Katabatic Winds, and Bibliography.* University of Alaska, Technical Report UAG-R-189: 125 pp.

Mather, K. B. & Goodspeed, M. J. 1959. Australian Antarctic ice thickness measurements and sastrugi observations, Mac-Robertson Land, 1957–58. *Polar Record* **9 (62)**: 436–445.

Matsuoka, Norikazu; Moriwaki, Kiichi & Hirakawa, Kazuomi 1996. Field experiments on physical weathering and wind erosion in an Antarctic cold desert. *Earth Surface Processes and Landforms* **21**: 687–699.

Matthes, François E. 1900. Glacial sculpture of the Bighorn Mountains, Wyoming. *21st Annual Report of the United States Geological Survey 1899–1900 (Part II)*: 167–190.

Mattsson, Jan O. 1973. Vidstjälpta stenar i sand – några fältiakttagelser och enkla försök. (Abstract: Wind-tilted pebbles in sand – some field observations and simple experiments.) *Lunds Universitets Naturgeografiska Institution, Rapporter och Notiser* **14**: 1–41.

Mawson, Douglas 1915. *The Home of the Blizzard. Being the Story of the Australasian Antarctic Expedition, 1911–1914.* Vol. I. William Heinemann, London. xxx and 349 pp.

Mawson's Antarctic Diaries. 1988. Jacka, Fred & Jacka, Eleanor (eds.). Unwin Hyman, London. 414 pp.

Maxwell, J. B. 1980. *The Climate of the Canadian Arctic Islands and Adjacent Waters.* Volume 1. Environment Canada, Atmospheric Environmental Service, Climatological Studies 30: 1–532.

McCabe, L. H. 1939. Nivation and corrie erosion in West Spitsbergen. *Geographical Journal* **94**: 447–465.

McEwan, I. K. 1993. Bagnold's kink: A physical feature of a wind velocity profile modified by blown sand? *Earth Surface Processes & Landforms* **18**: 145–156.

McEwan, I. K. & Willetts, B. B. 1993. Sand transport by wind: a review of the current conceptual model. *Geological Society Special Publication* **72**: 7–16.

McEwan, I. K.; Willetts, B. B. & Rice, M. A. 1992. The grain/bed collision in sand transport by wind. *Sedimentology* **39**: 971–981.

McGowan, H. A. 1994. *Thermal and Dynamic Influences on Alpine Dust Storms, Lake Tekapo, New Zealand.* Unpublished Ph.D. Thesis, University of Canterbury, Christchurch, New Zealand. 268 pp.

McGowan, Hamish A.; Sturman, Andrew P. & Owens, Ian F. 1995. Aeolian dust transport and deposition by foehn winds in an alpine environment, Lake Tekapo, New Zealand. *Geomorphology* **15**: 135–146.

McKay, G. A. & Gray, D. M. 1981. The distribution of snowcover. In: Gray, D. M. & Male, D. H. (eds.): *Handbook of Snow.* Pergamon Press, Toronto. 153–190.

McKay, G. A. & Thompson, H. A. 1968. Snowcover in the Prairie Provinces of Canada. *Transactions of the American Society of Agricultural Engineers* **11**: 812–815.

McKee, Edwin D. 1966. Structures of dunes at White Sands National Monument, New Mexico. *Sedimentology* **7**: 3–69.

— (ed.) 1979. *A Study of Global Sand Seas*. US Geological Survey Professional Paper 1052. Washington. 429 pp.

McKenna-Neuman, Cheryl 1989. Kinetic energy transfer through impact and its role in entrainment by wind of particles from frozen surfaces. *Sedimentology* **36**: 1007–1015.

— 1990a. Observations of winter aeolian transport and niveo-aeolian deposition at Crater Lake, Pangnirtung Pass, N.W.T., Canada. *Permafrost and Periglacial Processes* **1**: 235–247.

— 1990b. Role of sublimation in particle supply for aeolian transport in cold environments. *Geografiska Annaler* **72A**: 329–335.

— 1993. A review of aeolian transport processes in cold environments. *Progress in Physical Geography* **17(2)**: 137–155.

McKenna-Neuman, Cheryl & Gilbert, Robert 1986. Aeolian processes and landforms in glaciofluvial environments of southeastern Baffin Island, N.W.T., Canada. In: Nickling, William G. (ed.): *Aeolian Geomorphology. Proceedings of the 17th Annual Binghamton Geomorphology Symposium*. Allen & Unwin, Winchester. 213–235.

McKenna-Neuman, C. & Nickling, W. G. 1989. A theoretical and wind tunnel investigation of the effect of capillary water on the entrainment of sediment by wind. *Canadian Journal of Soil Science* **69**: 79–96.

— 1995. Aeolian sediment flux decay: non-linear behaviour on developing deflation lag surfaces. *Earth Surface Processes and Landforms* **20**: 423–435.

McKeon, J. B. 1989. Late glacial dunes, ventifacts and wind direction in west-central Maine. In: Tucker, R. D. & Marvinney, R. G. (eds.): *Studies in Maine Geology 6. Quaternary Geology*. Maine Geological Survey, Augusta. 89–101.

McKeown, S.; Clark, J. I. & Matheson D. 1988. Frost penetration and thermal regime in dry gravel. *Journal of Cold Regions Engineering* **2(3)**: 111–123.

Mears, Jr., Brainerd 1987. Late Pleistocene periglacial wedge sites in Wyoming: an illustrated compendium. *Geological Survey of Wyoming, Memoir* **3**: 1–77.

Meinardus, W. 1938. Klimakunde der Antarktis. In: Köppen, W. & Geiger, R. (eds.): *Handbuch der Klimatologie IV: U.* Gebrüder Borntraeger, Berlin. 133 pp.

Mellor, Malcolm 1960. Gauging Antarctic drift snow. In: *Antarctic Meteorology. Proceedings of the Symposium Held in Melbourne, February 1959*. Pergamon Press, London. 347–355.

— 1963. Polar snow – A summary of engineering properties. In: Kingery, W. D. (ed.): *Ice and Snow. Properties, Processes and Applications*. The M.I.T. Press, Cambridge, Massachusetts. 528–559.

— 1965. *Blowing Snow*. Cold Regions Science and Engineering CRREL monographs Part III, Section A3c. 79 pp.

— 1977. Engineering properties of snow. *Journal of Glaciology* **19(81)**: 15–66.

Mellor, M. & Radok, U. 1960. Some properties of drifting snow. In: *Antarctic Meteorology. Proceedings of the Symposium Held in Melbourne, February 1959*. Pergamon Press, London. 333–346.

Merrill, G. P. 1898. Desert varnish. *U.S. Geological Survey Bulletin* **150**: 389–391.

Middendorff, A. Th. v. 1867. *Reise in den äussersten Norden und Osten Sibiriens. Band IV, Übersicht Natur Nord- und Ost-Sibiriens. Theil I. Einleitung, Geographie,*

Hydrographie, Orographie, Geognosie, Klima und Gewächse. Kaiserlichen Akademie der Wissenschaften, St. Petersburg.

Migala, Krzysztof & Sobik, Mieczyslaw 1984. Deflation and nival eolian phenomena observed under conditions of congelation in the forefield of the Werenskiold glacier (SW Spitsbergen). *Zeitschrift für Gletscherkunde und Glazialgeologie* **20**: 197–206.

Miller, Maynard M. 1961. A distribution study of abandoned cirques in the Alaska–Canada boundary range. In *Geology of the Arctic*, University of Toronto Press. 833–847.

— 1967. Alaska's mighty rivers of ice. *National Geographic Magazine* **131(2)**:194–217.

Miotke, Franz-Dieter 1979. Die Formung und Formungsgeschwindigkeit von Windkantern in Victoria-Land, Antarktis. *Polarforschung* **49(1)**: 30–43.

— 1982. Formation and rate of formation of ventifacts in Victoria Land, Antarctica. *Polar Geography and Geology* **6(2)**: 98–113.

— 1983. Colder than Siberia, drier than the Sahara: the Antarctic – a total desert. *German Research – Reports of the DFG* **1/83**: 22–25.

Möller, Per; Hjort, Christian; Adrielsson, Lena & Salvigsen, Otto 1994. Glacial history of interior Jameson Land, East Greenland. *Boreas* **23**: 320–348.

Moore, T. R. 1978. Soil development in Arctic and Subarctic areas of Quebec and Baffin Island. In: Mahaney, W. C. (ed.): *Quaternary Soils*. Geo Abstracts, Norwich. 379–411.

Morgan, R. P. C. 1986. *Soil Erosion and Conservation*. Longman, Hong Kong. 298 pp.

Morin, Hubert & Payette, Serge 1988. Holocene gelifluction in a snowpatch environment at the Forest Tundra transition along the eastern Hudson Bay coast. *Boreas* **17**: 79–88.

Morris, Wesley R. & Peters, Norman L. 1960. IGY Meteorology in Antarctica. Inside Antarctica No. 5 - Byrd Station. *Weatherwise* **13**: 162–165.

Mrózek, Wladyslaw 1958. Wydmy Kotliny Torunsko – Bydgoskiej. *Wydmy Sródladowe Polski* **II**: 1–59.

Nakamori, Toru 1994. Meteorological observations in mountain area of cold snowy regions. *Fourth International Symposium on Cold Region Development (ISCORD '94), June 13–16, 1994, Finland. Extended Abstracts*. 308–309.

Nansen, Fridtjof 1920. *En Ferd til Spitsbergen*. Jacob Dybwads Forlag, Kristiania. 279 pp.

— 1922. *Spitzbergen*. F. U. Brockhaus, Leipzig. 327 pp.

Needham, C. E. 1937. Ventifacts from New Mexico. *Journal of Sedimentary Petrology* **7**: 31–33.

Nichols, Robert L. 1966. Geomorphology of Antarctica. In: Tedrow, J. C. F. (ed.): *Antarctic Soils and Soil-forming Processes*. American Geophysical Union Antarctic Research Series 8: 1–46.

Nichols, Robert L. 1969. Geomorphology of Inglefield Land, North Greenland. *Meddelelser om Grønland* **188(1)**:1–109.

Nickling, W. G. 1978. Eolian sediment transport during dust storms: Slims River Valley, Yukon Territory. *Canadian Journal of Earth Sciences* **15**: 1069–1084.

— 1984. The stabilizing role of bonding agents on the entrainment of sediment by wind. *Sedimentology* **31**: 111–117.

— 1988. The initiation of particle movement by wind. *Sedimentology* **35**: 499–511.

— 1994. Aeolian sediment transport and deposition. In: Pye, Kenneth (ed.): *Sediment Transport and Depositional Processes*. Blackwell Scientific Publications, Oxford. 293–350.

Nickling, W. G. & Brazel, A. J. 1985. Surface wind characteristics along the icefield ranges, Yukon Territory, Canada. *Arctic and Alpine Research* **17**: 125–134.

Nickling, W. G. & Ecclestone, M. 1981. The effects of soluble salts on the threshold shear velocity of fine sand. *Sedimentology* **28**: 505–510.

Nickling, W. G. & McKenna-Neuman, C. 1995. Development of deflation lag surfaces. *Sedimentology* **42**: 403–414.

Nielsen, Niels 1928. Landskabet Syd-Ost for Hofsjökull i det indre Island. *Geografisk Tidsskrift* **31**: 23–44.

— 1933. Contributions to the physiography of Iceland with particular reference to the highlands west of Vatnajökull. *Det kongelige danske Videnskabernes Selskabs Skrifter. Naturvidenskabelig og Mathematisk Afdeling* **9**. Raekke IV(5): 183–288.

Niessen, Augusta C. H. M.; Koster, Eduard A. & Galloway, John P. 1984. *Periglacial Sand Dunes and Eolian Sand Sheets. An Annotated Bibliography*. U.S. Geological Survey Open-file Report 84–167: 1–61.

Nitz, Bernhard 1965. Windgeschliffene Geschiebe und Steinsohlen zwischen Fläming und Pommerscher Eisrandlage. *Geologie* **14**: 686–698.

Nordenskjöld, Otto 1910. Från danska Sydvästgrönland. *Ymer* **30**: 17–46.

Nordenskjöld, Otto & Mecking, Ludwig 1928. *The Geography of the Polar Regions*. American Geographical Society. Special Publication 8. 359 pp.

Nordstrom, Karl; Psuty, Norbert & Carter, Bill (eds.) 1990. *Coastal Dunes: Form and Process*. John Wiley & Sons, Chichester. 392 pp.

Norrman, John O. 1981. Coastal dune systems. In: Bird, Eric C. F. & Koike, Kazuyuki (eds.): *Coastal Dynamics and Scientific Sites*. Reports on the projects of the I. G. U. Commission on the coastal environment 1976–1980. Department of Geography, Komazawa University, Tokyo, 119–157.

Norton, D. C.; Bolsenga, S. J. & Badarch, M. 1993. Snow depth mapping in Mongolia. *50th Eastern Snow Conference, 61th Western Snow Conference, Quebec City*. 381–387.

Nyberg, Rolf 1985. *Debris Flows and Slush Avalanches in Northern Swedish Lappland*. Ph.D. thesis. Lunds Universitets Geografiska Institution, Avhandlingar 97: 1–222.

— 1989. Observations of slushflows and their geomorphological effects in the Swedish mountain area. *Geografiska Annaler* **71A**: 185–198.

— 1991. Geomorphic processes at snowpatch sites in the Abisko mountains, northern Sweden. *Zeitschrift für Geomorphologie N.F.* **35**: 321–343.

Obleitner, F. 1994. Climatological features of glacier and valley winds at the Hintereisferner (Ötztal Alps, Austria). *Theoretical and Applied Climatology* **49(4)**: 225–239.

O'Brian, Morrough P. & Rindlaub, Bruce D. 1936. The transport of sand by wind. *Civil Engineering* **6(5)**: 325–327.

Odensky, Wm. 1958. U-shaped dunes and effective wind directions in Alberta. *Canadian Journal of Soil Science* **38**: 56–62.

Ohata, Tetsuo 1989a. Katabatic wind on melting snow and ice surfaces (I). Stationary glacier wind on a large maritime glacier. *Journal of the Meteorological Society of Japan* **67**: 99–112.

— 1989b. Katabatic wind on melting snow and ice surfaces (II). Application of a theoretical model. *Journal of the Meteorological Society of Japan* **67**: 113–120.

Ohlson, Birger 1957. Om flygsandfälten på Hietatievat i östra Enontekiö. (Summary: On the drift-sand formations at Hietatievat in eastern Enontekiö.) *Terra* **69**: 129–137.

— 1964. Frostaktivität, Verwitterung und Bodenbildung in den Fjeldgegenden von Enontekiö, Finnisch-Lappland. *Fennia* **89(3)**: 1–180.

Ohmura, Atsumu & Reeh, Niels 1991. New precipitation and accumulation maps for Greenland. *Journal of Glaciology* **37(125)**: 140–148.

Oksanen, Lauri & Virtanen, Risto 1995. Topographic, altitudinal and regional patterns in continental and suboceanic heath vegetation of northern Fennoscandia. *Acta Botanica Fennica* **153**: 1–80.

Oksanen, Lauri; Moen, Jon & Helle, Timo 1995. Timberline patterns in northernmost Fennoscandia. Relative importance of climate and grazing. *Acta Botanica Fennica* **153**: 93–105.

Oura, H.; Ishida, T.; Kobayashi, D.; Kobayashi, S. & Yamada, T. 1967. Studies on blowing snow II. In: Oura, H. (ed.): *Physics of Snow and Ice*, Part 2. Institute of Low Temperature Science, Sapporo: 1099–1117.

Owens, J. S. 1927. The movement of sand by wind. *The Engineer* **143**: 377.

Parish, Thomas R. & Wendler, Gerd 1991. The katabatic wind regime at Adelie Land, Antarctica. *International Journal of Climatology* **11**: 97–107.

Passarge, Siegfried 1921. *Vergleichende Landschaftskunde. Heft 2. Kältewüsten und Kältesteppen.* Dietrich Reimer/ Ernst Vohsen/ A.-G. Berlin. 1–163.

Paterson, W. S. B. 1981. *Physics of Glaciers*. 2nd edition. Pergamon Press, Oxford. 380 pp.

Payette, Serge & Filion, Louise 1993. Holocene water-level fluctuations of a subarctic lake at the tree-line in northern Québec. *Boreas* **22**: 7–14.

Payette, Serge; Morneau, Claude; Sirois, Luc & Desponts, Mireille 1989. Recent fire history of the northern Québec biomes. *Ecology* **70**: 656–673.

Peary, E. A. 1898. Journeys in North Greenland. *Geographical Journal* **11**: 213–240.

Peltier, Louis C. 1950. The geographic cycle in periglacial regions as it is related to climatic geomorphology. *Annals of the Association of American Geographers* **40(3)**: 214–236.

Penny, Cheryl E. & Pruitt, William O. 1984. Subnivean accumulation of CO_2 and its effects on winter distribution of small mammals. *Special Publication Carnegie Museum of Natural History* **10**: 373–380.

Pedgley, D. E. 1967. The shapes of snowdrifts. *Weather* **22**: 42–48.

Périard, Christophe & Pettré, Paul 1993. Some aspects of the climatology of Dumont d'Urville, Adélie Land, Antarctica. *International Journal of Climatology* **13**: 313–327.

Pernarowski, Leszek 1959. O procesie sortowania piasków eolicznych na przykładzie wydm okolic Rzędzowa. (Summary: Notes on sorting of aeolian sands.) *Czasopismo Geograficzne* **30**: 33–60.

— 1966. Glacjalny i postglacjalny cyrkulacja atmosfery w swietle kierunku wiatrów wydmotwórczych (Summary: Glacial and postglacial atmospheric circulation in the light of directions of dune-forming winds). *Czasopismo Geograficzne* **37**: 3–24.

Perrier, Alain 1982. Land surface processes: vegetation. In: Eagleson, P. S. (ed.): *Land Surface Processes in Atmospheric General Circulation Models*. Cambridge University Press, Cambridge. 395–448.

Peterson, K. M. & Billings, W. D. 1980. Tundra vegetational patterns and succession in relation to microtopography near Atkasook, Alaska. *Arctic and Alpine Research* **12**: 473–482.

Petterssen, Sverre 1950. Some aspects of the general circulation of the atmosphere. *Centenary Proceedings of the Royal Meteorological Society, London.* 120–155.

Pettijohn, F. J. 1975. *Sedimentary Rocks*. 3rd edition. Harper and Row, New York. 628 pp.

Pettijohn, F. J.; Potter, P. E. & Siever, R. 1987. *Sand and Sandstone*. Springer-Verlag, New York. Second edition. 553 pp.

Pettré, Paul & André, Jean-Claude 1991. Surface-pressure change through Loewe's phenomena and katabatic flow jumps: study of two cases in Adélie Land, Antarctica. *Journal of the Atmospheric Sciences* **48**: 557–571.

Péwé, Troy L. 1955. Origin of the upland silt near Fairbanks, Alaska. *Bulletin of the Geological Society of America* **66**: 699–724.

⊿ 1959. Sand-wedge polygons (tesselations) in the McMurdo Sound Region, Antarctica – A progress report. *American Journal of Science* **257**: 545–552.

— 1960. Multiple glaciation in the McMurdo Sound region, Antarctica – a progress report. *Journal of Geology* **68(5)**: 498–514.

— (ed.) 1965. *Guidebook to the Quaternary Geology. Central and South-Central Alaska*. 1965 INQUA Field Conference F.

— 1974. Geomorphic processes in polar deserts. In: Smiley, Terah L. & Zumberge, James H. (eds.): *Polar Deserts and Modern Man*. The University of Arizona Press, Tucson. 33–52.

Péwé, Troy L. & Journaux, André 1983. *Origin and Character of Loesslike Silt in Unglaciated South-Central Yakutia, Siberia, U.S.S.R.* U.S. Geological Survey, Professional Paper 1262: 1–46.

Péwé, T. L. & Reger, R. D. (eds.) 1983. *Guidebook to Permafrost and Quaternary Geology along the Richardson and Glenn Highways between Fairbanks and Anchorage, Alaska*. Fourth International Conference on Permafrost, July 18–22, 1983, University of Alaska, Fairbanks, Alaska, U.S.A. 263 pp.

Philip, Arne L. 1990. Ice-pushed boulders on the shores of Gotland, Sweden. *Journal of Coastal Research* **6**: 661–676.

Piotrowski, Andrzej 1983. Results of investigations over a magnitude of aeolian transport in the western part of Oscar II Land (NW Spitsbergen) during summer 1979. *Acta Universitatis Nicolai Copernici, Geografia XVIII*, **56**: 62–67.

Pissart, A. 1966. Le rôle géomorphologique du vent dans la région de Mould Bay (Ile Prince Patrick – N.W.T. – Canada). *Zeitschrift für Geomorphologie N.F.* **10**: 226–236.

Pissart, A.; Vincent, J.-S. & Edlund, S. A. 1977. Dépôts et phénomènes éoliens sur l'île de Banks, Territoires du Nord-Ouest, Canada. *Canadian Journal of Earth Sciences* **14**: 2462–2480.

Pluis, J. L. A. & De Winder, B. 1990. Natural stabilization. *Catena Supplement* **18**: 195–208.

Polar Regions Atlas. 1978. Central Intelligence Agency. Washington.

Polunin, Nicholas 1951. The real Arctic: suggestions for its delimitation, subdivision and characterization. *Journal of Ecology* **39**: 308–315.

Porsild, A. E. 1938. Earth mounds in unglaciated Arctic Northwestern America. *Geographical Review* **28**: 46–58.

Poser, H. 1932. Einige Untersuchungen zur Morphologie Ostgrönlands. *Meddelelser om Grønland* **84(5)**: 1–55.

— 1936. Talstudien aus Westspitzbergen und Ostgrönland. *Zeitschrift für Gletscherkunde* **24**: 43–98.

— 1948. Äolische Ablagerungen und Klima des Spätglazials in Mittel- und Westeuropa. *Die Naturwissenschaften* **35**: 269–276, 307–312.

— 1950. Zur Rekonstruktion der spätglazialen Luftdruckverhältnisse in Mittel- und Westeuropa auf Grund der vorzeitlichen Binnendünen. *Erdkunde* **4**: 81–88.

Potter, S. A. (ed.) 1987. *ANARE Antarctic Field Manual*. Third edition. Antarctic Division, Kingston, Tasmania. 140 pp.

Powers, W. E. 1936. The evidences of wind abrasion. *Journal of Geology* **44**: 214–219.

Price, W. Armstrong 1960. Barchans of southern Peru. *Geographical Review* **50**: 585–586.

— 1963. The oriented lakes of Arctic Alaska: A discussion. *Journal of Geology* **71**: 530–531.

— 1968a. Carolina Bays. In: Fairbridge, Rhodes W. (ed.): *The Encyclopedia of Geomorphology*. Encyclopedia of Earth Sciences Series Vol. III, Reinhold Book Corporation, New York. 102–109.

— 1968b. Oriented lakes. In: Fairbridge, Rhodes W. (ed.): *The Encyclopedia of Geomorphology*. Encyclopedia of Earth Sciences Series Vol. III, Reinhold Book Corporation, New York. 784–796.

Pruitt, Jr, William O. 1984. Snow and living things. In: Olson, Rod; Hastings, Ross & Geddes, Frank (eds.): *Northern Ecology and Resource Management*. The University of Alberta Press. Edmonton. 51–77.

Putnins, P. 1970. The climate of Greenland. In: Orvig. S. (ed.): *Climates of the Polar Regions*. World Survey of Climatology vol. 14. Elsevier, Amsterdam. 3–128.

Pye, Kenneth 1983. Coastal dunes. *Progress in Physical Geography* **7**: 531–557.

— 1984. Loess. *Progress in Physical Geography* **8**: 176–217.

— 1987. *Aeolian Dust and Dust Deposits*. Academic Press, London. 334 pp.

— 1992. Aeolian dust transport and deposition over Crete and adjacent parts of the Mediterranean Sea. *Earth Surface Processes and Landforms* **17**: 271–288.

Pye, Kenneth & Tsoar, Haim 1990. *Aeolian Sand and Sand Dunes*. Unwin Hyman, London. 396 pp.

Pyritz, Ewald 1972. Binnendünen und Flugsandebenen im niedersächsischen Tiefland. *Göttinger geographische Abhandlungen* **61**: 1–153.

Qingsong, Zhang 1983. Periglacial landforms in the Vestfold Hills, East Antarctica: Preliminary observations and measurements. In: Oliver, R. L.; James, P. R. & Jago, J. B. (eds.): *Antarctic Earth Science, Fourth International Symposium*. Cambridge University Press, New York. 478–481.

Radok, Uwe 1970. *Boundary Processes of Drifting Snow*. University of Melbourne, Meteorology Department Publication 13: I: 1–20.

— 1977. Snow drift. *Journal of Glaciology* **19**: 123–139.

Rapp, Anders 1959. Avalanche boulder tongues in Lappland. A description of little-known landforms of periglacial debris accumulation. *Geografiska Annaler* **41**: 34–48.

— 1960. Recent development of mountain slopes in Kärkevagge and surroundings, northern Scandinavia. *Geografiska Annaler* **42**: 73–200.

— 1982. Zonation of permafrost indicators in Swedish Lappland. *Geografisk Tidsskrift* **82**: 37–38.

— 1983. Impact of Nivation in Steep Slopes in Lappland and Scania, Sweden. *Abhandlungen der Akademie der Wissenschaften in Göttingen, Mathematisch-Physikalische Klasse, Dritte Folge* **35**: 97–115.

— 1986. Comparative studies of actual and fossil nivation in north and south Sweden. *Zeitschrift für Geomorphologie*, Supplementband **60**: 251–263.

Rapp, Anders; Nyberg, Rolf & Lindh, Lars 1986. Nivation and local glaciation in N. and S. Sweden. A progress report. *Geografiska Annaler* **68A**: 197–205.

Raudkivi, A. J. 1990. *Loose Boundary Hydraulics*. 3rd edition. Pergamon Press, Oxford. 538 pp.

Reed, Richard J. & Kunkel, Bruce A. 1960. The Arctic circulation in summer. *Journal of Meteorology* **17**: 489–506.

Reimer, A. 1980. The effect of wind on heat transfer in snow. *Cold Regions Science and Technology* **3**: 129–137.

Rex, R. W. 1961. Hydrodynamic analysis of circulation and orientation of lakes in northern Alaska. In: Raasch, Gilbert O. (ed.): *Geology of the Arctic*. University of Toronto Press vol. 2: 1021–1043.

Rickert, D. A. & Tedrow, J. C. F. 1967. Pedologic investigations of some aeolian deposits of northern Alaska. *Soil Science* **104**: 250–262.

Riezebos, P. A.; Boulton, G. S.; van der Meer, J. J. M.; Ruegg, G. H. J.; Beets, D. J.; Castel, I. I. Y.; Hart, J.; Quinn, I.; Thornton, M. & van der Wateren, F. M. 1986. Products and effects of modern eolian activity on a nineteenth-century glacier-pushed ridge in West Spitsbergen, Svalbard. *Arctic and Alpine Research* **18(4)**: 389–396.

Riordan, A. J. 1975. The climate of Vanda Station, Antarctica. In: Weller, Gunter & Bowling, Sue Ann (eds.): *Climate of the Arctic*. 24th Alaska Science Conference, Fairbanks. 268–275.

Ritchie, James C. 1993. Northern vegetation. In: French, Hugh M. & Slaymaker, Olav (eds.): *Canada's Cold Environments*. McGill-Queen's University Press, Montreal & Kingston. 93–116.

Ritchie, J. C. & Hare, F. K. 1971. Late Quaternary vegetation and climate near the arctic tree line of northwest North America. *Quaternary Research* **1**: 331–342.

Robertson-Rintoul, Maralyn J. 1990. A quantitative analysis of the near-surface wind flow pattern over coastal parabolic dunes. In: Nordstrom, Karl; Psuty, Norbert & Carter, Bill (eds.). *Coastal Dunes. Form and Process*. Wiley, Chichester. 57–78.

Robinson, David A.; Dewey, Kenneth F. & Heim, Richard R., Jr. 1993. Global snow cover monitoring: An update. *Bulletin of the American Meteorological Society* **74**: 1689–1696.

Rochette, Jean-Claude & Cailleux, André 1971. Dépôts nivéo-éoliens annuels à Poste-de-la-Baleine, Nouveau-Québec. *Revue de Géographie de Montréal* **25**: 35–41.

Rosen, Peter S. 1978. An efficient, low cost, aeolian sampling system. *Geological Survey of Canada, Current Research*, Paper 78–1A, 531–532.

— 1979. Boulder barricades in Central Labrador. *Journal of Sedimentary Petrology* **49**: 1113–1124.

Rosenfeld, G. A. & Hussey, K. M. 1958. A consideration of the problem of oriented lakes. *Iowa Academy of Science, Proceedings* **65**: 279–287.

Rouse, Wayne R. 1993. Northern climates. In: French, Hugh M. & Slaymaker, Olav (eds.): *Canada's Cold Environments*. McGill-Queen's University Press, Montreal & Kingston. 65–92.

Rowe, J. S. & Hermesh, R. 1974. Saskatchewan's Athabasca sand dunes. *Nature Canada* **3(3)**: 19–23.

Rubin, Morton J. & Giovinetto, Mario B. 1962. Snow accumulation in Central West Antarctica as related to atmospheric and topographic factors. *Journal of Geophysical Research* **67(13)**: 5163–5170.

Rudberg, Sten 1968. Wind erosion – preparation of maps showing the direction of eroding winds. *Biuletyn Peryglacjalny* **17**: 181–193.

— 1970. Naturgeografiska seminarieuppsatser vid Göteborgs universitet höstterminen 1959 – vårterminen 1969 (Seminar papers in physical geography at the University of Gothenburg 1959–1969 summarized and edited by Sten Rudberg). *Meddelanden från Geografiska Föreningen i Göteborg Gothia* **10**: 1–164.

— 1974. Some observations concerning nivation and snow melt in Swedish Lapland. *Abhandlungen der Akademie der Wissenschaften in Göttingen, Mathematisch-Physikalische Klasse, Dritte Folge* **29**: 263–273.

— 1994. Glacial cirques in Scandinavia. *Norsk Geografisk Tidsskrift* **48**: 179–197.

Ruegg, Gerard H. J. 1983. Periglacial eolian evenly laminated sandy deposits in the late Pleistocene of NW Europe, a facies unrecorded in modern sedimentological handbooks. In: Brookfield, M. E. & Ahlbrandt, T. S. (eds.): *Eolian Sediments and Processes*. Elsevier, Amsterdam. 455–482.

Russell, I. C. 1896–97. Glaciers of Mount Rainier. *U.S. Geological Survey 18th Annual Report, Part II*: 391.

Rutin, J. 1983. *Erosional Processes on a Coastal Sand Dune, De Blink Noordwijkerhout, The Netherlands.* Doctoral dissertation. Faculty of Science, University of Amsterdam. 144 pp.

Rutter, N. W.; Foscolos, A. E. & Hughes, O. L. 1978. Climatic trends during the Quaternary in central Yukon based upon pedological and geomorphological evidence. In: Mahaney, W. C. (ed.): *Quaternary Soils*. Geo Abstracts, Norwich. 309–359.

Ruz, Marie-Hélène 1993. Coastal dune development in a thermokarst environment: some implications for environmental reconstruction, Tuktoyaktuj Peninsula N.W.T. *Permafrost and Periglacial Processes* **4**: 255–264.

Ryan, William L. 1990. Surface water supplies. In: Ryan, William L. & Crissman, Randy D. (eds.): *Cold Regions Hydrology and Hydraulics*. American Society of Civil Engineers, New York. 301–316.

Salisbury, Frank B. 1985. Plant growth under snow. *Aquilo, Ser. Botanica* **23**: 1–7.

Samuelsson, Carl 1921. Till frågan om vinderosion i arktiska trakter med särskild hänsyn till de å Spetsbergen rådande förhållandena. *Ymer* **41**: 122–138.

— 1925. Några studier över erosionsföreteelserna på Island. *Ymer* **45**: 339–355.

— 1926. Studien über die Wirkungen des Windes in den kalten und gemässigten Erdteilen. *Bulletin of the Geological Institution of the University of Upsala* **20**: 57–230.

Samuelsson, Gunnar (1910). Scottish peat mosses. A contribution to the knowledge of the late-quaternary vegetation and climate of North Western Europe. *Bulletin of the Geological Institute of Upsala* **10**: 197–260.

Sanderson, Marie 1950. Is Canada's Northwest subhumid? *Canadian Geographical Journal* **31**: 142–146.

Sapper, Karl 1909. Die Bedeutung des Windes auf Island. *Aus der Natur (Leipzig)* **5(1)**: 43–51 and 77–84.

Sarre, R. D. 1987. Aeolian sand transport. *Progress in Physical Geography* **11**: 157–182.

Sater, John E. 1968. Arctic regions. In: Fairbridge, Rhodes W. (ed.): *The Encyclopedia of Geomorphology*. Encyclopedia of Earth Sciences Series Vol. III, Reinhold Book Corporation, New York. 22–28.

Savat, J. 1982. Common and uncommon selectivity in the process of fluid transportation: Field observations and laboratory experiments on bare surfaces. In Yaalon, Dan H. (ed.): *Aridic Soils and Geomorphic Processes*. Catena Supplement 1: 139–159.

Savile, D. B. O. 1963. Factors limiting the advance of spruce at Great Whale River, Quebec. *Canadian Field-Naturalist* **77**: 95–97.

Schemenauer, R. S.; Berry, M. O. & Maxwell, J. B. 1981. Snowfall formation. In: Gray, D. M. & Male, D. H. (eds.): *Handbook of Snow*. Pergamon Press, Toronto. 129–151.

Schlichting, Hermann 1979. *Boundary-layer Theory*. McGraw-Hill, New York. 7th edition. 817 pp.

Schlyter, Peter 1991. Recent and periglacial wind action in Scandia and adjacent areas of S Sweden. *Zeitschrift für Geomorphologie*, Supplement-Band **90**: 143–153.

— 1992. Large sorted stone polygons, and ventifact distribution, in the Syrkadal area, Scania, S. Sweden. *Geografiska Annaler* **74A**: 219–226.

— 1994. Paleo-periglacial ventifact formation by suspended silt or snow – site studies in South Sweden. *Geografiska Annaler* **76A**: 187–201.

— 1995a. *Palaeo-wind Abrasion in Southern Scandinavia. Field and laboratory studies*. Meddelanden från Lunds Universitets Geografiska Institutioner, Avhandlingar **122**: 1–116.

— 1995b. Ventifacts as palaeo-wind indicators in southern Scandinavia. *Permafrost and Periglacial Processes* **6**: 207–219.

Schlyter, P.; Maeno, N.; Nishimura, K. & Kosugi, K. 1995. Wind abrasion of rock – Laboratory experiments with quartz and snow particles. In: Schlyter, Peter: *Palaeo-wind Abrasion in Southern Sweden. Field and Laboratory Studies*. Meddelanden från Lunds Universitets Geografiska Institutioner, Avhandlingar **122**: 101–116.

Schmidt, R. A. 1980. Threshold wind-speeds and elastic impact in snow transport. *Journal of Glaciology* **26**: 453–467.

— 1982. Vertical profiles of wind speed, snow concentration and humidity in blowing snow. *Boundary Layer Meteorology* **23**: 223–246.

Schoewe, Walter H. 1932. Experiments on the formation of wind-faceted pebbles. *American Journal of Science*, 5th series, **24**: 111–134.

Schönhage, W. 1969. Note on the ventifacts in The Netherlands. *Biuletyn Peryglacjalny* **20**: 355–360.

Schumacher, R. 1988. Aschenaggregate in vulkaniklastischen Transportsystemen. Ph.D. Thesis, University of Bochum, Germany. 152 pp.

Schunke, Ekkehard 1986. Periglazialformen und Morphodynamik im südlichen Jameson-Land, Ost-Grönland. *Abhandlungen der Akademie der Wissenschaften in Göttingen, Mathematisch-Physikalische Klasse*, Dritte Folge **36**: 1–142.

Schutz, Charles & Bregman, L. D. 1988. *Global Annual Snow Accumulation by Months*. A RAND Note N-2687-RC: 1–85.

Schütz, Lothar; Jaenicke, Ruprecht & Pietrer, Horst 1981. Saharan dust transport over the North Atlantic Ocean. *Geological Society of America, Special Paper* **186**: 87–100.

Schwan, J. 1988. The structure and genesis of Weichselian to early Holocene aeolian sand sheets in western Europe. *Sedimentary Geology* **55**: 197–232.

Schwerdtfeger, W. 1970. The climate of the Antarctic. In: Orvig, S. (ed.): *Climates of the Polar Regions*. World Survey of Climatology vol. 14. Elsevier, Amsterdam. 253–355.

Schytt, V. 1959. The glaciers of the Kebnekajse-Massif. *Geografiska Annaler* **41**: 213–227.

Scoresby, W. (1820). *An Account of the Arctic Regions, with a History and Description of the Northern Whale-fishery*. Vol. I. Archibald Constable and Co., Edinburgh. xx and 551 pp.

Sekyra, Josef 1969. Periglacial phenomena in the oases and the mountains of the Enderby Land and the Dronning Maud Land (East Antarctica). *Biuletyn Peryglacjalny* **19**: 277–289.

Selby, M. J. 1977. Transverse erosional marks on ventifacts from Antarctica. *New Zealand Journal of Geology and Geophysics* **20**: 949–969.

Selby, M. J.; Palmer, R. W. P.; Smith, C. J. R. & Rains, R. B. 1973. Ventifact distributions and wind directions in the Victoria Valley, Antarctica. *New Zealand Journal of Geology and Geophysics* **16**: 303–306.

Selby, M. J.; Rains, R. B. & Palmer, R. W. P. 1974. Eolian deposits of the ice-free Victoria Valley, Southern Victoria Land and Antarctica. *New Zealand Journal of Geology and Geophysics* **17**: 543–562.

Seligman, G. 1980. *Snow Structure and Ski Fields.* 3rd edition. First published in 1936. International Glaciological Society, Cambridge. 555 pp.

Sellman, Paul V.; Carey, Kevin L.; Keeler, Charles & Hartwell, Allan, D. 1972. Terrain and coastal conditions on the Arctic Alaskan coastal plain. Arctic environmental data package. Supplement 1. *CRREL Special Report* **165**: 1–72.

Seppälä, Matti 1966. Recent ice-wedge polygons in eastern Enontekiö, northernmost Finland. *Reports from the Kevo Subarctic Research Station* **3**: 274–287.

— 1968. Etelä-Unkarin deflaatiojärvien erikoipiirteitä. (Summary: Features of the deflation lakes in southern Hungary). *Terra* **80**: 53–58.

— 1969. On the grain size and roundness of wind-blown sands in Finland as compared with some Central European samples. *Bulletin of the Geological Society of Finland* **41**: 165–181.

— 1971a. Evolution of eolian relief of the Kaamasjoki- Kiellajoki river basin in Finnish Lapland. *Fennia* **104**: 1–88.

— 1971b. Stratigraphy and material of the loess layers at Mende, Hungary. Appendix by Punakivi, Kalevi & Seppälä, Matti: Mineral analyses of the loess samples. *Bulletin of the Geological Society of Finland* **43**: 109–123.

— 1972a. Location, morphology and orientation of inland dunes in northern Sweden. *Geografiska Annaler* **54A**: 85–104.

— 1972b. Peat at the top of Ruohttir fell, Finnish Lapland. *Reports from the Kevo Subarctic Research Station* **9**: 1–6.

— 1972c. Some remarks on the formation of scratch circles on wind-blown sand. *Bulletin of the Geological Society of Finland* **44**: 131–132.

— 1972d. The term 'palsa'. *Zeitschrift für Geomorphologie N.F.* **16**: 463.

— 1973a. On the formation of periglacial sand dunes in northern Fennoscandia. *Ninth Congress International Union for Quaternary Research (INQUA) Abstracts*, Christchurch. 318–319.

— 1973b. On the formation of small marginal lakes on the Juneau Icefield, south-eastern Alaska, U.S.A. *Journal of Glaciology* **12**: 267–273.

— 1974. Some quantitative measurements of the present-day deflation on Hietatievat, Finnish Lapland. *Abhandlungen der Akademie der Wissenschaften in Göttingen, Mathematisch-Physikalische Klasse, Dritte Folge* **29**: 208–220.

— 1975a. Influence of rock jointing on the asymmetric form of the Ptarmigan Glacier valley, south-eastern Alaska. *Bulletin of the Geological Society of Finland* **47**: 33–44.

— 1975b. International research programme for periglacial sand dune studies – A recommendation. *Quaestiones Geographicae* **2**: 139–149.

— 1976a. Periglacial character of the climate of the Kevo region (Finnish Lapland) on the basis of meteorological observations 1962–71. *Reports from the Kevo Subarctic Research Station* **13**: 1–11.

— 1976b. Seasonal thawing of a palsa at Enontekiö, Finnish Lapland, in 1974. *Biuletyn Peryglacjalny* **26**: 17–24.

— 1977. Frequency isopleth diagram to illustrate wind observations. *Weather* **32**: 171–175.

— 1981 Forest fires as activator of geomorphic processes in Kuttanen esker-dune region, northernmost Finland. *Fennia* **159**: 221–228.

— 1982a. An experimental study of the formation of palsas. *Proceedings Fourth Canadian Permafrost Conference, Calgary, Alberta 1981.* National Research Council of Canada, Ottawa. 36–42.

— 1982b. Present-day periglacial phenomena in northern Finland. *Biuletyn Peryglacjalny* **29**: 231–243.

— 1983a. Palsasuon talvilämpötiloista Utsjoella. (Summary: Winter temperatures of a palsa bog in Finnish Lapland). *Oulanka Reports* **4**: 20–24.

— 1983b. Seasonal thawing of palsas in Finnish Lapland. *Permafrost Fourth International Conference, July 17–22, 1983, Proceedings.* National Academy Press, Washington, D.C. 1127–1132.

— 1984. Deflation measurements on Hietatievat, Finnish Lapland, 1974–77. In: Olson, Rod; Geddes, F. & Hastings, R. (eds.): *Northern Ecology and Resource Management.* The University of Alberta Press, Edmonton. 39–49.

— 1986. The origin of palsas. *Geografiska Annaler* **68A**: 141–147.

— 1987a. Nordenskiöldhuset (Svenskehuset) och dess omgivningar på Spetsbergen. *Nordenskiöld-samfundets Tidskrift* **47**: 3–34.

— 1987b. Periglacial phenomena of northern Fennoscandia. In: Boardman, John (ed.): *Periglacial Processes and Landforms in Britain and Ireland.* Cambridge University Press, Cambridge. 45–55.

— 1988. Palsas and related forms. In: Clark, M. J. (ed.): *Advances in Periglacial Geomorphology.* John Wiley & Sons, Chichester. 247–278.

— 1990. Depth of snow and frost on a palsa mire, Finnish Lapland. *Geografiska Annaler* **72A**: 191–201.

— 1991. *Rinodina endophragmia*, a lichen from the Vestfjella, Dronning Maud Land, Antarctica. *Annales Botanici Fennici* **28**: 193–196.

— 1992. Stabilization of snow temperature in Dronning Maud Land, Antarctica, January 1989. *Geografiska Annaler* **74A**: 227–230.

— 1993. Climbing and falling sand dunes in Finnish Lapland. In: Pye, K. (ed.): *The Dynamics and Environmental Context of Aeolian Sedimentary Systems.* Geological Society Special Publication 72: 269–274.

— 1994. Snow depth controls palsa growth. *Permafrost and Periglacial Processes* **5**: 283–288.

— 1995a. Deflation and redeposition of sand dunes in Finnish Lapland. *Quaternary Science Reviews* **14**: 799–809.

— 1995b. How to make a palsa: a field experiment on permafrost formation. *Zeitschrift für Geomorphologie N.F.*, Supplement-Band **99**: 91–96.

— 1997a. Distribution of permafrost in Finland. *Bulletin of the Geological Society of Finland* **69**: 87–96.

— 1997b. Introduction to the periglacial environment in Finland. *Bulletin of the Geological Society of Finland* **69**: 73–86.

— 1997c. Piping causing thermokarst in permafrost, Ungava Peninsula, Quebec, Canada. *Geomorphology* **20**: 313–319.

— 1998. New permafrost formed in peat hummocks (pounus), Finnish Lapland. *Permafrost and Periglacial Processes* **9**: 367–373.

— 1999. Geomorphological aspects of road construction in a cold environment, Finland. *Geomorphology* **31**: 65–91.

— 2001. Strong deflation on palsas in Finnish Lapland. *Transactions, Japanese Geomorphological Union* **22(4)**: C-216.

— 2002. Relief control of summer wind direction and velocity: a case study from Finnish Lapland. *Norsk Geografisk Tidsskrift* **56**: 117–121.

Seppälä, Matti; Gray, James & Ricard, Jean 1991. Development of low-centred ice-wedge polygons in the northernmost Ungava Peninsula, Québec, Canada. *Boreas* **20**: 259–282.

Seppälä, Matti & Koutaniemi, Leo 1985. Formation of a string and pool topography as expressed by morphology, stratigraphy and current processes on a mire in Kuusamo, Finland. *Boreas* **14**: 287–309.

Seppälä, Matti & Lindé, Krister 1978. Wind tunnel studies of ripple formation. *Geografiska Annaler* **60A**: 29–42.

Seppälä, Matti & Rastas, Jukka 1980. Vegetation map of northernmost Finland with special reference to subarctic forest limits and natural hazards. *Fennia* **158**: 41–61.

Sharp, Robert P. 1949. Pleistocene ventifacts east of the Big Horn Mountains, Wyoming. *Journal of Geology* **57**: 175–195.

— 1963. Wind ripples. *Journal of Geology* **71**: 617–636.

— 1964. Wind-driven sand in Coachella Valley, California. *Geological Society of America Bulletin* **75**: 785–804.

Sherman, Douglas J. & Hotta, Shintaro 1990. Aeolian sediment transport: theory and measurement. In: Nordstrom, Karl; Psuty, Norbert & Carter, Bill (eds.): *Coastal Dunes: Form and Process*. John Wiley & Sons, Chichester. 17–37.

Sirén, Gustav 1961. Taka-Lapin metsien historiasta ja ilmastosta historiallisena aikana. (Summary: On the history and climate of forest in Northern Lapland during the historical time.) *Lapin Tutkimusseuran Vuosikirja* **2**: 29–47.

Smalley, I. J. 1966. The properties of glacial loess and the formation of loess deposits. *Journal of Sedimentary Petrology* **36**: 669–676.

Smalley, Ian J. & Smalley, Valerie 1983. Loess material and loess deposits: formation, distribution and consequences. In: Brookfield, M. E. & Ahlbrandt, T. S. (eds.): *Eolian Sediments and Processes*. Elsevier, Amsterdam. 51–68.

Smiley, Terah L. & Zumberge, James H. (eds.) 1974. *Polar Deserts and Modern Man*. The University of Arizona Press, Tucson. 173 pp.

Smith, B. J.; Wright, J. S. & Whalley, W. B. 1991. Simulated aeolian abrasion of Pannonian sands and its implications for the origins of Hungarian loess. *Earth Surface Processes and Landforms* **16**: 745–752.

Smith, H. T. U. 1949a. Periglacial features in the driftless area of southern Wisconsin. *Journal of Geology* **57**: 196–215.

— 1949b. Physical effects of Pleistocene climatic changes in nonglaciated areas: eolian phenomena, frost action, and stream terracing. *Bulletin of the Geological Society of America* **60**: 1485–1516.

— 1964. Periglacial eolian phenomena in the United States. *Sixth INQUA Congress, Warsaw 1961, Report* **4**: 177–186.

— 1965. Dune morphology and chronology in central and western Nebraska. *Journal of Geology* **73**: 557–578.

— 1966. Wind-formed pebble ripples in Antarctica. *Geological Society of America, Special Paper* (abstracts for 1965) **87**: 160.

Sokolów, N. A. 1894. *Die Dünen. Bildung, Entwicklung und innerer Bau*. Verlag von Julius Springer, Berlin. 298 pp.

Solger, F. 1910. Studien über Nordostdeutsche Inlanddünen. *Forschungen zur deutschen Landes- und Volkeskunde* **19(1)**: 1–89.

Sollid, J. L.; Andersen, S.; Hamre, N.; Kjeldsen, O.; Salvigsen, O.; Sturød, S.; Tveitå, T. & Wilhelmsen, A. 1973. Deglaciation of Finnmark, North Norway. *Norsk geografisk Tidsskrift* **27**: 233–325.

Solopov, A. V. 1967. *Oases in Antarctica*. Academy of Sciences of the USSR, Interdepartmental Geophysical Committee, Meteorology Number 14. Israel Program for Scientific Translations, Jerusalem, 1969. 146 pp.

Sorenson, Curtis J.; Knox, James C.; Larsen, James A. & Bryson, Reid A. 1971. Paleosols and the forest bonder in Keewatin, N.W.T. *Quaternary Research* **1**: 468–473.

Stäblein, Gerhard 1983. Zur arktisch-periglazialen Talformung Ost-Grönlands. *Abhandlungen der Akademie der Wissenschaften in Göttingen, Mathematisch-Physikalische Klasse, Dritte Folge* 35: 281–293.

Stankowski, Wojciech 1963. Rzezba eoliczna Polski Pólnocno-Zachodniej na podstawie wybranych obszrów (Summary: Eolian relief of North-West Poland on the ground of chosen regions). *The Poznan Society of Friends of Sciences* **IV(1)**: 1–147.

Steenstrup, K. J. V. 1893. Bliver isen saa haard som staal ved höje kuldegrader? *Geologiska Föreningen i Stockholm Förhandlingar* **15**:119–120.

Steidtmann, James R. 1973. Ice and snow in eolian sand dunes of southwestern Wyoming. *Science* **179**: 796–798.

Stengel, Ingrid 1992. Zur äolische Morphodynamik von Dünen und Sandoberflächen. *Würzburger Geographische Arbeiten* **83**: 1–363.

St-Onge, Dennis 1965. La géomorphologie de l'Île Ellef Ringnes, Territoires du Nord-Ouest, Canada. Ministère des Mines et des Relevés techniques, Ottawa, Direction de la Géographie, *Etude Géographique* **38**: 1–46.

Strzemski, M. 1957. Efekty erozji wietrznej gleb na terenie poludniowo-wschodniej Polski w lutym 1956 r. (Summary: The effects of aeolian erosion of soils in southeastern Poland in February 1956). *Przeglad Geografiszny* **29**: 371–374.

Stull, Roland 2000. *Meteorology for Scientists and Engineers*. Second edition. Brook/Cole Thomson Learning, Pacific Grove, CA. 502 pp.

Sugden, David 1982. *Arctic and Antarctic. A Modern Geographical Synthesis*. Basil Blackwell, Oxford. 472 pp.

Sundborg, Åke 1955. Meteorological and climatological conditions for the genesis of aeolian sediments. *Geografiska Annaler* **37**: 94–111.

Suzuki, Takasuke & Takahashi, Ken'ichi 1981. An experimental study of wind abrasion. *Journal of Geology* **89**: 23–36.

Svensson, Harald 1969. Open fissures in a polygonal net on the Norwegian Arctic coast. *Biuletyn Peryglacjalny* **19**: 389–398.

— 1977. Observations on polygonal fissuring in non-permafrost areas of the Norden countries. *Abhandlungen der Akademie der Wissenschaften in Göttingen, Mathematisch-Physikalische Klasse, Dritte Folge* **31**: 63–76.

— 1981. Vinderosion i bergblock. (Abstract: A boulder ventifact.) *Svensk Geografisk Årsbok* 1981: 200–208.

— 1991. Vindabrasion i fast berg under högsta kustlinjen i södra Halland. (Abstract: Wind abrasion in solid rocks below the highest (Late Weichselian) shoreline in southern Halland, the Swedish west coast.) *Svensk Geografisk Årsbok* 1991: 157–167.

Sverdrup, H. U. 1938. Notes on erosion by drifting snow and transport of solid material by sea ice. *American Journal of Science* **35**: 370–373.

— 1957. The stress of the wind on the ice of the Polar Sea. *Norsk Polarinstitutt Skrifter* **111**: 1–11.

Svoboda, J. 1989. Vegetation of the Soviet polar deserts by V. D. Aleksandrova. A book review. *Arctic and Alpine Research* **21**: 316.

Swan, L. W. 1963. Aeolian zone. *Science* **140**: 77–78.

Swan, L. W. 1967. Alpine and aeolian regions of the world. In Wright, H. E. & Osburn, W. H. (eds.) *Arctic and Alpine Environments*. Indiana University Press, Bloomington. 29–54.

Swan, Lawrence W. 1992. The aeolian biome. Ecosystems of the earth's extremes. *Bioscience* **42**: 262–273.

Swett, Keene & Mann, Keith 1986. Terrace scarp deflation as a renewable source for eolian sediments in an arctic periglacial setting. *Polar Research* **5**, 45–52.

Taber, Stephen 1943. Perennially frozen ground in Alaska: Its origin and history. *Bulletin of the Geological Society of America* **54**: 1433–1548.

Tabler, Ronald D. 1980a. Geometry and density of drifts formed by snow fences. *Journal of Glaciology* **26**: 405–419.

— 1980b. Self-similarity of wind profiles in blowing snow allows outdoor modeling. *Journal of Glaciology* **26**: 421–434.

Tabuchi, H. & Hara, Y. 1992. Block fields and sorted polygons in Finnish Lapland (In Japanese, Summary in English). *Geographical Review Japan* **65(2)**: 105–113.

Takahashi, Shuhei; Ohmae, Hirokazu; Ishikawa, Masao; Katsushima, Takayoshi & Nishio, Fumihiko 1984a. Observation of snow drift flux at Mizuho Station, East Antarctica, 1982. *Memoirs of National Institute of Polar Research, Special Issue* **34**: 113–121.

— 1984b. Snow surface features of the Shirase Glacier drainage basin, Antarctica (Abstracts). *Memoirs of National Institute of Polar Research, Special Issue* **34**: 234.

Takeuchi, Masao 1980. Vertical profile and horizontal increase of drift-snow transport. *Journal of Glaciology* **26**: 481–492.

Tarnocai, C. 1978. Genesis of organic soils in Manitoba and the Northwest Territories. In: Mahaney, W. C. (ed.): *Quaternary Soils*. Geo Abstracts, Norwich. 453–470.

Täubert, Heinrich 1956. Sowjetische Untersuchungen in der Bunger-"Oase" (Antarktika). *Petermanns Geographische Mitteilungen* **100**: 329–334.

Tedrow, J. C. F. 1966. Polar desert soils. *Soil Science Society of America Proceedings* **30**: 381–388.

— 1969. Thaw lakes, thaw sinks, and soils in northern Alaska. *Biuletyn Peryglacjalny* **20**: 337–344.

— 1970. Soil investigations in Inglefield Land, Greenland. *Meddelelser om Grønland* **188(3)**: 1–93.

— 1977. *Soils of the Polar Landscapes*. Rutgers University Press, New Brunswick, New Jersey. 638 pp.

— 1978. Development of polar desert soils. In: Mahaney, W. C. (ed.): *Quaternary Soils*. Geo Abstracts, Norwich. 413–425.

Tedrow, J. C. F.; Bruggemann, P. F. & Walton, G. F. 1968. *Soils of Prince Patrick Island*. Arctic Institute of North America Research Paper 44: 1–82.

Teeri, J. A. & Barrett, P. E. 1975. Detritus transport by wind in the High Arctic terrestrial ecosystem. *Arctic and Alpine Research* **7**: 387–391.

Teichert, Curt 1939. Corrasion by wind-blown snow in polar regions. *American Journal of Science* **237**: 146–148.

— 1948. Corrasion by drifting snow. *Journal of Glaciology* **1**: 145.

Thiesmeyer, Lincoln R. 1942. Wind-worn stones in glacial deposits of the Middle West. *Science* **96**: 242–244.

Thomas, David S. G. (ed.) 1989. *Arid Zone Geomorphology*. Belhaven Press, London. 372 pp.

— (ed.) 1997. *Arid Zone Geomorphology. Process, form and change in drylands*. 2nd edition. John Wiley & Sons, Chichester. 713 pp.

Thorarinsson, Sigurdur 1962. L'érosion éolienne en Islande a la lumière des études téphrochronologiques. *Revue de Géomorphologie Dynamique* **XIII(7–9)**: 107–124.

Thorn, Colin E. 1975. Influence of late-lying snow on rock-weathering rinds. *Arctic and Alpine Research* **7**: 373–378.

— 1976. Quantitative evaluation of nivation in the Colorado Front Range. *Geological Society of America, Bulletin* **87**: 1169–1178.

— 1978. The geomorphic role of snow. *Annals of the Association of American Geographers* **68(3)**: 414–425.

— 1979a. Bedrock freeze-thaw weathering regime in an alpine environment, Colorado Front Range. *Earth Surface Processes* **4**: 211–228.

— 1979b. Ground temperatures and surficial transport in colluvium during snowpatch meltout; Colorado Front Range. *Arctic and Alpine Research* **11**: 41–52.

— 1988. Nivation: A geomorphic chimera. In: Clark, M. J. (ed.): *Advances in Periglacial Geomorphology*. John Wiley & Sons, Chichester. 3–31.

Thorn, Colin E. & Hall, Kevin 1980. Nivation: an arctic-alpine comparison and reappraisal. *Journal of Glaciology* **25**: 109–124.

Thornbury, William D. 1954. *Principles of Geomorphology*. John Wiley & Sons, New York & London. 618 pp.

Thorson, Robert M. & Schile, Carol Ann 1995. Deglacial eolian regimes in New England. *Geological Society of America, Bulletin* **107**: 751–761.

Tikkanen, Matti & Heikkinen, Olavi 1995. Aeolian landforms and processes in the timberline region of northern Finnish Lapland. *Zeszyty Naukowe Universytetu Jagiellonskiego Prace Geograficzne* **98**: 67–90.

Tobiasson, Wayne 1988. Buildings and utilities in very cold regions: Overview and research needs. *The Northern Engineer* **20(3–4)**: 4–11.

Tobolski, Kazimierz 1975. Succession of vegetation on drifting sands of Finnish Lapland dunes. *Quaestiones Geographicae* **2**: 157–168.

Trainer, Frank W. 1961. Eolian deposits of the Matanuska Valley agricultural area Alaska. *U.S. Geological Survey Bulletin* **1121-C**: 1–35.

Trautz, Max 1919. Am Nordrand des Vatnajökull im Hochland von Island. *Dr. A. Petermanns Mitteilungen aus Justus Perthes' geographischer Anstalt* **65**: 121–126.

Treshnikov, A. F. (ed.) 1985. *Atlas Arktiki*. Glavnoye Upravleniye Geodezii i Kartografii pri Sovete Ministrov SSSR, Moskva.

Tricart, Jean 1967. Le modéle des périglaciaires. In: Tricart, J. & Cailleux, A.: *Traité de Géomorphologie 2*. SEDES, Paris. 512 pp.

— 1970. *Geomorphology of Cold Environments*. Macmillan, London. 320 pp.

Troelsen, J. C. 1949. Contributions to the geology of the area round Jørgen Brønlunds Fjord, Peary Land, North Greenland. *Meddelelser om Grønland* **149(2)**: 1–29.

— 1952. An experiment on the nature of wind erosion conducted in Peary Land, North Greenland. *Meddelelser fra Dansk Geologisk Forening* **12**: 221–222.

Troll, C. 1944. Strukturböden, Solifluktion und Frostklimate der Erde. *Geologische Rundschau* **34**: 545–694.

— 1948. Der subnivale oder periglaziale Zyklus der Denudation. *Erdkunde* **2**: 1–21.

— 1973. Rasenabschälung (turf exfoliation) als periglaziales Phänomen der subpolaren Zonen und der Hochgebirge. *Zeitschrift für Geomorphologie N.F.* Supplement **17**:1–32.

Tsoar, Haim 1983. Wind tunnel modeling of echo and climbing dunes. In: Brookfield, M. E. & Ahlbrandt, T. S. (eds.): *Eolian Sediments and Processes*. Elsevier, Amsterdam. 247–259.

Tsoar, Haim & Yaalon, Dan H. 1983. Deflection of sand movement on a sinuous longitudinal (seif) dune: use of fluorescent dye as tracer. *Sedimentary Geology* **36**: 25–39.

Tuhkanen, Sakari 1980. Climatic parameters and indices in plant geography. *Acta Phytogeographica Suecica* **67**: 1–110.

— 1984. A circumboreal system of climatic-phytogeographical regions. *Acta Botanica Fennica* **127**: 1–50.

Ugolini, F. C.; Bockheim, J. G. & Anderson, Duwayne M. 1973. Soil development and patterned ground evolution in Beacon Valley, Antarctica. *North American Contribution, Permafrost Second International Conference*, National Academy of Science, Washington. 246–254.

Urbaniac, Urszula 1969. Les sables de coauertive, les cryoturbations et les fractures dans les dunes du bassin de plock. *Biuletyn Peryglacjalny* **19**: 399–422.

van Dieren, J. W. 1934. *Organogene Dünenbildung. Eine geomorphologische Analyse der Dünenlandschaft der West-Friesischen Insel Terschelling mit pflanzensoziologischen Methoden*. Nijhoff, Amsterdam. 304 pp.

van Vliet-Lanoë, Brigitte & Seppälä, Matti 2002. Stratigraphy, age and formation of peaty earth hummocks (pounus), Finnish Lapland. *The Holocene* **12**: 187–199.

van Vliet-Lanoë, Brigitte; Seppälä, Matti & Käyhkö, Jukka 1993. Dune dynamics and cryoturbation features controlled by Holocene water level change, Hietatievat, Finnish Lapland. *Geologie en Mijnbouw* **72**: 211–224.

Vieira, Goncalo Teles 1999. Coarse sand accumulations in granite mountains: the case-studies of the Serra do Gerês and Serra da Estrela (Portugal). *Zeitschrift für Geomorphologie N.F.*, Supplement-Band **119**: 105–118.

Viereck, L. A. 1965. Relationship of white spruce to lenses of perennially frozen ground, Mount McKinley National Park, Alaska. *Arctic* **18**: 262–267.

Vierhuff, Hellmut 1967. Untersuchungen zur Stratigraphie und Genese der Sandlössvorkommen in Niedersachsen. *Mitteilungen aus dem Geologischen Institut der Technischen Hochschule Hannover* **5**: 1–99.

Vilborg, Lennart 1977. The cirque forms of Swedish Lapland. *Geografiska Annaler* **59A**: 89–260.

— 1984. The cirque forms of Central Sweden. *Geografiska Annaler* **66A**: 41–77.

von Zahn, G. W. 1930. Wüstenrinden am Rand der Gletscher. *Chemie der Erde* **4**: 145–156.

Vowinckel, E. & Orvig, S. 1970. The climate of the north polar basin. In: Orvig, S. (ed.): *Climates of the Polar Regions*. World Survey of Climatology vol. 14. Elsevier, Amsterdam. 129–252.

Wade, F. Alton 1945. The geology of the Rockefeller Mountains, King Edward VII Land, Antarctica. Reports on scientific results of the US Antarctic Service Expedition, 1939–1941. *Proceedings of the American Philosophical Society* **89(1)**: 67–77.

Wahrhaftig, Clyde 1965. *Physiographic Division of Alaska*. U.S. Geological Survey, Professional Paper 482: 1–52.

Walker, H. J. 1967. Riverbank dunes in the Colville Delta, Alaska. *Lousiana State University Studies, Coastal Studies Bulletin* **1**: 7–17.

Walker, H. Jesse 1993. Lakes and permafrost in the Colville River Delta, Alaska. *Permafrost Sixth International Conference, Beijing*, Vol. 2: 1345.

Wallace, Robert E. 1948. Cave-in lakes in Nabesna, Chisana, and Tanana river valleys, eastern Alaska. *Journal of Geology* **56**: 171–181.

Walter, Heinrich & Leith, Helmut 1960. *Klimadiagramm-Weltatlas*. Gustav Fischer, Jena.

Wangstrom, Per 1989. Collector snow fences in the Arctic. *The Northern Engineer* **21(1–2)**: 13–19.

Warren Wilson, J. 1958. Dirt on snow patches. *Journal of Ecology* **46**: 191–198.

Washburn, A. L. 1953. Geography and Arctic lands. In: Taylor, Griffith (ed.): *Geography in the Twentieth Century*. Philosophical Library, New York. 267–287.

— 1969. Weathering, frost action and patterned ground in the Mesters Vig District, Northeast Greenland. *Meddelelser om Grønland* **176 (4)**: 1–303.

— 1970. An approach to a genetic classification of patterned ground. *Acta Geografica Lodziana* **24**: 437–446.

— 1973. *Periglacial Processes and Environments*. Arnold, London. 320 pp.

— 1979. *Geocryology. A Survey of Periglacial Processes and Environments*. Arnold, London. 406 pp.

Wasylikowa, Kystyna 1964. Pollen analysis of the late-glacial sediments in Witow near Leczyca, Middle Poland. *Report of the VIth INQUA Congress II*: 497–502.

Watson, Andrew 1989. Desert crusts and rock varnish. In: Thomas, David S. G. (ed.): *Arid Zone Geomorphology*. Belhaven Press, London. 25–55.

Webb, P. N. & McKelvey, B. C. 1959. Geological investigations in South Victoria Land, Antarctica. *New Zealand Journal of Geology and Geophysics* **2**: 120–136.

Webber, Patrick J. 1974. Tundra primary production. In: Ives, Jack D. & Barry, Roger G. (eds.): *Arctic and Alpine Environments*. Methuen, London. 445–473.

Wegener, Alfred 1911. Meteorologische Terminsbeobachtungen am Danmarks-Havn. *Meddelanden om Grønlund* **42**: 125–355.

Wendler, Gerd 1987. Blowing snow in eastern Antarctica. *Antarctic Journal of the United States* **22(5)**: 264–265.

Wendler, Gerd; Kodama, Yuji & Eaton, Frank 1983. Strong winds at Delta. *The Northern Engineer* **15(3)**: 15–19.

Wentworth, C. K. & Dickey, R. I. 1935. Ventifact localities in the United States. *Journal of Geology* **43**: 97–104.

Werner, B. T.; Haff, P. K.; Livi, R. P. & Anderson, R. S. 1986. Measurement of eolian ripple cross-sectional shapes. *Geology* **14**: 743–745.

Westgate, John A.; Stemper, Becky A. & Péwé, Troy L. 1990. A 3 m.y. record of Pliocene-Pleistocene loess in interior Alaska. *Geology* **18**: 858–861.

Weyant, William S. 1966. The Antarctic climate. In: Tedrow, J. C. F. (ed.): *Antarctic Soils and Soil-forming Processes*. American Geophysical Union Antarctic Research Series 8: 47–59.

— 1967. *The Antarctic Atmosphere: Climatology of the Surface Environment*. Antarctic map folio series. American Geographical Society. Folio 8: 1–4 and 13 plates.

Whalley, W. B.; Marshall, J. R. & Smith, B. J. 1982. Origin of desert loess from some experimental observations. *Nature* **300 (5891)**: 433–434.

Whalley, W. B.; Smith, B. J.; McAlister, J. J. & Edwards, A. J. 1987. Aeolian abrasion of quartz particles and the production of silt-size fragments: preliminary results. In: Frostick, L. & Reid, I. (eds.): *Desert Sediments: Ancient and Modern*. Geological Society Special Publication **35**: 129–138.

White, Sidney E. 1976. Is frost action really only hydration shattering? *Arctic and Alpine Research* **8**: 1–6.

Whitney, Marion I. 1978. The role of vorticity in developing lineation by wind erosion. *Geological Society of America Bulletin* **89**: 1–18.

— 1983. Eolian features shaped by aerodynamic and vorticity processes. In: Brookfield M. E. & Ahlbrandt, T. S. (eds.): *Eolian Sediments and Processes.* Elsevier, Amsterdam. Developments in Sedimentology **38**: 223–245.

Whitney, Marion I. & Dietrich, Richard V. 1973. Ventifact sculpture by windblown dust. *Geological Society of America, Bulletin* **84**: 2561–2582.

Willetts, B. B. & Rice, M. A. 1983. Practical representation of characteristic grain shape of sands: a comparison of methods. *Sedimentology* **30**: 557–565.

Williams, Joseph E. 1949. Chemical weathering at low temperatures. *Geographical Review* **39**: 129–135.

Williams, Peter J. 1961. Climatic factors controlling the distribution of certain frozen ground phenomena. *Geografiska Annaler* **43**: 339–347.

— 1991/1992. Thermal properties and the nature of freezing soils. *The Northern Engineer* **23(4)/24(1)**: 46–52.

Wilson, R. C. L.; Drury, S. A. & Chapman, J. L. 2000. *The Great Ice Age: Climate Change and Life.* Routledge, London. 267 pp.

Winkelmolen, A. M. 1969. The rollability apparatus. *Sedimentology* **13**: 291–305.

Wishart, E. R. 1970. *Electrification of Antarctic Drifting Snow.* University of Melbourne, Meteorology Department Publication 13: III: 1–27.

Woo, Ming-Ko 1993. Northern hydrology. In: French, Hugh M. & Slaymaker, Olav (eds.): *Canada's Cold Environments.* McGill-Queen's University Press, Montreal & Kingston. 117–142.

Woodworth, J. B. 1894. Postglacial eolian action in southern New England. *American Journal of Science* **47**: 63–71.

Wright, H. E. Jr.; Almendinger, J. C. & Grüger, J. 1985. Pollen diagram from the Nebraska Sandhills and the age of the dunes. *Quaternary Research* **24**: 115–120.

Yatsu, Eiju 1988. *The Nature of Weathering. An Introduction.* Sozosha, Tokyo. 624 pp.

Yen, Yin-Chao; Cheng, K. C. & Fukusako, S. 1991/1992. A review of intrinsic thermophysical properties of snow, ice, sea ice, and frost. *The Northern Engineer* **23(4)/24(1)**: 53–74.

Yosida, Zyungo 1963. Physical properties of snow. In: Kingery, W. D. (ed.): *Ice and Snow. Properties, Processes and Applications.* The M.I.T. Press, Cambridge, Massachusetts. 485–527.

Yosida, Zyungo and Colleagues 1955. Physical studies on deposited snow. I. Thermal properties. *Contributions from the Institute of Low Temperature Science, Hokkaido University, Sapporo,* **7**: 19–74.

Young, Kathy L. & Lewkowicz, Antoni G. 1988. Measurement of outflow from a snowbank with basal ice. *Journal of Glaciology* **34**: 358–362.

Yurtsev, Boris A. 1994. Floristic division of the Arctic. *Journal of Vegetation Science* **5**: 765–776.

Zav'yalova, I. N. 1987. Humidity of air in Antarctica. In: Dolgin, I. M. (ed.): *Climate of Antarctica.* A. A. Balkema, Rotterdam. 92–101.

Zingg, A. W. 1953. Wind-tunnel studies of the movement of sedimentary material. *Proceedings of the Fifth Hydraulics Conference.* State University of Iowa, Studies in Engineering, Bulletin **34**: 111–135.

Zhang, T; Osterkamp, T. E. & Stamnes, K. 1996. Some characteristics of the climate in Northern Alaska, U.S.A. *Arctic and Alpine Research* **28**: 509–518.

Zotov, V. D. 1938. Survey of the tussock-grasslands of the South Island, New Zealand. *The New Zealand Journal of Science and Technology* **20(4A)**: 212A–244A.

— 1940. Certain types of soil erosion and resultant relief features on the higher mountains of New Zealand. *The New Zealand Journal of Science and Technology* **21(5B)**: 256B–262B.

Index